M. B. Pepys (Ed.)

Acute Phase Proteins in the Acute Phase Response

With 45 figures

Springer-Verlag
London Berlin Heidelberg New York
Paris Tokyo Hong Kong

M. B. Pepys, MA, MD, PhD, FRCP, MCRPath
Immunological Medicine Unit,
Royal Postgraduate Medical School,
Hammersmith Hospital, Du Cane Road, London W12 0NN, UK

ISBN-13:978-1-4471-1741-4 e-ISBN-13:978-1-4471-1739-1
DOI: 10.1007/978-1-4471-1739-1

Cover: Fig. 10.6. One subunit of an SAP molecule shows extended regions of density to which a polypeptide fragment could be built.

British Cataloguing in Publication Data
The acute phase proteins in the acute phase response.
I. Pepys, B. (Mark B.)
599'.01'13
ISBN-13:978-1-4471-1741-4

Library of Congress Cataloging-in-Publication Data
Acute phase proteins in the acute phase response/M.B. Pepys (ed.).
p. cm.
Proceedings of a 1988 symposium.
ISBN-13:978-1-4471-1741-4(U.S.)
1. Acute phase reaction—Congresses. 2 Acute phase proteins—Congresses.
I. Pepys, M.B. (Mark B.)
[DNLM: 1. Acute Phase Proteins—congresses. 2. Inflammation—congresses.
OZ 150 A189 1988]
RB131.A27 1989
616.07'9—dc20
DNLM/DLC
for Library of Congress 89–21716
CIP

Softcover reprint of the hardcover 1st edition 1989

Typeset by Wilmaset, Birkenhead, Wirral

2128/3916–543210 Printed on acid-free paper

Preface

The discovery of C-reactive protein in the laboratory of O.T. Avery at Rockefeller University in 1929–30 was the first specific observation of the acute phase plasma protein response (Tillett and Francis 1930). This was one of three contributions of fundamental importance which emerged from that laboratory, the other two being the recognition that polysaccharides could act as antigens and that DNA transmits genetic information. In the course of characterization of pneumococcal carbohydrate antigens, a somatic polysaccharide common to all R and S forms of pneumococci was identified and designated Fraction "C" (Tillet et al. 1930). Testing of sera from patients with pneumococcal infection revealed the presence of material which precipitated with the C-polysaccharide but which differed from antibody in that calcium was required for the reaction. Furthermore, the amount of reactive material was greatest when patients were acutely ill and decreased in the convalescent phase, the precise opposite of specific anti-pneumococcal antibodies. Subsequently, the C-reactive material was shown to be a protein and to be present in the sera of individuals who were acutely ill with other, non-pneumococcal infections and tissue damaging conditions, hence Avery coined the term "acute phase" and called the protein "acute phase protein" (Abernethy and Avery 1941; MacLeod and Avery 1941). At that time methods were too insensitive to detect C-reative protein (CRP) in sera of healthy subjects and it was considered to be a pathological product. However, CRP is always present in the plasma and no individual lacking CRP has yet been described.

In response to almost all forms of tissue damage, infection and inflammation, the rate of synthesis and secretion of CRP by the liver, which is the sole significant site of production, increases dramatically, leading to greatly raised plasma levels. A large number of other diverse plasma proteins also participate in this acute phase response to "injury" and it is clearly an important pathophysiological process present in all warm blooded species. The nature of the acute phase response, the structure and functions of some acute phase proteins, the regulation and control mechanisms of their expression, and the clinical implications and significance of these phenomena were the subject of the 1988 Symposium of the Princesse Liliane Cardiology Foundation, the proceedings of which form the contents of the present volume.

Included among the acute phase proteins are proteinase inhibitors, coagulation factors, complement components, transport proteins and a number of proteins of unknown function. In selecting topics for a meeting which of necessity was limited in size and scope, the focus was on areas of particular basic and clinical interest in which there is currently active progress. One such area concerns the stimuli and mediator mechanisms which trigger the adaptive regulation of gene expression underlying the acute phase response. A closely related area is the molecular genetic organization of specific acute phase proteins. In both these fields there have been many exciting new developments and these are both reviewed and reported here. Another growing field concerns the structure and structure–function relationships of the acute phase proteins themselves, particularly CRP, the prototype acute phase reactant, the related protein, serum amyloid P component (SAP), serine proteinase inhibitors (serpins) and serum amyloid A protein (SAA).

Although the functions of neither CRP nor SAA are precisely known, they are of major importance because of their clinical relevance and because of the speed and dynamic range of their incremental change in the acute phase response, making them outstanding models for experimental investigation.

The acute phase response in general, and that of CRP in particular, is non-specific and clinical measurement of serum CRP therefore provides an extremely sensitive and valuable screening test for organic disease. Furthermore, once a diagnosis is established serial measurements of CRP generally provide the best means of objective monitoring of disease activity and response to treatment in most forms of bacterial infection, non-infective inflammatory disease and trauma, whether accidental or surgical. Finally, CRP measurement is a sensitive test for the development of intercurrent infection in patients at risk by virtue of some other primary disease, immunosuppressive medication, or invasive medical intervention.

SAA is the only other plasma protein which parallels CRP in the rate and magnitude of its dynamic response to acute phase stimuli. Unlike CRP it is, however, a poor antigen and no routine methods for its assay are commercially available. Its clinical importance lies in the fact that SAA is the precursor of AA protein, a partial degradation product which polymerizes in vivo to form AA amyloid fibrils. These are laid down as amyloid deposits in the tissues and organs of a certain proportion of individuals suffering from chronic active infections and inflammatory diseases, leading to the condition known as reactive systemic amyloidosis. Once deposited these fibrils persist and seem not to be cleared away and their continued accumulation leads in most patients to progressive functional deterioration in affected organs, morbidity and ultimately death. Since there is no specific treatment which promotes resolution of amyloid deposits, this generally fatal complication of otherwise non-fatal primary conditions is of considerable clinical significance. In addition to the amyloid fibrils themselves, amyloid deposits also always contain a non-fibrillar glycoprotein, amyloid P component (AP), derived from circulating SAP. The function of SAP is not yet known but nevertheless it is of major importance in this field, firstly because in the mouse it is a dramatic acute phase reactant and hence a very useful model for experimental study, and secondly because radiolabelled human SAP can now be used for the safe, non-invasive in vivo diagnosis, localization and quantitation of amyloid deposits. All these areas were discussed at the meeting and are reviewed in this volume.

London, June 1989 M. B. Pepys

References

Abernethy T J, Avery O T (1941) The occurrence during acute infections of a protein not normally present in the blood. I. Distribution of the reactive protein in patient's sera and the effect of calcium on the flocculation reaction with C polysaccharide of Pneumococcus. J Exp Med 73:173–182

MacLeod C M, Avery O T (1941) The occurrence during acute infections of a protein not normally present in the blood. II. Isolation and properties of the reactive protein. J Exp Med 73:183–190

Tillett W S, Francis T (1930) Serological reactions in pneumonia with a non-protein somatic fraction of pneumococcus. J Exp Med 52:561–571

Tillett W S, Goebel W F, Avery O T (1930) Chemical and immunological properties of a species-specific carbohydrate of pneumococci. J Exp Med 52:895–900

Contents

List of Contributors xv

1 **Interleukin-1 in the Acute Phase Response**
J. Gauldie
Introduction 1
The Acute Phase Response 2
Structure of IL-1 2
Synthesis and Regulation of IL-1 3
IL-1 Receptor 3
Pleotropic Actions of IL-1 4
IL-1 and Acute Phase Reactions 6
Fever 6
Leukocytosis 7
Acute Phase Proteins 7
IL-1 and Other Cytokines 11
Synergy of Action 11
Induction of Other Cytokines by IL-1 11
IL-1 and Corticosteroid 12
In Vivo Action of IL-1 13
Summary 14

2 **Regulation of Human SAA Gene Expression by Cytokines**
M. R. Edbrooke and Patricia Woo
Introduction 21
In Vitro Expression of Human SAA 21
Cis-acting Sequences Responsible for PMA Induction of SAA 22
Cytokine Control of SAA Expression 23

3 **Transcriptional Regulation of Acute Phase Response Genes with Emphasis on the Human C-reactive Protein Gene**
G. Ciliberto
Introduction 29
The Physiological Role of the Acute Phase Response 30
Systems for the Study of the Acute Phase Response . 31

Monokines Responsible for the Modulation of Liver-specific Gene Expression 32
Transcriptional Regulation – the Mechanism 35
Liver-specific Promoters and Enhancers 35
Promoters Induced During the Acute Phase 37
Conclusions and Perspectives 42

4 Organization, Structure and Expression of Pentraxin Genes
A. S. Whitehead
Introduction 47
The Structure of Pentraxin Proteins 48
The Structure of Pentraxin Genes 51
Heat Shock Elements 52
The Purine–Pyrimidine Repeat Region and Oligo-A Stretch in the CRP Intron 53
The 3′ Untranslated Region 53
Elements Responding to Cytokines 54
Genetics 54

5 ApoSSA: Structure, Tissue Expression and Possible Functions
E. P. Benditt, R. L. Meek and N. Eriksen
Introduction 59
Background 59
SAA is an Apolipoprotein 60
Where are SAAs Made and Where do They go? 61
Patterns of Cell and Tissue Expression of SAA mRNAs in Mice 62
Rats Have SAA-related Genes 63
Summary 64

6 Regulation of Biosynthesis and Secretion of Human C-reactive Protein and Serum Amyloid A
I. Kushner, M. K. Ganapathi and S. S. Macintyre
Introduction 69
Regulation of Biosynthesis 71
Role of Cytokines 71
Role of Cofactors: Signal Transduction 75
Heterogeneity in the Acute Phase Response 77
Regulation of CRP Secretion 77
Dynamics of CRP Secretion by Rabbit Hepatocytes 78
Subcellular Localization of the Intracellular Pool of CRP 79
CRP is Specifically Retained Within the Endoplasmic Reticulum 80
Conclusions 81

7 **Molecular Regulation of the Acute Phase Complement Proteins**
H. R. Colten and J. D. Gitlin
Introduction 85
Factor B 87
The Second Component, C2 89
Constitutive Expression of Factor B and C2 89
Regulated Gene Expression 90
Interleukin-1 and Interferon-γ 90
Tumor Necrosis Factor and Interleukin-6 93
Summary 93

8 **Biosynthesis of Acute Phase Proteins by the Liver Cells**
G. Feldmann, D. Bernuau, J. Y. Scoazec and M. Maurice
Introduction 97
Role of Hepatocytes in the In Vivo Biosynthesis of APR 98
Expression of APR in the Normal Unstimulated Liver 98
Expression of APR During the AIR 101
Role of Sinusoidal Cells in the In Vivo Biosynthesis of APR 103
Summary 103

9 **The Plasma Serine Protease Inhibitors (Serpins): Structural Modifications in Inflammation**
P. E. Stein and R. W. Carrell
The Serpins 107
Structure of the Serpins 107
Synthesis of Serpins in Inflammation 109
Antitrypsin 111
Antitrypsin Deficiency 111
The Reactive Centre 112
Mechanism of Protease Inhibition 112
Specificity of Inhibition 112
Reactive Centre Oxidation of Antitrypsin 113
The SR Conformational Change 114
Physiological Significance of the SR Change 115
Pathological Significance of the SR Change 116
Evolutionary Loss of the SR Change 118
Other Structural Modifications in Inflammation 118
N-terminal Cleavage: Angiotensinogen 118
Modification of Carbohydrate 119
Conclusions 119

10 The Three Dimensional Structure of SAP
H. E. White, B. P. O'Hara, G. Oliva, T. L. Blundell M. B. Pepys and S. P. Wood 123

11 Structure, Metabolism and Function of Acute Phase High Density Lipoprotein
F. C. de Beer, A. F. Strachan and E. G. Shephard
Introduction 137
Apo-SAA in Plasma 138
Structure of Apo-SAA 138
Structure of Acute Phase HDL 140
Metabolic Function of Normal HDL 142
Synthesis of Apo-SAA 143
Plasma Clearance of Apo-SAA 144
Cellular Association and Degradation of Apo-SAA . 145
Functions of Apo-SAA 147
Concluding Remarks 148

12 Clinical Measurement of Acute Phase Proteins to Detect and Monitor Infectious Diseases
P. E. Laurent
Introduction 151
Microbial Growth and AP Protein Response 152
General Pattern of AP Protein Response in Infection 154
Monitoring Disease Activity Using AP Protein Response 155
AP Protein Response in Viral Infection 155
Immunomodulation and AP Protein Response 156

13 C-reactive Protein: Clinical Aspects
M. H. van Rijswijk and M. A. van Leeuwen
Introduction 161
The Acute Phase Proteins 161
The Erythrocyte Sedimentation Rate as an Indirect Indicator for the Acute Phase Reaction 162
C-reactive Protein as a Direct Indicator of the Acute Phase Reaction 163
Assays for CRP 163
CRP Levels in Different Diseases 163
Functions of CRP 165
Conclusion 166

14 Pathogenesis of AA Amyloidosis
G. Husby
Definition and Classification of Amyloidosis 169
Serum and Tissue Amyloid A Proteins 171
Structure of SAA and AA 171
Induction and Production of the Acute Phase Protein SAA 172

Apo SAA and "Apo AA" 175
Displacement of SAA from HDL–SAA Complexes by Apo AI and Apo AII 176
Formation and Deposition of AA Amyloid 177
Incomplete Degradation of SAA 178
Amyloid Enhancing Factor and Glycosaminoglycans 178
Protein AP: The Amyloid "P Component" 179
Summary of Some Current Hypotheses Regarding AA Amyloid Formation 180

15 Serum Amyloid P Component: A Specific Molecular Targeting Vehicle in Amyloidosis
P. N. Hawkins and M. B. Pepys
Introduction 187
Serum Amyloid P Component 188
Ligand Binding by SAP/AP 188
Binding of SAP to Chromatin 189
SAP and Amyloidosis 190
Tissue Amyloid P Component 191
SAP as a Targeting Vehicle in Amyloidosis 191
Scintigraphic Imaging of Amyloid Deposits In Vivo . 191
Radiolabelled-SAP Studies in Experimental Murine Amyloidosis 192
Mouse Imaging Studies Using ^{123}I-Human SAP 192
Mouse Imaging Studies Using ^{123}I-Mouse SAP 195
Localization of ^{125}I-Pentraxins in AEF-Induced Amyloid 196
^{125}I-SAP Localization as a Method of Quantitating Murine Amyloid Deposits 197
Radiolabelled SAP Studies in Man 198
Preparation of ^{123}I-Human SAP for Clinical Studies 198
Imaging of Human Amyloid Deposits 199
Human SAP Turnover Studies 199
Conclusions 203

Subject Index 207

Contributors

E. P. Benditt,
Department of Pathology, School of Medicine, University of Washington, Seattle, WA 98195, USA

D. Bernuau,
Laboratoire de Biologie Cellulaire, INSERM U24, Faculté de Médecine Xavier-Bichat, 16 rue Henri Huchard, 75018 Paris, France

T. L. Blundell,
Laboratory of Molecular Biology, Department of Crystallography, Birkbeck College, Malet Street, London WC1E 7HX, UK

R. W. Carrell,
Department of Haematological Medicine, University of Cambridge Clinical School, Hills Road, Cambridge CB2 2QL, UK

G. Ciliberto,
Instituto di Scienze Biochimiche, Il Facoltà di Medicina e Chirurgia, Università di Napoli, Via S. Pansini 5, 80131 Napoli, Italy

H. R. Colten,
Professor of Pediatrics, St Louis Children's Hospital, 400 South Kingshighway Blvd., St Louis, MI 63110, USA

F. C. de Beer,
Department of Medicine, University of Kentucky, Medical Center MN614, 800 Rose Street, Lexington, KY 40475–0084, USA

M. R. Edbrooke,
Section of Molecular Rheumatology, MRC Clinical Research Centre, Harrow, Middlesex HA1 3UJ, UK

N. Eriksen,
Department of Pathology, University of Washington, Seattle, WA 98195, USA

G. Feldmann,
Laboratoire de Biologie Cellulaire, INSERM U24, Faculté de Médecine Xavier-Bichat, 16 rue Henri Huchard, 75018 Paris, France

J. Gauldie,
Department of Pathology, McMaster University, 1200 Main Street West, Hamilton, Ontario, Canada L8N 3Z5

M. K. Ganapathi,
Department of Medicine, Case Western Reserve University at Cleveland Metropolitan General Hospital, 3395 Scranton Road, Cleveland, OH 44109, USA

J. D. Gitlin,
The Edward Mallinckrodt Department of Pediatrics, Washington University School of Medicine, 400 South Kingshighway Blvd., St Louis, MI 63110, USA

P. N. Hawkins,
Immunological Medicine Unit, Department of Medicine, Royal Postgraduate Medical School, Hammersmith Hospital, Du Cane Road, London, W12 0NN, UK

G. Husby,
Department of Rheumatology, 9012 Regional Hospital of Tromsø, Tromsø, Norway

I. Kushner,
Professor of Medicine, Cleveland Metropolitan General Hospital, 3395 Scranton Road, Cleveland, OH 44109, USA

P. E. Laurent,
Institut Pasteur de Lyon, 77 rue Pasteur, 69365 Lyon Cedex 07, France

S. S. Macintyre,
Department of Medicine, Case Western Reserve University at Cleveland Metropolitan General Hospital, 3395 Scranton Road, Cleveland, OH 44109, USA

M. Maurice,
Laboratoire de Biologie Cellulaire, INSERM U24, Faculté de Médecine Xavier-Bichat, 16 rue Henri Huchard, 75018 Paris, France

R. L. Meek,
Department of Pathology, University of Washington, Seattle, WA 98195, USA

B. P. O'Hara,
Laboratory of Molecular Biology, Department of Crystallography, Birkbeck College, Malet Street, London WC1E 7HX, UK

G. Oliva,
Laboratory of Molecular Biology, Department of Crystallography, Birkbeck College, Malet Street, London WC1E 7HX, UK

M. B. Pepys,
Immunological Medicine Unit, Department of Medicine, Royal Postgraduate Medical School, Hammersmith Hospital, Du Cane Road, London W12 0NN, UK

J. Y. Scoazec,
Laboratoire de Biologie Cellulaire, INSERM U24, Faculté de Médecine Xavier-Bichat, 16 rue Henri Huchard, 75018 Paris, France

E. G. Shephard,
Department of Internal Medicine, University of Stellenbosch Medical School, PO Box 63, Tygerberg 7505, Republic of South Africa

P. E. Stein,
Department of Haematological Medicine, University of Cambridge Clinical School, Hills Road, Cambridge CB2 2QL, UK

A. F. Strachan,
Department of Internal Medicine, University of Stellenbosch Medical School, PO Box 63, Tygerberg 7505, Republic of South Africa

M. A. van Leeuwen,
Division of Rheumatology, Department of Medicine, State University, 59 Oostersingel, 9713 EZ Groningen, The Netherlands

M. H. van Rijswijk,
Division of Rheumatology, Department of Medicine, State University, 59 Oostersingel, 9713 EZ Groningen, The Netherlands

H. E. White,
Laboratory of Molecular Biology, Department of Crystallography, Birkbeck College, Malet Street, London, WC1E 7HX, UK

A. S. Whitehead,
Children's Hospital, 300 Longwood Avenue, Boston, MA 02115, USA

Patricia Woo,
Division of Rheumatology, MRC Clinical Research Centre, Watford Road, Harrow, Middlesex HA1 3UJ, UK

S. P. Wood,
Laboratory of Molecular Biology, Department of Crystallography, Birkbeck College, Malet Street, London WC1E 7HX, UK

Chapter 1

Interleukin-1 in the Acute Phase Response

J. Gauldie

Introduction

The body responds to injury in an orderly and orchestrated fashion to arrest the process of injury, provide protection for the rest of the organism against further injury and begin the repair process aimed at returning the body to homeostasis. These responses are both local and systemic. Immediate local responses include alterations in vascular permeability, and the release of active components from cells, including lysosomal enzymes, vasoactive amines and arachidonic acid metabolites. Systemic responses involve the activation of phagocytic cells, further generation of arachidonate metabolites, including prostaglandins and leukotrienes, and the release of polypeptide hormones or cytokines from various cells. The systemic reaction includes fever, an increase in the circulating granulocyte pool, changes in the metabolic activity of the liver and other tissues, with the induction of synthesis and secretion by the liver of a number of plasma proteins termed the "acute phase proteins" or "acute phase reactants" (APR).

The cytokine which appears to play a pivotal role in the induction of the acute phase response has been identified as interleukin-1 (IL-1) and over the past eight years, the molecule has been isolated and sequenced, cDNAs have been cloned and expressed and recombinant-derived molecules from man and rodents have been isolated and examined. There has been a veritable explosion of information as to the function of this particular polypeptide family. It is not the intention of this review to examine all of the data surrounding the characterization and the structure of the molecule, since this information and many of the biological functions of the molecule have been summarized in recent detailed reviews (Dinarello 1984, 1985, 1987; Dinarello and Mier 1987; Oppenheim et al. 1986). Rather, I shall concentrate on a more detailed examination of the involvement of IL-1 in the acute inflammatory response with particular reference to the induction of acute phase proteins. I will also examine some very recent data involving a comparison of in vivo and in vitro administration of recombinant derived IL-1 as well as the evidence of significant interaction of IL-1 and other peptide cytokines in describing a complex array of hormone interaction in the initiation and control of the acute phase response.

The Acute Phase Response

The tracing of the history and identification of IL-1 as being the molecule previously known as lymphocyte activating factor, endogenous pyrogen or leukocyte endogenous mediator is already well documented (Kushner 1982; Dinarello 1984; Koj 1986). Thus, the recognition that leukocyte endogenous mediator could elicit the acute phase induction of fibrinogen in vivo (Kampschmidt et al. 1982; Bornstein 1982) and that lymphocyte activating factor could elicit the acute phase synthesis of serum amyloid A and haptoglobin in mice in vivo and in vitro (Sipe et al. 1979, 1982; Selinger et al. 1980; Sztein et al. 1981; Hooper et al. 1981; McAdam et al. 1982) opened the way for speculation that in addition to the ability of IL-1 to mediate the febrile component of the acute phase response, it was also responsible for the initiation of the hepatic acute phase response in inflammation.

Since inflammation which occurs at a distant tissue site gives rise to metabolic changes in the liver and induces acute phase protein synthesis, this implies the presence of a blood-borne mediator. Since this mediator arises at one tissue site and causes the induction of cellular changes at a secondary organ, it can be classified as a hormone, even though many diverse and dispersed cells can release such mediators. In addition, to be classed as a hepatocyte stimulating hormone, the factor must interact specifically with the hepatocyte via a recognizable receptor and cause the induction of a spectrum of plasma proteins that are characteristic of an hepatic acute phase reaction. Several recent reviews highlight the involvement of such mediators in the control of the acute phase liver response (Gauldie et al. 1985; Koj 1986; Fey and Fuller 1987).

Structure of IL-1

There are two distinct IL-1 molecules: an acidic form with pI around 5 (IL-1α) and a neutral-to-basic form with a pI around 7 (IL-1β). Over the past several years, both forms have been cloned in the human (Auron et al. 1984; March et al. 1985), in the mouse (Lomedico et al. 1984; Gray et al. 1986) and more recently in the rat (Nishida et al. 1987). Between the α and β forms there is approximately 25% homology, but between the murine and human isoforms there is between 71% and 76% homology (March et al. 1985; Gray et al. 1986). In the human the gene indicates the presence of seven exons for each form and the gene is located on chromosome 2 (Dinarello 1987). Both molecules are translated initially as a 31 000 Da precursor polypeptide with the mature forms having similar 17 500 Da molecular sizes. There are both soluble and membrane-bound forms of the polypeptides which have measurable activity in a number of assays and there may be some smaller fragments of IL-1 which also exhibit biological activity. Both forms interact with the same receptor on the cell surface, and the receptor on the mouse lymphocyte has recently been cloned (Bird and Saklatvala 1986; Dower et al. 1986; Bird et al. 1988; Sims et al. 1988). It is not yet clear whether the same receptor exists on all cell types for both forms of IL-1. In using the lymphocyte activating factor co-mitogenic assay, the recombinant molecules are reported to have activities of 10^6–10^7 U mg^{-1}, depending on the method of production and isolation. In the human, IL-1β is the predominant molecule released by monocytes, while in the mouse the IL-1α form is predominant.

Synthesis and Regulation of IL-1

Originally described as being derived from leukocytes, the monocyte has been recognized as the most significant source of IL-1. However, many other cell types have been shown to express IL-1 including astrocytes, endothelial cells, keratinocytes and some B lymphocytes. In addition, recent evidence indicates that the neutrophil, once thought to be negative for IL-1, can be shown to express the mRNA and synthesize the protein for IL-1 upon appropriate activation by growth factors (Lindemann et al. 1988). Since the monocyte/macrophage is the cell most likely to initiate the inflammatory response upon contact with pathogen, it represents the most important source of IL-1 in the induction of acute inflammation. This cell has been reported to make significant quantities of IL-1 (100 fg per cell per 24 h) (Dinarello 1988). There are, however, reports of heterogeneity in the monocyte/macrophage population in the ability to produce IL-1. The peripheral blood monocyte appears to be the cell most capable of making IL-1 and maturation in vitro is associated with a decrease in this ability. In addition, human alveolar macrophages have been shown by immunohistochemistry and *in situ* hybridization to display heterogeneity in synthesizing IL-1. This probably accounts for the lowered ability of macrophage populations to make IL-1 compared to peripheral blood monocytes.

Once triggered by factors such as endotoxin (LPS) or phorbol-ester (TPA), the accumulation of mRNA coding for IL-1 is transcriptionally regulated (Fenton et al. 1987) and recent work with the IL-1β gene 5′ coding sequence implies the presence of *cis*-acting elements which convey tissue specificity in the expression of IL-1 (Clark et al. 1988). While the molecule does not appear to have a recognized leader sequence, there are obviously adequate control elements in the gene to allow initiation of transcription, intracellular processing and finally export of the mature 17 kDa polypeptide.

IL-1 Receptor

Using primarily cells of the lymphoid class, a number of studies have demonstrated that IL-1α and IL-1β bind to the same receptor on the cell with the same affinity. This in itself is notable given the lack of homology between the two forms (Bird and Saklatvala 1986, 1987; Bird et al. 1988; Urdal et al. 1988). Recently, the cDNA for a high affinity receptor on the murine lymphoid cell EL4 has been cloned showing the receptor has a molecule weight of 80 000 Da and belongs to a class of molecules known as the immunoglobulin supergene family (Sims et al. 1988). The availability of this clonally-derived material and the appropriate cDNA probes will shortly allow the confirmation of the disperse nature of the IL-1 receptor as well as the modifications to the density of the receptor that can occur during the inflammatory response, such as has been shown recently by Akahoshi et al. (1988) for the induction of the IL-1 receptor by glucocorticoid.

While the IL-1 receptor on the murine EL4 T lymphoma cell line has been shown to have a molecular weight of 80 000 Da, there is a suggestion of heterogeneity either between different cell types or cross-species since Horuk et al. (1988) have shown that the receptor isolated from the human Raji B lymphoma cell has an apparent molecular weight of 95 000 Da. These discrepancies await the examination of cDNA sequences for the IL-1 receptors to determine if these differences are real. In addition, it should be recognized that there are several instances in which it has been demonstrated that IL-1 can mediate cellular responses in the absence of detectable IL-1 receptors on the cells (Rosoff et al. 1988; Baumann et al. 1988). These examples of activation of T cells and hepatomas need to be investigated further and are likely to show exceptionally low levels of IL-1 receptors on such cells.

However, given the pleotropic nature of IL-1, and the diverse nature of the cell actions that it causes, it would not be surprising to find that some modifications or even other receptors may be present in different cell types such as the hepatocyte.

Pleotropic Actions of IL-1

Given that many diverse cell types are capable of making IL-1 in vitro, it is surprising that a broad spectrum of cell types are affected by IL-1. Table 1.1 summarizes the very many broad actions of IL-1 in the diverse cell types that it affects. Rather than reiterate what has already been well documented in recent reviews (Dinarello 1984, 1987; Oppenheim et al. 1986), only the most recent developments will be mentioned here. Those novel aspects involving interaction of IL-1 with other cytokines will be dealt with in the following sections.

Table 1.1. Multiple systemic effects of IL-1

Proliferation/activation of T and B lymphocytes and induction of IL-2 and IL-2 receptors
Proliferation of fibroblasts and keratinocytes and release of PGE_2
Induction of neutrophilia
Activation of chondrocytes and release of proteases and PGE_2
Activation of endothelial cells and release of plasminogen activator inhibitors
Cytotoxic to tumor cells and β cells of the pancreas
Induction of multiple cytokines from many cells
Induction of fever and glucocorticoid synthesis
Induction of synthesis of some acute phase proteins

From Table 1.1, it is obvious that IL-1 has the aspects of being an alarm polypeptide. It can initiate many of the body's responses to trauma and infection, but in so doing it contributes to the inflammation, particularly when chronic stimulation ensues. Since IL-1 has been involved in thermal regulation, much controversy has surrounded the issue of whether IL-1 plays a role in neuronal modulation. Recent data indicating that substance P can stimulate the production of IL-1 (Kimball et al. 1988) coupled with some elegant immunohistochemistry studies by Breder et al. (1987) lead to the speculation that since IL-1 appears to be present in the brain, that astrocytes may be seen to produce IL-1 (Fontana et al. 1982), and the neuropeptide substance P may be involved in regulating IL-1, it is possible that IL-1 behaves as a neuromodulator and may control many other functions yet to be established in neuronal regulation. Developments in this area await the sensitive detection systems of *in situ* hybridization to identify the potentially large number of cell types in the brain which may participate in the production and utilization of IL-1. With the recent cloning of the IL-1 receptor, the possibility of detecting cells responding to IL-1 in the brain becomes more realistic.

A second area with recent development involves the demonstration by Vlassara et al. (1988) that macrophages can be stimulated to release IL-1 upon uptake of glycosylated proteins. Glycosylation of proteins is enhanced under conditions of hyperglycaemia. This brings to importance the finding that IL-1 is directly toxic to β cells of the pancreas (Bendtzen et al. 1986) and suggests a possible mechanism for the continual destruction of the β cells in diabetes. These findings, however, must be contrasted with recent data by Hill et al. (1988) indicating that IL-1 decreases intracellular glucocorticoid receptors, resulting in decreased induction of PEPCK enzyme and subsequent reduced gluconeogenesis and plasma glucose. Thus, normal inflammation is associated with an apparent inhibition of hepatic glucose synthesis which would reduce glycosylated protein, but it would appear that IL-1 can also modulate the other signals for mediating blood glucose levels.

A third area concerns the involvement of polymorphonuclear neutrophils in inflammation. While it was previously shown that IL-1 induced neutrophilia in vivo, it is now apparent, using recombinant-derived molecules, that IL-1 is not directly chemotactic for neutrophils but can apparently enhance the action of other chemotactic agents and thereby contribute to the neutrophilia in inflammation. More importantly, the neutrophil, long thought to be devoid of IL-1 activity, has recently been shown to be able to secrete IL-1 upon appropriate stimulation, that being by another cytokine – granulocyte/macrophage colony-stimulating factor (GMCSF) (Lindemann et al. 1988). In addition, IL-1 can be shown to stimulate the release of other cytokines including GMCSF from stromal cells (see below), and this might represent an auxiliary pathway to amplification of the local release of IL-1 from inflammatory tissue sites.

The fourth area involves examination of intracellular events controlling the release of IL-1 as well as events occurring after the interaction of cells with IL-1. Recently, the cloning of the IL-1 genes (Clark et al. 1986; Telford et al. 1986; Furutani et al. 1986) has allowed an examination of the *cis*-acting elements 5′ upstream of the coding region, which control the expression of the IL-1 molecule (Clark et al. 1988). In addition, studies using antibodies to recombinant molecules and suitable molecular probes demonstrate that the synthesis and secretion of IL-1 differs from other polypeptide cytokines and that the differences may extend to significant alterations in the mechanisms of secretion of IL-1α and IL-1β. These data imply that further investigations in this area may well lead to a clearer understanding of the physiologic roles for the α and β forms of this molecule. Further, demonstration of significant phosphorylation events occurring in cells after IL-1 stimulation (Matsushima et al. 1988) allows a more suitable examination of the mechanism of signal transduction in various cells after IL-1 interacts with its receptor.

A fifth recent development involves the action of IL-1 on the induction of a cell-surface glycoprotein (ICAM-1), which is thought to be necessary for effective cellular interactions in the immune response, through the binding of molecules such as ICAM-1 to a series of adhesion molecules on macrophage and lymphocytes including the lymphocyte function associated antigen-3 (LFA-3). ICAM-1 belongs to the immunoglobulin supergene family, a characteristic similar to the IL-1 receptor. It is believed that the interaction of all activated leukocytes with ICAM-1 is a crucial step in the induction and development of an inflammatory response and the ability of IL-1 to induce expression of ICAM-1 is obviously important in the initiation of inflammation. It will be interesting to determine the effect of IL-1 on the induction of other cell-surface binding molecules in other situations (Rothlein et al. 1988; Dustin et al. 1988).

Finally, it should be noted that there have been several descriptions of the use of molecular probes in *in situ* hybridization to detect the presence of IL-1 gene expression in tissue. These have included Takacs et al. (1988) in the C57/B16 mouse, Bernaudin et al. (1988) in human alveolar macrophages and peripheral blood monocytes and Duff et al. (1988) demonstrating the presence of IL-1-containing cells in the synovium of individuals with arthritis. These developments indicate that very shortly there will be available a considerable body of literature defining the localization and state of activation of various cells and the prevalence of either spontaneous or induced IL-1 gene expression in normal physiology and during an inflammatory process.

IL-1 and Acute Phase Reactions

Fever

Since the recognition that IL-1 is identical to endogenous pyrogen (EP) it is clear that IL-1 contributes to thermal regulation and is responsible in part for the alteration of the temperature set-point of the body. EP or IL-1 was shown to induce fever via a prostaglandin E_2-dependent mechanism with an increased level of PGE_2 occurring in or near the preoptic-anterior hypothalamus (Dinarello 1984). While it is now clear that IL-1 is not the only cytokine capable of inducing the febrile response (IL-6 has recently been shown to stimulate fever, Helle et al. 1988), it appears logical that IL-1 contributes to the induction of fever since it has also been shown that IL-1 appears to be more active at an elevated temperature in the induction of thymocyte proliferation (Duff and Durum 1982; Hanset et al. 1983). With the recent observation that IL-1 may be an integral component in the brain (Breder et al. 1988) and earlier evidence that the central nervous system is probably involved in acute phase responses to IL-1 (Turchik and Bornstein 1980), there is speculation that IL-1 from the blood may enter the brain to mediate these changes directly, though there is still no direct evidence to show the point of entry, and this speculation remains controversial.

Leukocytosis

Injection of recombinant-derived IL-1 into tissue sites results in the rapid accumulation of polymorphonuclear leukocytes. The known association of leukocytosis with inflammation and the ability of IL-1 to induce this phenomenon indicate that, directly or indirectly, IL-1 mediates granulopoiesis and the tissue accumulation of neutrophils, presumably by chemotactic responses (Granstein et al. 1985; Cybulsky et al. 1986). Indeed, IL-1 has been shown to mediate the local Shwartzman reaction and mediates changes in vascular permeability (Beck et al. 1986). While early reports implicated IL-1 as being directly chemotactic for neutrophils, examination of highly purified preparations of recombinant IL-1 (rIL-1) showed no chemotactic activity for human neutrophils (Yoshimura et al. 1987). Regardless of whether IL-1 acts via a direct or indirect mechanism, injection of the cytokine into the peritoneal cavity of mice results in a dramatic infiltration of neutrophils, and Sayers et al. (1988) have shown that the induction of neutrophilia is independent of the action of mediators of the arachidonic pathway. Moreover, using a potent neutrophilic chemotactic factor, LTB4, did not result in neutrophil accumulation in the peritoneal cavity. Taken together, these data imply that IL-1 causes neutrophil accumulation via an indirect route or via its action as a haemopoietic factor, but still that IL-1 is an important modulator of the leukocytosis often associated with the acute phase response. This fact was dramatically shown by the elegant work of Dunn et al. (1988) who used a slow releasing copolymer containing IL-1 to demonstrate that chronic local release of IL-1 induces an inflammatory granuloma consisting of dense monocyte infiltrates mixed with scattered polymorphonuclear leukocytes and neovascularization.

Acute Phase Proteins

The best data documenting that IL-1 is involved in the induction of acute phase proteins comes from studies using isolated primary hepatocytes and hepatoma cells. While the early investigation of IL-1 involvement in acute phase protein induction made use of purified material, most of those earlier studies have since been confirmed with the use of purified recombinant IL-1. As will become clear in this and in later articles, IL-1 is only one of the hormones that regulate the acute phase protein response, and to talk about it in isolation negates the understanding that most of the hepatic reaction is controlled by combinations of several hormones. We will deal here only with IL-1 mechanisms. However, it is obvious that IL-1 and the recently described hepatocyte-stimulating factor (HSF) or IL-6 (Ritchie and Fuller 1983; Gauldie et al. 1987a) and tumor necrosis factor (TNF), along with glucocorticoid, can account for the full regulation of liver acute phase genes in inflammation.

By itself, rIL-1 has been shown to induce serum amyloid A (SAA) in vitro in primary mouse hepatocytes at both the proteins and mRNA level (Ramadori et al. 1985). While this was true for mouse SAA, the same could not be seen in human hepatoma cells (Perlmutter et al. 1986a; Ganapathi et al. 1988), which highlights an important consideration in dealing with isolated cells. Not only are there species differences in the liver's response to cytokine challenge (α2-macroglobulin [α2-M] and cysteine proteinase inhibitor [CPI] are acute phase proteins in the rat but not in the mouse or man), but there are differences in how primary cells and hepatoma cells may respond and it is now clear that even subclones of hepatoma cells may respond differently to these cytokines (H. Baumann, personal communication). Thus, evidence for the lack of stimulation of a particular acute phase protein gene by a cytokine should be taken in the context of this variability and confirmed only when it has been examined in several systems with different cell types.

In the case of SAA, rIL-1 in vivo causes induction of plasma SAA within a very short period of time (4–12 hours) while the maximum response to stimulation by endotoxin in vivo occurs between 16 and 36 hours after administration (Sipe et al. 1988; Sipe and Ramadori 1986). This indicates not only a direct action of IL-1 on the liver, as confirmed by the in vitro experiments, but also indicates that SAA production is probably directly stimulated by IL-1. The notable difference was that the magnitude of induction of the SAA gene by IL-1 in vitro was less than expected from the in vivo experiments. This was also true for the murine SAA gene transfected and expressed in cultured cells, where maximal induction was only achieved by combinations of both IL-1 and HSF/IL-6 (W. Liao et al., unpublished data), similar to recent results seen for the acidic form of SAA in *Mus caroli* primary hepatocytes (Prowse and Baumann 1989).

In addition to SAA, rIL-1 increased the expression of complement factor B and C3, the third component of complement, in mouse hepatocytes and human hepatoma cells (Perlmutter et al. 1986a,b; Ramadori et al. 1985). A further acute phase protein, serum amyloid P component (SAP), was also induced in mouse hepatocytes by purified natural IL-1 (Le and Mortensen 1984), and we and others have shown that one of the most prominent acute phase proteins of man, α1-acid glycoprotein (α1-AGP), was induced by both purified and recombinant IL-1 in human, rat and mouse hepatocytes and in hepatoma cells (Gauldie et al. 1987a,b; Koj et al. 1987; Baumann et al. 1987a,b; Geiger et al. 1988). While the action on the positive acute phase proteins is the induction of the message and protein synthesis, IL-1 also has a down-regulating action on the negative acute phase protein albumin at the RNA and protein levels.

Again, injection of IL-1 in vivo causes changes in the same acute phase proteins as indicated above. In addition, recombinant IL-1 and purified natural IL-1 administered in vivo cause the induction of fibrinogen in all species tested (Kampschmidt and Mesecher 1985; Dinarello 1987). However, in contrast to the previous proteins, at either the mRNA or protein synthesis level, IL-1 has no stimulating effect on the expression of the fibrinogen gene in vitro in rat, rabbit, mouse or human hepatocytes or hepatoma cells (Woloski and Fuller 1985; Darlington et al. 1986; Evans et al. 1987; Koj et al. 1987; Baumann et al. 1987a). On the contrary, rIL-1 acts in a negative manner causing suppression of induction of some of the acute phase proteins (Darlington et al. 1986; Baumann et al. 1987a; Koj et al. 1987; Gauldie et al. 1987a, 1988; Andus et al. 1988; Marinkovic et al. 1989). The proteins negatively affected by IL-1 include fibrinogen in rat and human and cysteine proteinase inhibitor or thiostatin as well as α2-M in the rat. Thus, IL-1 has a positive regulatory function on some of the acute phase protein genes and negative effects on others. The most striking regulation by IL-1 is seen with α1-AGP. Here IL-1 acts in synergy with HSF/IL-6 in the presence of glucocorticoid to cause maximum stimulation (Baumann et al. 1987a, Gauldie et al. 1988).

While there has been no full detailed study reported as yet, it appears that IL-1α and IL-1β are equipotent in inducing the various acute phase proteins, and even membrane-bound IL-1α on mouse macrophages apears to result in the induction of C3 and the reduction of albumin in hepatoma cells (Beuscher et al. 1987). A recent suggestion by Geiger et al. (1988) that there may be differences in the ability of IL-1α and IL-1β to induce α1-AGP would appear to be the only evidence that there may be differences in the way the two molecules act. However, since they act at the same receptor, perhaps minor differences in specific activity of the molecules used in this particular instance could account for the differences seen.

In the light of the above data, particularly the negative effect on fibrinogen, it is now necessary to postulate a different mechanism of induction when IL-1 is administered in vivo, since the outcome is the induction of the protein. We have recently shown that IL-6 is the major regulator of fibrinogen synthesis (Gauldie et al. 1987a). Since one of the most potent inducers of IL-6 from stromal cells in vitro is IL-1 (Sehgal et al. 1987), it is probable that the administration of IL-1 in vivo causes the release of IL-6 from stromal cells resulting in the stimulation of the fibrinogen gene product and the demonstration of the full hepatic acute phase response.

By comparing the action of recombinant IL-1, TNF and IL-6 in the presence of glucocorticoid on a number of primary hepatocyte and hepatoma cells in mouse, rat and human, we were able to determine that IL-1 induces a subset of the acute phase protein genes and that it has additive, negative and synergistic actions with the other cytokines and with glucocorticoid, but that all three or all four hormones are necessary to induce the full hepatic acute phase protein response (Gauldie et al. 1988). Table 1.2 summarizes these findings. It is worth noting that haptoglobin exhibits marked species difference in that it appears to be IL-6-dependent in human hepatoma cells (Baumann et al. 1987a; Morrone et al. 1988), yet is IL-1- and IL-6-dependent in the rat (Baumann et al. 1987b). Thus, while in the early 1980s, IL-1 was considered the major regulator of the acute phase response (Dinarello 1987), it is clear that IL-1 is not the main direct hepatic mediator but rather a modulator of the IL-6-induced hepatocyte response and acts in both a direct and indirect manner in the induction of acute phase proteins.

Table 1.2. Acute phase protein genes regulated by inflammatory cytokines

Induced by IL-1 and IL-6	Induced by IL-6 only	Synergy IL-1 and IL-6	Inhibition of IL-6 by IL-1
α1 acid glycoprotein	α2-macroglobulin	α1 acid glycoprotein	Fibrinogen
C3	Fibrinogen	C3	α2-macroglobulin
Factor B	α1-proteinase inhibitor	Haptoglobin (rat)	Cystein proteinase inhibitor
SAA	Haemopexin		
SAP (mouse)	Haptoglobin (human)		
Haptoglobin (rat)	α1-antichymotrypsin		
C-reactive protein	Ceruloplasmin		
	Cystein proteinase inhibitor		
	C1 esterase inhibitor		

It is possible that IL-1 exerts its negative effect on IL-6-mediated induction of fibrinogen by causing a reduction in cytosolic glucocorticoid binding in the hepatocyte as has recently been shown by Hill et al. (1988) using the Reuber rat hepatoma cell line. Interestingly, the induction of the PEPCK enzyme was also inhibited by IL-1, suggesting PEPCK, which is important as the rate-limiting enzyme in gluconeogenesis, may come under the same influences for metabolic control as do a number of the acute phase proteins.

Recently several of the acute phase protein genes have been cloned. The 5′ region of the gene presumably contains *cis*-acting elements which are regulatory regions controlling tissue specificity of expression as well as inducibility during inflammation. The regions appear to be complex and while there is good evidence indicating tissue-specific elements (Grayson et al. 1988; De Simone et al. 1987; Courtois et al. 1987), the data regarding specific IL-1 or IL-6 regulatory elements are not yet fully detailed. However, Prowse and Baumann (1988) recently showed that there was a regulatory region responding to combinations of IL-1, IL-6 and glucocorticoid in the 5′ region of the rat α1-AGP gene. Oliviero et al. (1987) with the haptoglobin gene and Ciliberto et al. (1987) with the CRP gene indicated that there were responsive elements on these genes corresponding to inducibility by monocyte conditioned medium which they recently implicated as being an IL-6-dependent response (Morrone et al. 1988). To date, no consensus sequence of IL-1-dependent inducible *cis*- or *trans*-acting element has been determined.

An additional approach has been the construction of transgenic mice in which the human α1-AGP gene, SAA gene, α1-antitrypsin gene and CRP gene have been inserted in the murine genome. These chimerics have been used to study the tissue specificity of expression and inducibility during inflammation in the intact animal (Dente et al. 1987; Sifers et al. 1987; Ciliberto et al. 1987). Using this approach, an important finding was made relating to species differences in acute phase protein responses. The human CRP gene transfected into the mouse genome maintained its response to acute phase signals while the murine CRP was not induced, suggesting that it is the *cis*-control elements of the murine CRP gene that renders it non-responsive to acute signals rather than some aspect of receptor and signal transduction in the mouse hepatocyte.

The mechanism of signal transduction in the hepatocyte after IL-1 stimulation has been recently assessed. Baumann et al. (1988) demonstrated that both IL-1 and IL-6 act via mechanisms which can be shown to be independent of protein kinase-C (inducible by phorbol ester-TPA) and also independent of intracellular Ca^{2+} changes. Treatment with TPA inhibits the effect of IL-1, but probably does so by down-regulation of IL-1 receptor function and decreased binding of IL-1 by the hepatocyte. Definitive proof will await the generation of appropriate cDNA probes and antibodies for these receptors.

For a full review of the cytokines involved in the regulation of the acute phase protein genes and the response of the liver in inflammation, the readers are referred to reviews by Fey and Fuller (1987) and Fey and Gauldie (1989).

Given that IL-1 induces a subset of the acute phase proteins and may down-regulate the expression of others, it is interesting to note the recent description of a plasma binding factor for IL-1 which was shown to be α2-M (Teodorescu et al. 1988). It is unclear whether α2-M is providing a carrier function, a protective function or an absorptive and removal function for this cytokine. Since α2-M has also been shown to carry PDGF and TGFβ as well as a recent report that it carries

IL-6 (O'Connor-McCourt and Wakefield 1987; Matsuda et al. 1988), this suggests a general role of cytokine transport for one of the major antiproteases involved in inflammation.

IL-1 and Other Cytokines

The most interesting recent developments have occurred in broadening our understanding of how these various protein hormones interact with cells. It is obvious that they do not act alone; rather they mediate cell function by cascade events or in concert and synergy. The physiological response is defined by multiple cytokine interactions. It is probable that the final understanding of the role each cytokine plays will only come with the ability to understand the context of the language in which the cells interact with multiple cytokines. While we have been restricted to simplistic approaches, recent evidence implies that we must investigate more fully cytokine interactions in our experimental models.

Synergy of Action

IL-1 clearly synergizes with IL-6 as shown above on the induction of acute phase proteins such as α1-AGP. The same two cytokines act in synergy on the induction of thymocyte proliferation when the in vitro experimental conditions are appropriate (Houssiau et al. 1988). In conjunction with another inflammatory cytokine, tumor necrosis factor (TNF), it acts synergistically on a number of cells including the pituitary cell, bone, endothelial cells and fibroblasts to modify many of their functions, including proliferation, cytotoxicity and the generation of PGE_2 (Mandrup-Poulsen et al. 1988; Gowen 1988; Dinarello and Mier 1987; Ruggiero and Baglioni 1987; Stashenko et al. 1987). In addition, IL-1 as haemopoietin-1 acts in synergy with a number of the colony-stimulating factors (CSF) in haemopoiesis and radioprotection (Broxmeyer et al. 1987; Neta 1988; Moreb et al. 1988; Castelli et al. 1988). Finally, Kimball (1988) has also shown that IL-1 causes a synergistic enhancement of the action of neuropeptide substance P on the modulation of fibroblast proliferation.

It is probable that there will be many more examples of how IL-1 acts with the other inflammatory hormones in a synergistic manner and highlights the complex nature of these interactions.

Induction of Other Cytokines by IL-1

It is obvious that IL-1 is recognized as the "alarm" polypeptide hormone. It appears to be the major signal for the generation of a multitude of other cytokines from many different cell types. Thus, IL-1 causes the release from activated T-cells of IL-2, the IL-2 receptor, TGFβ, IL-3, IL-4, IL-6 and interferon-γ (IFN-γ) (Hagiwara et al. 1987; Dinarello 1988), as well as TNF from endothelial cells (Philip and Epstein 1986), and it is a strong inducer of IL-6 release from stromal cells (Sehgal et al. 1987). Moreover, it can induce a number of the CSFs including GM-CSF and G-CSF from fibroblasts, endothelial cells and T-cells (Hermann et al. 1988; Zsebo et al. 1988; Sieff et al. 1987; Kaushansky et al. 1988; Zucali et al. 1986). Whether IL-1 will remain as the most potent inducer of these various cytokines remains to be confirmed, but the role of many of these other hormones in cellular proliferation and differentiation, functions which are important in healing and repair mechanisms, implies that they are likely to be involved in recovery from inflammation.

Recently it was shown that IL-1 can induce its own expression in smooth muscle cells, monocytes and endothelial cells (Warner et al. 1987; Dinarello et al. 1987). While these reports raise the issue that IL-1 may be involved in its own self-amplification mechanism in some diseases, a follow-up to that study indicated that this induction was inhibited by the presence of interferon-γ even though previous studies showed that IFN-γ enhanced rather than inhibited the endotoxin-induced IL-1 production. The inhibition of this IL-1-mediated induction of IL-1 synthesis was comparable to that observed from 10^{-7} M dexamethasone, a level consistent with an inflammatory response. In addition to IL-1 induction by IL-1, other cytokines, including TNF and neuropeptide substance P, have been shown to induce the expression of IL-1 in several cell types (Ghezzi and Dinarello 1988). Once again, the complex nature of this interaction is shown, with autocrine, paracrine and exocrine actions of the hormones being important in the final outcome of the event.

IL-1 and Corticosteroid

It is thought that the increase in corticosteroid seen in inflammation is part of the homeostatic response and results in anti-inflammatory activity. However, it must be remembered that the acute phase protein response of the liver is dependent on the presence of glucocorticoid, even though glucocorticoid by itself is not specifically stimulatory to acute phase protein synthesis. Thus it is not surprising that IL-1 may play a role in the induction of the changes seen in glucocorticoid levels and inflammation. This would imply that IL-1 manages to directly or indirectly bridge the blood–brain barrier and manifest its effect at either the hypothalamus or pituitary gland. There is dispute regarding whether IL-1 can directly cause release of ACTH from corticotroph cells dispersed from the interior pituitary glands of rats (Bernton et al. 1987). Sapolski et al. (1987) and Berkenbosch et al. (1987) have found no such ACTH-releasing activity of IL-1 on pituitary cells in vitro. Recent demonstrations by Breeder et al. (1987) that IL-1 fibres can be shown to exist in the brain by immunohistochemistry indicates that IL-1 is probably effective in vivo in raising ACTH and subsequently glucocorticoid levels during inflammation. The result of the raised glucocorticoid is that in synergistic action with IL-1 and IL-6, the acute phase proteins are induced and conversely, in an inhibitory fashion, glucocorticoid inhibits the action of IL-1 on the induction of other cytokines such as IL-6 from fibroblasts (Kohase et al. 1987). Moreover, glucocorticoid inhibits the production of IL-1 from monocytes (Kern et al. 1988; Young et al. 1988) and can also inhibit the action of IL-1 on T-cell activation (Tracey et al. 1988). These in vitro experiments are consistent with recent in vivo data showing that administration of recombinant IL-1 causes a three-fold increase in serum corticosterone levels which peaked approximately 3 hours after IL-1 administration. Thus, there is evidence for IL-1 stimulation of glucocorticoid production, glucocorticoid synergism with IL-1 action, glucocorticoid inhibition of IL-1 action and glucocorticoid inhibition of IL-1 production. While most of these actions can fit the proposed scheme of inflammatory reaction, recent work by Akahoshi et al. (1988) showing the induction of high affinity IL-1 receptors on human lymphocytes by exposure to glucocorticoids indicates that the final pathways of interaction between glucocorticoid and IL-1 may indeed be much more complex than so far documented. Nonetheless, there is solid evidence that IL-1 both modifies and interacts with glucocorticoid in the induction and modulation of the acute phase response.

In Vivo Action of IL-1

Considering the fact that IL-1 is active in vitro in the 1 pM–1nM range (17 pg–17ng ml^{-1}), it is not surprising that difficulties are encountered in detecting biologically active serum levels of IL-1 during inflammation. In addition, there have been several inhibitors of IL-1 (LAF) activity shown in plasma (Oppenheim et al. 1986). Only with the development of sensitive immunoassays will the real fluctuation in plasma levels be detected. Such findings are beginning to be reported for synovial fluid and serum in acute and chronic disorders (Cannon et al. 1988; Reuter et al. 1988; Duff et al. 1988).

There have been many postulates that IL-1 is involved in a number of the acute and chronic inflammatory disorders (Miossec 1987). However, apart from most of the in vitro data which supports the hypothesis that IL-1 is an inflammatory mediator, there are only a few recent experiments implicating IL-1 in inflammation and in haemopoiesis. The first documents the use of rIL-1 in the modulation of bone marrow cells which appear to be stimulated into cell cycle and this stimulation may be responsible for the reconstitution of haemopoietic and immune systems after sublethal radiation (Neta et al. 1987; Morrissey et al. 1988; Castelli et al. 1988). Increased survival may be due to the release of colony-stimulating factors which may help to minimize the myelosuppressive effects of DNA-damaging agents such as cyclophosphamide or γ-irradiation. The second involves the demonstration that recombinant IL-1 can be used to induce the local Shwartzman reaction and in addition is also pro-inflammatory in a chronic way as it accelerates the development of collagen-induced arthritis in mice (Beck et al. 1986; Hom et al. 1988). In contrast to these pro-inflammatory responses to administering IL-1 in vivo is recent work by Sauder and colleagues using topically administered IL-1 to improve dermal wound healing (Mertz et al. 1988). Thus, it appears that like many of the in vitro findings, there will be opposite and contrary data on the outcome of utilization of IL-1 in vivo; protective and reparative processes may be enhanced, while inflammatory and destructive processes may be worsened. It is obvious that the complex nature of how IL-1 interacts in the acute phase response is manifest in the many other systemic actions initiated by IL-1.

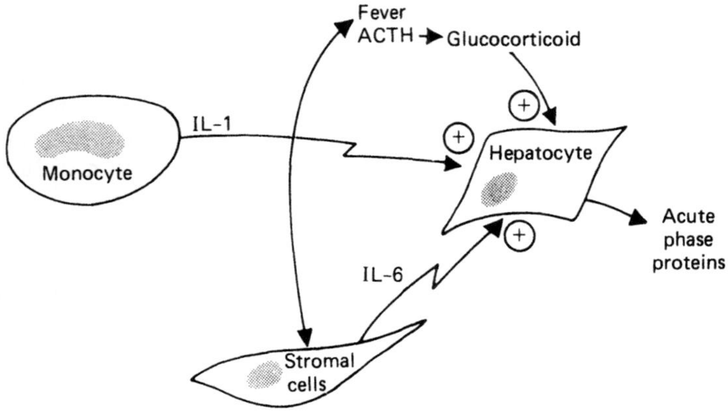

Fig. 1.1. IL-1 in acute phase response.

Summary

While IL-1 is one of the most potent immunoregulatory molecules produced by macrophages in the initiation of the acute phase response, its role should be seen as that of "alarm" cytokine. It initiates many of the immediate inflammatory processes, as well as participating in some of the repair processes. It stimulates the release of many other physiological cytokines such as growth factors and interleukins, some of which appear to have potent anti-inflammatory activities. Figure 1.1 summarizes the ways in which IL-1 is involved in the acute phase response and the induction of acute phase proteins from the liver. The development of new immunoassays and documentation of altered levels of IL-1 in the circulation will lead to a better understanding of the fluctuations and changes in the levels of this hormone throughout and immediately after an inflammatory response. It is obvious that the molecule does not work alone, rather that it participates in the overall orchestration of many cytokine messages and it will be important to examine these cytokine networks both in vivo and in vitro with the aid of newly available recombinant molecules before we can fully understand the role of IL-1 and its involvement in the acute phase response.

Acknowledgements. Work carried out in the author's laboratory was supported by the Medical Research Council of Canada. Helpful discussion and collaboration with Drs Baumann, Fey and Koj are gratefully acknowledged, as is the secretarial assistance of Janice Butera and Joan Hickey.

References

Akahoshi T, Oppenheim JJ, Matsushima K (1988) Induction of high-affinity IL-1 receptors on human pBL by glucocorticoid hormones. J Exp Med 167:924–936

Andus T, Geiger T, Hirano T, Kishimoto T, Tran-Thi TA, Kecker K, Heinrich PC (1988) Regulation of synthesis and secretion of major rat acute-phase proteins by recombinant human interleukin-6 (BSF-2/IL-6) in hepatocyte primary cultures. Eur J Biochem 173:287–293

Auron PE, Webb AC, Rosenwasser LJ, Mucci SF, Rich A, Wolff SM, Dinarello CA (1984) Nucleotide sequence of human monocyte interleukin-1 precursor cDNA. Proc Natl Acad Sci USA 81:7907–7911

Baumann H, Richards C, Gauldie J (1987a) Interaction among hepatocyte-stimulating factors, interleukin-1, and glucocorticoids for regulation of acute phase plasma proteins in human hepatoma (HepG2) cells. J Immunol 139:4122–4128

Baumann H, Onorato V, Gauldie J, Jahreis GP (1987b) Distinct sets of acute phase plasma proteins are stimulated by separate human hepatocyte-stimulating factors and monokines in rat hepatoma cells. J Biol Chem 262:9756–9768

Baumann H, Isseroff H, Latimer JJ, Jahreis GP (1988) Phorbol ester modulates interleukin-6 and interleukin-1 — regulated expression of acute phase plasma proteins in hepatoma cells. J Biol Chem 263: 17390–17396

Beck G, Habicht GS, Benach JL, Miller F (1986) Interleukin-1: a common endogenous mediator of inflammation and the local Shwartzman reaction. J Immunol 136:3025–3031

Bendtzen K, Mandrup-Poulsen T, Nerup J, Nielsen JH, Dinarello CA, Svenson M (1986) Cytotoxicity of human pI 7 interleukin-1 for pancreatic islets of Langerhans. Science 232:1545–1547

Berkenbosch F, vanOers J, DelRey A, Tilders F, Besedovsky H (1987) Corticotropin-releasing factor-producing neurons in the rat activated by interleukin-1. Science 238:524

Bernaudin JF, Yamauchi K, Wewers MD, Tocci MJ, Ferrans VJ, Crystal RG (1988) Demonstration by in situ hybridization of dissimilar IL-1β gene expression in human alveolar macrophages and blood monocytes in response to lipopolysaccharide. J Immunol 140:3822–3829

Bernton EW, Beach JE, Holaday JW, Smallridge R, Fein HG (1987) Release of multiple hormones by a direct action of interleukin-1 on pituitary cells. Science 238:652

Beuscher HU, Fallon RJ, Colten HR (1987) Macrophage membrane interleukin-1 regulates the expression of acute phase proteins in human hepatoma Hep 3B cells. J Immunol 139:1896–1901

Bird TA, Saklatvala J (1986) Identification of a common class of high affinity receptors for both types of porcine interleukin-1 on connective tissue cells. Nature 324:263–266

Bird TA, Saklatvala J (1987) Studies on the fate of receptor-bound ^{125}-I-interleukin-1β in porcine synovial fibroblasts. J Immunol 139:92–97

Bird TA, Gearing AJH, Saklatvala J (1988) Murine interleukin-1 receptor — Direct identification by ligand blotting and purification to homogeneity of an interleukin-1-binding glycoprotein. J Biol Chem 263:12063–12069

Bornstein DL (1982) Leukocytic pyrogen: a major mediator of the acute phase reaction. Ann NY Acad Sci 389:323–337

Breder CD, Dinarello CA, Saper CB (1987) Interleukin-1 immunoreactive innervation of the human hypothalamus. Science 240:321–324

Broxmeyer HE, Williams DE, Hagoc G, Cooper S, Gillis S, Shadduck RK, Bicknell DC (1987) Synergistic myelopoietic actions in vivo after administration to mice of combinations of purified natural murine colony-stimulating factor 1, recombinant murine interleukin-3, and recombinant murine granulocyte/macrophage colony-stimulating factor. Proc Natl Acad Sci USA 84:3871

Cannon JG, van der Meer JWM, Endres S, Lonnemann G, Dinarello CA (1988) Interleukin-1 in human blood: Chloroform extraction and radioimmunoassay. In: Powanda MC, Oppenheim JJ, Kluger MJ, Dinarello CA (eds) Monokines and other non-lymphocytic cytokines. Allan R. Liss, New York, pp 373–376

Castelli MP, Black PL, Schneider M, Pennington R, Abe F, Talmadge JE (1988) Protective, restorative and therapeutic properties of recombinant human IL-1 in rodent models. J Immunol 140:3830–3837

Ciliberto G, Arcone R, Wagner EF, Ruther U (1987) Inducible and tissue-specific expression of human C-reactive protein in transgenic mice. EMBO J 6:4017–4022

Clark BD, Collins KL, Gandy MS, Webb AC, Auron PE (1986) Genomic sequence for human prointerleukin 1 beta: possible evolution from a reverse transcribed prointerleukin 1 alpha gene. Nucl Acids Res 14:7897–7914

Clark BD, Fenton MJ, Rey HL, Webb AC, Auron PE (1988) Characterization of *cis* and *trans* acting elements involved in human proIL-1β gene expression. In: Powanda MC, Oppenheim JJ, Kluger MJ, Dinarello CA (eds) Progress in leukocyte biology, vol 6. Alan R. Liss, New York, pp 47–53

Courtois G, Morgan JG, Campbell LA, Fourel G, Crabtree GR (1987) Interactive of a liver-specific nuclear factor with the fibrinogen and α-1-antitrypsin promoters. Science 238:688–692

Cybulsky MI, Colditz IG, Movat HZ (1986) The role of interleukin-1 in neutrophil leukocyte emigration induced by endotoxin. Am J Pathol 124:367–372
Darlington GJ, Wilson DR, Lachman LB (1986) Monocyte-conditioned medium, interleukin-1, and tumor necrosis factor stimulate the acute phase response in human hepatoma cells in vitro. J Cell Biol 103:787–793
Dente L, Ruther U, Tripodi M, Wagner EF, Cortese R (1987) Expression of human α 1-acid glycoprotein genes in cultured cells and in transgenic mice. Genes Dev pp 259–266
De Simone V, Ciliberto G, Hardon E, Paonessa G, Palla F, Lundberg L, Cortese R (1987) *Cis*- and *trans*-acting elements responsible for the cell-specific expression of the human α 1-antitrypsin gene. EMBO J 67:2759–2766
Dinarello CA (1984) Interleukin-1. Rev Infect Dis 6:51–95
Dinarello CA (1985) An update on human interleukin-1: from molecular biology to clinical relevance. J Clin Immunol 5:287–297
Dinarello CA (1987) The biology of interleukin-1 and comparison to tumor necrosis factor. Immunol Lett 16:227–232
Dinarello CA, Mier JW (1987) Lymphokines. N Engl J Med 317:940–945
Dinarello CA, Ikejima T, Warner SJC, Orencole SF, Lonnemann G, Cannon JG, Libby P (1987) Interleukin-1 induces interleukin-1: Induction of circulating interleukin-1 in rabbits in vivo and in human mononuclear cell in vitro. J Immunol 139:1902–1910
Dower SK, Kronheim SR, Hopp TP, Cantrell M, Deeley M, Gillis S, Henney CS, Urdal DL (1986) The cell surface receptors for interleukin-1α and interleukin-1β are identical. Nature 324:266–268
Duff GW, Durum SK (1982) Fever and immunoregulation: hyperthermia, interleukin-1 and -2 and T-cell proliferation. Yale J Biol Med 55:437–446
Duff GW, Dickens E, Wood N, Manson J, Symons J, Poole S, di Giovine F (1988) Immunoassay, bioassay and in situ hybridization of monokines in human arthritis. In: Powanda MC, Oppenheim JJ, Kluger MJ, Dinarello CA (eds) Monokines and other non-lymphocytic cytokines. Alan R. Liss, New York, pp 387–392
Dunn CJ, Hardee MM, Gibbons AJ, Staite ND, Richard KA (1988) Interleukin-1 induces chronic granulomatous inflammation. In: Powanda MC, Oppenheim JJ, Kluger MJ, Dinarello CA (eds) Progress in leukocyte biology, vol 6. Alan R. Liss, New York, pp 329–335
Dustin ML, Staunton DE, Springer TA (1988) Supergene families meet in the immune system. Immunol Today 9:213–215
Evans E, Courtois GM, Kilian PL, Fuller GM, Crabtree GR (1987) Induction of fibrinogen and a subset of acute phase response genes involves a novel monokine which is mimicked by phorbol esters. J Biol Chem 262:10850–10854
Fenton MT, Clark BD, Collins KL, Webb AC, Rich A, Auron PE (1987) Transcriptional regulation of the human pro-interleukin-1β gene. J Immunol 138:3972–3979
Fey GH, Gauldie J (1989) The acute phase response of the liver in inflammation. In: Popper H, Schaffner F (eds) Progress in liver diseases, vol 9. WB Saunders (in press)
Fey GH, Fuller GM (1987) Regulation of acute phase gene expression by inflammatory mediators. Mol Biol Med 4:323–338
Fontana A, Kristensen F, Dubs R, Gemsa D, Weber E (1982) Production of PGE and an interleukin-1-like factor by cultured astrocytes and C_6 glioma cells. J Immunol 129:2413–2419
Furutani Y, Notake M, Fukui T, Ohue M, Nomura H, Yamada M, Nakamura S (1986) Complete nucleotide sequence of the gene for human interleukin-1 alpha. Nucl Acids Res 14: 3167–3179
Ganapathi MK, Schultz D, Mackiewicz A, Samols D, Hu S-I, Brabenec A, MacIntyre SS, Kushner I (1988) Heterogeneous nature of the acute phase response. Differential regulation of human serum amyloid A, C-reactive protein, and other acute phase proteins by cytokines in Hep 3B cells. J Immunol 141:564–569
Gauldie J, Lamontagne L, Stadnyk A (1985) Acute phase response in infectious disease. Surv Synth Pathol Res 4:126–151
Gauldie J, Richards C, Harnish D, Lansdorp P, Baumann H (1987a) Inteferon β2/B-cell stimulatory factor type 2 shares identity with monocyte-derived hepatocyte-stimulating factor and regulates the major acute phase protein response in liver cells. Proc Natl Acad Sci USA 84:7251–7255
Gauldie J, Sauder DN, McAdam KPWJ, Dinarello CA (1987b) Purified interleukin-1 (IL-1) from human monocytes stimulates acute-phase protein synthesis by rodent hepatocytes in vitro. Immunol 60:203–207
Gauldie J, Richards C, Harnish D, Baumann H (1988) Interferon β-2 is identical to monocytic HSF and regulates the full acute phase protein response in liver cells. In: Powanda MC, Oppenheim JJ, Kluger MJ, Dinarello CA (eds) Monokines and other non-lymphocytic cytokines. Alan R. Liss, New York, pp 15–20

Geiger T, Andus T, Klapproth J, Northof H, Heinrich PC (1988) Induction of α 1-acid glycoprotein by recombinant human interleukin-1 rat hepatoma cells. J Biol Chem 263:7141–7146
Ghezzi P, Dinarello CA (1988) IL-1 induces IL-1 III. Specific inhibition of IL-1 production by IFN-τ^1. J Immunol 140:4238–4244
Gowen M (1988) Actions of Il-1 and TNF on human osteoblast-like cells: similarities and synergism. In: Powanda MC, Oppenheim JJ, Kluger MJ, Dinarello CA (eds) Monokines and other non-lymphocytic cytokines. Alan R. Liss, New York, pp 261–266
Granstein RD, Margolis R, Mizel SB, Sauder DN (1985) In vivo inflammatory activity of epidermal cell-derived thymocyte activating factor and recombinant interleukin-1 in the mouse. J Clin Invest 77:1020
Gray PW, Glaister D, Chen E, Goeddel DV, Pennica D (1986) Two interleukin 11 genes in the mouse: Cloning and expression of the cDNA for murine interleukin 1β. J Immunol 137:3644–3648
Grayson DR, Costa RH, Xanthopoulos KG, Darnell JE Jr (1988) A cell-specific enhancer of the mouse α 1-antitrypsin gene has multiple functional regions and corresponding protein-binding sites. Mol Cell Biol 8:1055–1066
Hagiwara H, Huang HS, Arai N et al. (1987) Interleukin-1 modulates RNA levels of lymphokines and of other molecules associated with T cell activation in the T cell lymphoma LBRM 33-1A5. J Immunol 138:2514–2519
Hansen DF, Murphy PA, Silicano R, Shin HS (1983) The effect of temperature on the activation of thymocytes by interleukin 1 and 2. J Immunol 130:216–221
Helle M, Brakenhoff JPJ, DeGroot ER, Aarden LA (1988) Interleukin 6 is involved in interleukin 1-induced activities. Eur J Immunol 18:957–959
Herrmann F, Oster W, Meuer SC, Klein K, Lindemann A, Mertelsmann R (1988) IL-1 stimulates T-lymphocytes to produce GM-CSF. J Clin Invest 81:1415–1418
Hill MR, Stith RD, McCallum RE (1988) Human recombinant IL-1 alters glucocorticoid receptor function in reuber hepatoma cells. J Immunol 141:1522–1528
Hom JT, Bendele AM, Carlson DG (1988) In vivo administration with IL-1 accelerates the development of collagen-induced arthritis in mice. J Immunol 141:834–841
Hooper T, Andus T, Klapproth J, Northoff H, Heinrich PC (1988) Induction of alpha 1-acid glycoprotein by recombinant human interleukin-1 rat hepatoma cells. J Biol Chem 263:7141–7146
Horuk R, Huang JJ, Covington M, Newton RC (1988) Evidence for differences in the molecular properties of interleukin-1 receptors. In: Powanda MC, Oppenheim JJ, Kluger MJ, Dinarello CA (eds) Monokines and other non-lymphocytic cytokines. Alan R. Liss, New York, pp 179–184
Houssiau FA, Coulie PG, Olive D, Van Snick J (1988) Synergistic activation of human T cells by interleukin-1 and interleukin-6. Eur J Immunol 18:653–665
Kampschmidt RF, Mesecher M (1985) Interleukin-1 from P388D: effects upon neutrophils, plasma iron, and fibrinogen in rats, mice, and rabbits. Proc Soc Exp Biol Med 179:197–200
Kampschmidt RF, Upchurch HF, Pulliam LA (1982) Characterization of a leukocyte-derived endogenous mediator responsible for increased plasma fibrinogen. Ann NY Acad Sci 389:338–353
Kaushansky K, Lin N, Adamson JW (1988) Interleukin-1 stimulates fibroblasts to synthesize granulocyte-macrophage and granulocyte colony-stimulating factors. J Clin Invest 81:92–97
Kern JA, Lamb RJ, Reed JC, Daniele RP, Nowell PC (1988) Dexamethasone inhibition of IL-1β production by human monocytes post-transcriptional mechanisms. J Clin Invest 8:237–244
Kimball ES, Vaught JL, Persico FJ, Fisher MC (1988) Generation of interleukin-1 in a macrophage cell line and enhancement of interleukin-1 fibroblast proliferative activity by substance P. In: Powanda MC, Oppenheim JJ, Kluger MJ, Dinarello CA (eds) Monokines and other non-lymphocytic cytokines. Alan R. Liss, New York, pp 131–126
Kohase M, May LT, Tamm I, Vilcek J, Sehgal PB (1987) A cytokine network in human kiploid fibroblasts: interactions of beta interferons, tumor necrosis factor, platelet-derived growth factor and interleukin-1. Mol Cell Biol 7:273–280
Koj A (1986) Biological functions of acute phase proteins and the cytokines involved in their induced synthesis. In: Reutter W, Popper H, et al. (eds) Modulation of liver cell expression, Falk Symposium 43. MTP Press, Lancaster, pp 331–342
Koj A, Kurdowska A, Magielska-Zero D, Rokita H, Sipe JD, Dayer JM, Denczuk S, Gauldie J (1987) Limited effects of recombinant human and murine interleukin-1 and tumour necrosis factor on production of acute phase proteins by cultured rat hepatocytes. Biochem Int 14:553–560
Kushner I (1982) The phenomenon of the acute phase response. Ann NY Acad Sci 389:39–48
Le PT, Mortensen RF (1984) In vitro induction of hepatocyte synthesis of the acute phase reactant mouse serum amyloid P-component by macrophages and IL-1. J Leuk Biol 35:587–603

Lindemann A, Riedel D, Oster W, Meuer SC, Blohm D, Mertelsmann RH, Herrmann F (1988) Granulocyte/macrophage colony-stimulating factor induces interleukin-1 production by human polymorphonuclear neutrophils. J Immunol 140:837–839

Lomedico PT, Gubler V, Hellmann CP, Dukovich M, Giri JG, Pan Y-CE, Collier K, Semionow R, Chua AO, Mizel SB (1984) Cloning and expression of murine interleukin-1 cDNA in *Escherichia coli*. Nature 312:458–461

Mandrup-Poulsen T, Bendtzen K, Dinarello CA, Nerup J (1988) Potentiating effects of tumor necrosis factor on interleukin-1 mediated pancreatic β-cell toxicity. In: Powanda MC, Oppenheim JJ, Kluger MJ, Dinarello CA (eds) Monokines and other non-lymphocytic cytokines. Alan R. Liss, New York, pp 291–295

March CJ, Mosley B, Larsen A, Cerretti DP, Braedt G, Price V, Gillis S, Henney CS, Kronheim SR, Grabstein K, Conlon PJ, Hopp TP, Cosman D (1985) Cloning, sequence and expression of two distinct human interleukin-1 complementary DNAs. Nature 315:641–647

Marinkovic S, Jahreis GP, Wong GG, Baumann H (1989) Interleukin-6 modulates the synthesis of a specific set of acute phase plasma proteins in vivo. J Immunol 142:808–812

Matsuda T, Hirano T, Nagasawa S et al. (1988) α_2-macroglobulin as a carrier protein for IL-6. Abstract No. 414408, 61st congress of the Japanese Biochemical Society, Tokyo, October 1988. Abstract Volume p 915

Matsushima K, Shiroo M, Lew W, Kobayashi Y, Akahoshi T, Oppenheim JJ (1988) Detection of protein phosphorylation in human peripheral blood mononuclear cells in response to interleukin-1 (IL-1). In: Powanda MC, Oppenheim JJ, Kluger MJ, Dinarello CA (eds) Monokines and other non-lymphocytic cytokines. Alan R. Liss, New York, pp 229–234

McAdam KPWJ, Li J, Knowles J, Foss NT, Dinarello CA, Rosenwasser LJ, Selinger MJ, Kaplan MM and Goodman R (1982) The biology of SAA: identification of the inducer, in vitro synthesis, and heterogeneity demonstrated with monoclonal antibodies. Ann NY Acad Sci 389:126–136

Mertz PM, Davis SC, Kilian P, Sauder DN (1988) The effect of interleukin-1 on the epidermal healing rate of partial thickness wounds. Clin Res 36:378A

Miossec P (1987) The role of interleukin-1 in the pathogenesis of rheumatoid arthritis. Clin Exp Rheumatol 5:305–308

Moreb J, Zucali J, Gross MA, Dinarello CA (1988) Influence of interleukin-1 on human bone marrow blast progenitor cells in vitro. In: Powanda MC, Oppenheim JJ, Kluger MJ, Dinarello CA (eds) Monokines and other non-lymphocytic cytokines. Alan R. Liss, New York, pp 297–300

Morrissey P, Charrier K, Bressler L, Alpert A (1988) The influence of IL-1 treatment on the reconstitution of the hemopoietic and immune systems after sublethal radiation. J Immunol 140:4204–4210

Morrone G, Ciliberto G, Oliviero S, Arcone R, Dente L, Content J, Cortese R (1988) Recombinant interleukin 6 regulates the transcriptional activation of a set of human acute phase genes. J Biol Chem 263:12554–12558

Neta R (1988) Synergistic activity of IL-1 with TNF and IL-1 with CSF in radioprotection. In: Powanda MC, Oppenheim JJ, Kluger MJ, Dinarello CA (eds) Monokines and other non-lymphocytic cytokines. Alan R. Liss, New York, pp 359–364

Neta R, Sztein MB, Oppenheim JJ, Gillis S, Douches SD (1987) The in vivo effects of interleukin-1 I. Bone marrow cells are induced to cycle after administration of interleukin-1. J Immunol 139:1861–1866

Nishida T, Nishino N, Takano M, Kawai K, Bando K, Masui Y, Nakai S, Hirai Y (1987) cDNA cloning of IL-1α and IL-1β from mRNA of U937 cell line. Biochem Biophys Res Commun 143:345–352

O'Connor-McCourt MD, Wakefield LM (1987) Latent transforming growth factor β in serum: a specific complex with α_2-macroglobulin. J Biol Chem 262:14090–14099

Oliviero S, Morrone G, Cortese R (1987) The human haptoglobin gene: transcriptional regulation during development and acute phase induction. EMBO J 6:1905–1912

Oppenheim JJ, Kovacs EJ, Matsushima K, Durum SK (1986) There is more than one interleukin-1. Immunol Today 7:45–55

Perlmutter DH, Dinarello CA, Punsal PI, Colten HR (1986a) Cachetin/tumor necrosis factor regulates hepatic acute phase gene expression. J Clin Invest 78:1349

Perlmutter DH, Strunk RC, Goldberger G, Cole FS (1986b) Regulation of complement proteins C2 and factor B by interleukin-1 and interferon-gamma acting on transfected L cells. Mol Immunol 23:1263–1266

Philip R, Epstein LB (1986) Tumour necrosis factor as immunomodulatory and mediator of monocyte cytotoxicity induced by itself, τ-interferon and interleukin-1. Nature 323:86

Prowse KR, Baumann H (1988) Hepatocyte-stimulating factor, $\beta 2$ interferon, and interleukin-1 enhance expression of the rat alpha 1-acid glycoprotein gene via a distal upstream regulatory region. Mol Cell Biol 8:a–i

Prowse KR, Baumann H (1989) Interleukin-1 and interleukin-6 stimulate acute phase protein production in primary mouse hepatocytes. J Leuk Biol 45:55–61

Ramadori G, Sipe JD, Dinarello CA, Mizel SB, Colten HR (1985) Pretranslational modulation of acute phase hepatic protein synthesis by murine recombinant interleukin-1 (IL-1) and purified human IL-1. J Exp Med 162:930–942

Reuter A, Bernier J, Gysen P, Gevaert Y, Gathy R, Lopez M, Dupont G, Damas P, Franchimont P (1988) A RIA for tumor necrosis factor (TNFα) and interleukin 1β (IL-1β) and their direct determination in serum. In: Powanda MC, Oppenheim JJ, Kluger MJ, Dinarello CA (eds) Monokines and other non-lymphocytic cytokines. Alan R. Liss, New York, pp 377–381

Ritchie DG, Fuller GM (1983) Hepatocyte-stimulating factor: a monocyte-derived acute-phase regulatory protein. Ann NY Acad Sci 408:490–502

Rosoff PM, Savage N, Dinarello CA (1988) Interleukin-1 stimulates diacylglycerol production in T lymphocytes by a novel mechanism. Cell 54:73–81

Rothlein R, Czajkowski M, O'Neill MM, Marlin SD, Mainolfi E, Merluzzi VJ (1988) Induction of intercellular adhesion molecule 1 on primary and continuous cell lines by pro-inflammatory cytokines—regulation by pharmacologic agents and neutralizing antibodies. J Immunol 141:1665–1669

Ruggiero V, Baglioni C (1987) Synergistic anti-proliferative activity of interleukin-1 and tumor necrosis factor. J Immunol 138:661–663

Sapolsky R, Rivier C, Yamamoto G, Plotsky P, Vale W (1987) Interleukin-1 stimulates the secretion of hypothalamic corticotropin releasing factor. Science 238:522

Sayers TJ, Wiltrout TA, Bull CA, Denn AC III, Pilaro AM, Lokesh B (1988) Effect of cytokines on polymorphonuclear neutrophil infiltration in the mouse. J Immunol 141:1670–1677

Sehgal PB, May LT, Tamm I, Vilcek J (1987) Human β_2 interferon and B-cell differentiation factor BSF-2 are identical. Science 235:731–732

Selinger MJ, McAdam KPWJ, Kaplan MM, Sipe JD, Vogel SN and Rosenstreich DL (1980) Monokine-induced synthesis of serum amyloid A protein by hepatocytes. Nature 285:498–500

Sieff CA, Tsai S, Faller DV (1987) Interleukin-1 induces cultured human endothelial cell production of granulocyte-macrophage colony-stimulating factor. J Clin Invest 79:48–51

Sifers RN, Carlson JA, Clift SM, DeMayo FJ, Bullock DW, Woo SLC (1987) Tissue specific expression of the human $\alpha 1$ antitrypsin gene in transgenic mice. Nucl Acids Res 15:1459–1475

Sims JE, March CJ, Cosman D, Widmer MB, MacDonald HR, McMaham CJ, Grubin CE, Wignall JM, Jackson JL, Call SM, Friend D, Alpert AR, Gillis S, Urdal DL, Dower SK (1988) cDNA expression of cloning of the IL-1 receptor, a member of the immunoglobulin superfamily. Science 241:585–589

Sipe JD, Ramadori G (1986) In: Marrink J, Van Rijswijk MH (eds) Amyloidosis. Martinus Nijhoff, Dordrecht, pp 319–328

Sipe JD, Vogel SN, Ryan JL, McAdam KPWJ, Rosenstreich DL (1979) Detection of a mediator derived from endotoxin-stimulated macrophages that induce the acute phase serum amyloid A response in mice. J Exp Med 150:597–606

Sipe JD, Vogel SN, Sztein MB, Skinner M, Cohen AS (1982) The role of interleukin-1 in acute phase serum amyloid A (SAA) and serum amyloid P (SAP) biosynthesis. Ann NY Acad Sci 389:137–150

Sipe JD, Johns MA, Ghezzi P, Knapschaefer G (1989) Modulation of serum amyloid A gene expression by cytokines and bacterial cell wall components. In: Malmendier CL (ed) Eicosanoids, apolipoproteins, lipoprotein particles, and atherosclerosis. Plenum, New York (in press)

Stashenko P, Dewhirst FE, Peros WJ, Kent RL, Ago JM (1987) Synergistic interactions between interleukin-1, tumor necrosis factor, and lymphotoxin in bone resorption. J Immunol 138:1464–1468

Sztein MB, Luger TA, Oppenheim JJ (1982) An epidermal cell-derived cytokine triggers the in vivo synthesis of serum amyloid A by hepatocytes. J Immunol 129:87–90

Takacs L, Kovacs EJ, Smith MR, Young HA, Durum SK (1988) Detection of IL-1 gene expression by in-situ hybridization histochemistry: tissue localization of IL-1 mRNA in the normal C57BL/6 mouse. In: Powanda MC, Oppenheim JJ, Kluger MJ, Dinarello CA (eds) Monokines and other non-lymphocytic cytokines. Alan R. Liss, New York, pp 89–93

Telford JL, Macchia G, Massone A, Carinci V, Melli M (1986) The murine interleukin 1β gene: structure and evolution. Nucl Acids Res 14:9955–9963

Teodorescu M, Skosey JL, Schlesinger C, Wallman J (1988) Covalent disulfide binding of IL-1 to α_2-macroglobulin. In: Powanda MC, Oppenheim JJ, Kluger MJ, Dinarello CA (eds) Monokines and other non-lymphocytic cytokines. Alan R. Liss, New York, pp 209–212

Tracey DE, Hardee, MM, Richard KA, Paslay JW (1988) Pharmacological inhibition of IL-1 activity on T-cells by hydrocortisone, cyclosporin, prostaglandins and cyclic nucleotides. Immunopharmacol 15:47–52

Turchik JB, Bornstein DL (1980) Role of the central nervous system in acute-phase responses to leukocytic pyrogen. Infect Immun 30:439–444

Urdal DL, Call SM, Jackson JL, Dower SK (1988) Affinity purification and chemical analysis of the interleukin-1 receptor. J Biol Chem 263:2870–2877

Vlassara H, Brownlee M, Manogue K, Pasagian A, Dinarello CA, Cerami A (1988) Cachectin/TNF and IL-1 synthesis and secretion are induced by glucose-modified protein binding to high-affinity macrophage receptor. In: Powanda MC, Oppenheim JJ, Kluger MJ, Dinarello CA (eds) Monokines and other non-lymphocytic cytokines. Alan R. Liss, New York, pp 145–151

Warner SJC, Auger KR, Libby P (1987) Interleukin-1 induces interleukin-1: recombinant human interleukin-1 induces interleukin-1 production by adult human vascular endothelial cells. J Immunol 139:1911–1917

Woloski BMRNJ, Fuller GM (1985) Identification and partial characterization of hepatocyte-stimulating factor from leukemia cell lines: comparison with interleukin-1. Proc Natl Acad Sci 82:1443–1447

Yoshimura T, Matsushima K, Oppenheim JJ, Leonard EJ (1987) Neutrophil chemotactic factor produced by lipopolysaccharide (LPS)-stimulated human blood mononuclear leukocytes: partial characterization and separation from interleukin-1 (IL-1). J Immunol 139:788–793

Young PR, Hazuda DJ, Connor JR, Dalton BJ (1988) Transcription and translation of IL-1α and IL-1β genes in the presence of the glucocorticoid hormone dexamethasone. In: Powanda MC, Oppenheim JJ, Kluger MJ, Dinarello CA (eds) Monokines and other non-lymphocytic cytokines. Alan R. Liss, New York, pp 83–88

Zsebo KM, Yuschenkoff VN, Schiffer S, Chang D, McCall E, Dinarello CA, Brown MA, Altrock B, Bagby GC Jr (1988) Vascular endothelial cells and granulopoiesis: interleukin-1 stimulates release of G-CSF and GM-CSF. Blood 71:99–103

Zucali JR, Dinarello CA, Oblon DJ, Gross MA, Anderson L, Weiner RS (1986) Interleukin-1 stimulates fibroblasts to produce granulocyte-macrophage colony-stimulating activity and prostaglandin E_2. J Clin Invest 77:1857–1863

Chapter 2

Regulation of Human SAA Gene Expression by Cytokines

M. R. Edbrooke and Patricia Woo

Introduction

Serum amyloid A (SAA) protein is one of the major human acute phase reactants. Its serum concentration can increase by a thousandfold in acute and chronic inflammation, similar to C-reactive protein (CRP). The function of SAA, like CRP, is still unknown. However, prolonged elevation of SAA in the serum can lead to a life-threatening complication: reactive amyloidosis.

There are multiple murine SAA genes, consisting of three active genes and one pseudogene, and the regulation of each gene in primary hepatocyte cultures appears to be mainly at the transcriptional level (Lowell et al. 1986a,b). In contrast, the regulation of murine SAA3 expression in a liver cell line appears to be post-transcriptional (Rienhoff and Groudine 1988). Further elucidation of the differential expression of each gene in different tissues may have important implications in the depositions of amyloid fibrils. From Southern analysis of population genomic DNA, it is evident that there are mutiple human SAA genes as well. Results from Sack (1988) and our own data have indicated the presence of at least three genes.

In Vitro Expression of Human SAA

We have previously characterized a human SAA gene, SAA_g9, and have demonstrated that expression of this gene can be induced by both recombinant interleukin-1β (IL-1) and tumor necrosis factor (TNF) in transfected mouse L-cells (Woo et al. 1987). Transfectants were used since the human hepatoma lines

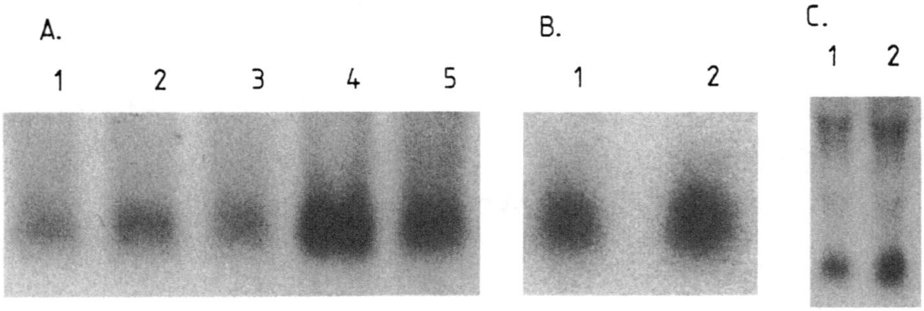

Fig. 2.1. Cytokine control of SAA gene expression. Northern analysis of total RNA isolated from L-cells transfected with human SAA_g9, after treatment with:

A Lane 1: IL-1, 10 ng ml^{-1}; lane 2: IL-1, 10 ng ml^{-1} and γ-IFN, 100 units ml^{-1}; lane 3: γ-IFN, 100 units ml^{-1}; lane 4: IL-1, 10 ng ml^{-1} and γ-IFN, 1000 units ml^{-1}; lane 5: γ-IFN, 1000 units ml^{-1}. Incubation time for all the above experiments was 18 hours.

B Lane 1: saline control; lane 2: IL-6, 10 units ml^{-1} for 24 hours.

C Lane 1: saline control; lane 2: PMA, 50 ng ml^{-1} for 4 hours.

(HepG2 and Hep3B) do not express SAA specific mRNA detectable by Northern analysis. The increase in specific mRNA is at least 50-fold by Northern analysis of total RNA extracted from stimulated transfectants. These results are similar to those described for murine hepatocyte cultures by Ramadori et al. (1985).

In view of the positive results obtained above, we decided to use this in vitro model to observe the effect of other cytokines on the expression of this particular human SAA gene. Figure 2.1 shows that γ-interferon(γ-IFN), recombinant interleukin-6 (IL-6), and the phorbol ester, phorbol 12-myristate 13-acetate (PMA), all produce an increase in specific SAA mRNA. The size of the increase of SAA mRNA is between 5- and 10-fold when the cells are stimulated maximally with IL-6, γ-IFN or PMA. This increase is less than that produced by IL-1 (>50-fold).

Cis-acting Sequences Responsible for PMA Induction of SAA

In order to study the mechanism of the regulation of human SAA gene expression, we have characterized the 5′ flanking region of the gene SAA_g9 by DNA sequence analysis (Fig. 2.2). Using DNA constructs containing 265 bp of the 5′ flanking region of SAA upstream of the reporter gene chloramphenicol acetyl transferase (CAT), we have identified a phorbol ester inducible enhancer region (Fig 2.3) (Edbrooke et al. 1989).

Since PMA is known to induce nuclear DNA-binding proteins, we performed gel retardation assays to see if PMA also induces a nuclear DNA-binding protein in our system (Edbrooke et al. 1989). Figure 2.4a demonstrates the induction of nuclear protein from HeLa cells (known to possess PMA responsive nuclear factor, NFκB), and from transfected L-cells, binding to the 265 bp promoter fragment of SAA. Moreover, footprint analysis (Fig. 2.4b) showed that the DNA

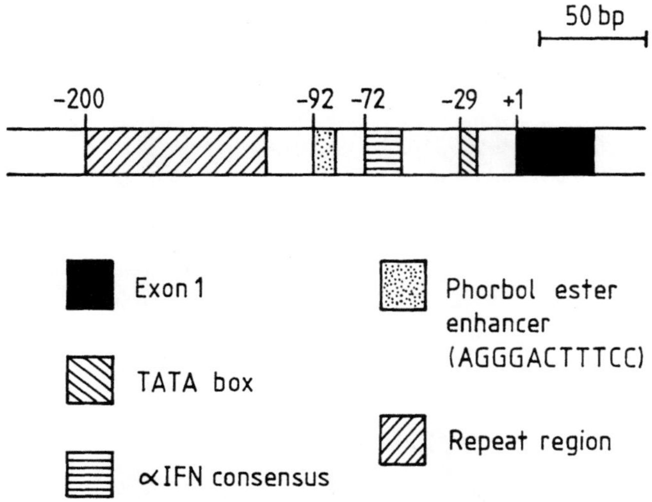

Fig. 2.2. Potential regulatory regions of the SAA_g9 gene. Diagrammatic representation of the 5′ flanking region of SAA_g9, indicating sequence homologies to other 5′ promoter/enhancer sequences.

sequence this nuclear protein binds to is GGGACTTTCC, a sequence previously described by Sen and Baltimore (1986) as the binding site for the nuclear transcription factor, NFκB. Furthermore, we were able to demonstrate that competition with a double-stranded oligonucleotide (containing this sequence) abolished the CAT activity induced by PMA, in co-transfection experiments. Thus this upstream sequence of human SAA_g9 enhances transcription of a heterologous gene, and is regulated by PMA via a transcription factor. Experiments are in progress to see whether PMA will induce transcription of the whole gene via this mechanism.

Cytokine Control of SAA Expression

The acute phase response results in a dramatic change in the expression of a group of serum proteins. These changes can be induced in vivo and in vitro by treatment with individual or combinations of cytokines. Examples include the induction of SAA (Ramadori et al. 1985; Woo et al. 1987), complement Factor B and C3 (Perlmutter et al. 1986), α1-acid glycoprotein (AGP) and haptoglobin by IL-1 (Gauldie et al. 1987), induction of AGP, fibrinogen, α1-antichymotrypsin, haptoglobin, cysteine protease inhibitor and α2-macroglobulin by IL-6 (Gauldie et al. 1987), the induction of SAA (Woo et al. 1987), complement C3, haptoglobin and AGP (Gauldie et al. 1987; Koj et al. 1987) by TNF and induction of complement Factor B and C2 by γ-IFN (Strunk et al. 1985).

5′ *cis*-acting control regions have been identified in a number of cytokine responsive acute phase genes. These include the IL-1 responsive Factor B, AGP, haptoglobin and stromelysin genes (Frisch et al. 1987; Oliviero et al. 1987; Prowse et al. 1988; H. R. Colten, personal communication) and the IL-6

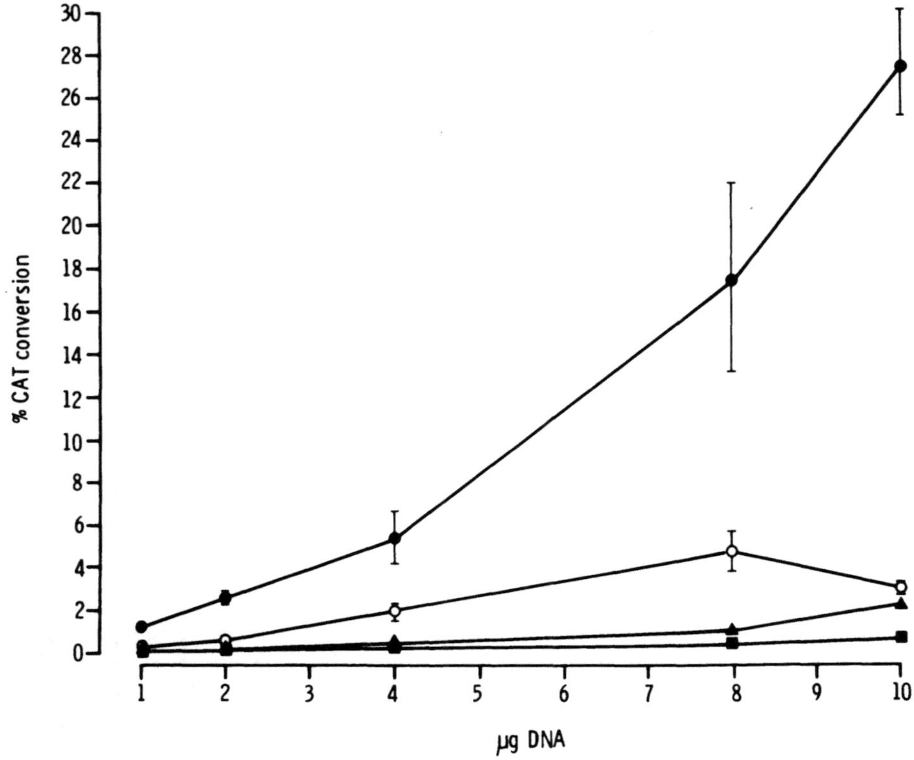

Fig. 2.3. Transcriptional induction of CAT gene activity with phorbol ester by a 5′ DNA fragment of the SAA gene. HeLa cells were transfected with varying amounts of CAT gene DNA (0CAT) or a construct containing 265 bp of the 5′ flanking region of SAA placed upstream of the CAT gene (0CAT/265), and then treated with 50 ng ml^{-1} of PMA for 4 hours. CAT activity was assayed, and the percentage CAT conversion measured. ■ 0CAT, −PMA; ▲ 0CAT, +PMA; ○ 0CAT/265, −PMA; ● 0CAT/265, +PMA.

responsive CRP gene (Arcone et al. 1988). We have characterized the 5′ flanking region of an SAA gene and have identified a very short homology between the IL-1 responsive SAA, stromelysin and Factor B genes. Within the 5′ region, we have identified a phorbol ester responsive enhancer sequence. It is not yet clear which cytokine(s) act via this enhancer region. Preliminary experiments using the construct containing 265 bp of the SAA promoter linked to the CAT gene showed increased CAT activity when transfectants are stimulated by recombinant IL-1, or γ-IFN, i.e. both cytokines can activate transcription via this region. In support of this hypothesis is the observation that PMA can replace IL-1 in lymphocyte comitogenesis assays (Williams et al. 1985). However, our experiments also demonstrate that PMA induces a moderate 5-fold increase in SAA mRNA whereas IL-1 produces an increase of more than 50-fold. Therefore, PMA can only mimic part of the action of IL-1, if at all, in this system.

IL-1 was recently shown to activate a novel mechanism to stimulate production of diacylglycerol (DG) that does not involve increased PI turnover, or binding of IL-1 to its receptor (Rosoff et al. 1988). This pathway involves DG and phosphorylcholine production from phosphatidylcholine (PC). Since PC hydrolysis is also stimulated by phorbol esters, this may explain the similarities between IL-1 and phorbol esters. In contrast, α-, β and γ-fibrinogen, α2-macroglobulin and albumin are inducible by a monokine which was not IL-1 or TNF, but whose activity could be mimicked by phorbol ester (Evans et al. 1987). One explanation of these observations is that PMA has multiple intracellular targets. A number of phorbol ester inducible nuclear proteins have been identified. These include AP-1, AP-2, AP-3, and NFκB. NFκB has yet to be demonstrated in tissues that normally express SAA. Factors in cell-types other than B-, T- and HeLa cells have been identified that possess very similar DNA-binding properties to NFκB, but have higher affinity to slightly different sequences. Moreover, the PMA inducible factor AP-3 is able to bind an identical sequence in SV40, but has slightly different binding characteristics from NFκB (Chiu et al. 1987). Thus, a series of closely related protein factors are able to bind to subtly different 5′ enhancer sequences, and any of these may be the true physiological activator for SAA gene transcription.

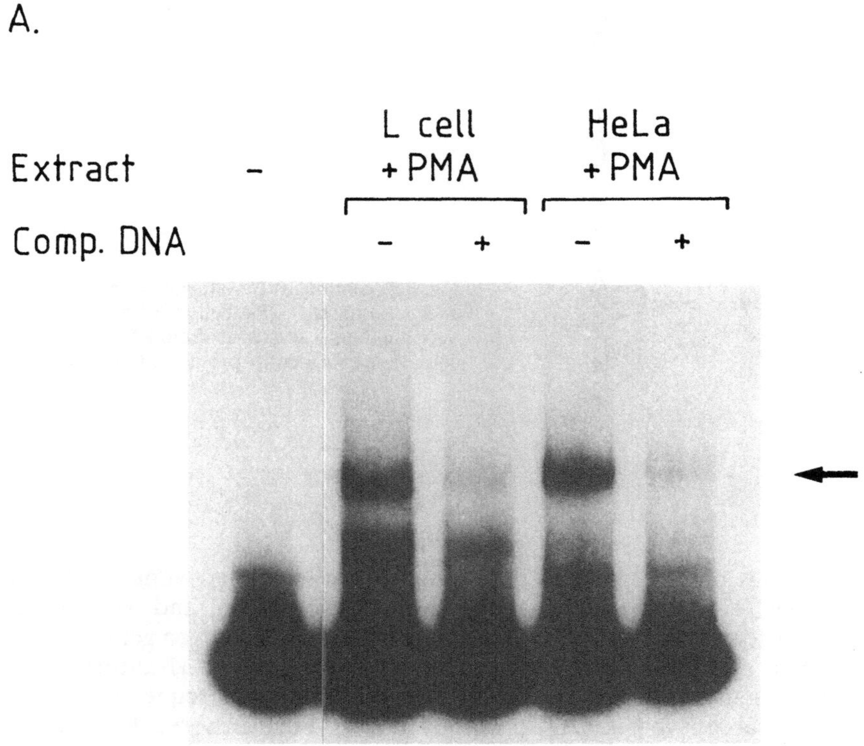

Fig. 2.4. Binding of a PMA induced nuclear transcription factor to the 5′ region of the SAA gene. Nuclear extracts were isolated from HeLa cells and transfected L-cells after PMA treatment (50 ng ml^{-1} for 4 hours. **A** Gel retardation assays were performed using the 265 bp 5′ fragment of SAA_g9, radiolabelled with 32P. Sequence specific binding was demonstrated by competition with 100 ng of a fragment of the HIV-LTR, containing two copies of the sequence GGGACTTTCC, known to bind the transcription factor NFκB. (Figure continued on p. 26.)

B.

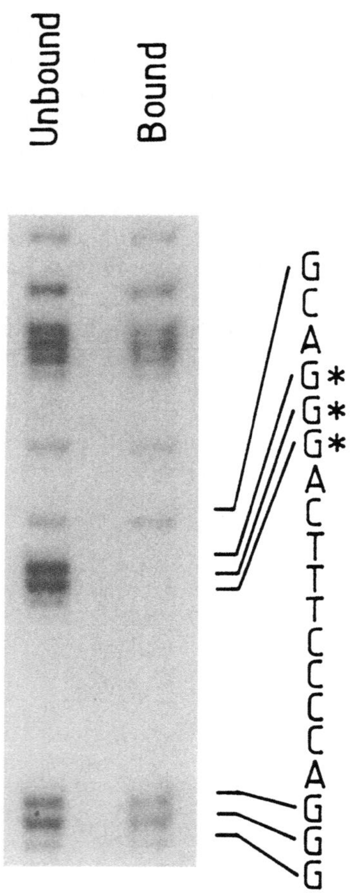

Fig. 2.4 (continued) **B** Methylation interference analysis using the 32P-labelled 265 bp fragment of SAA, and nuclear extracts from PMA treated cells. *Asterisks* identify protected G residues.

Other DNA binding proteins have been shown to activate some other acute phase genes, e.g. HNF1 activates the α- and β-fibrinogen, and α1-antitrypsin genes (Courtois et al. 1987), MLTF activates the rat γ-fibrinogen gene (Chodosh et al. 1987), and LF-A1 and LF-B1 (or HNF1) activate the α1-antitrypsin and haptoglobin genes (Hardon et al. 1988). However, binding sequences for these proteins are absent in the SAA promoter region. Furthermore, the SAA gene does not contain the heat shock consensus sequence found in the promotor regions of CRP and Factor B genes. Thus, a complex pattern of interactions between different, readily activatable nuclear factors and *cis*-acting DNA sequences of different acute phase genes is now emerging. Future comparison of wild-type and mutant SAA genes should test directly the importance of these factors in the physiology of the acute phase response.

References

Arcone R, Gualandi G, Ciliberto G (1988) Identification of sequences responsible for acute-phase induction of human C-reactive protein. Nucl Acids Res 16:3195–3207

Chiu RM, Imagawa M, Imbra RJ, Bockoven JR, Karin M (1987) Multiple *cis*- and *trans*-acting elements mediate the transcriptional response to phorbol esters. Nature 329:648–651

Chodosh LA, Carthew RW, Morgan JG, Crabtree GR, Sharp PA (1987) The adenovirus major late transcription factor activates the rat γ-fibrinogen promoter. Science 238:684–688

Courtois G, Morgan JG, Campbell LA, Fourel G, Crabtree GR (1987) Interaction of a liver-specific nuclear factor with the fibrinogen and α1-antitrypsin promoters. Science 238:688–692

Edbrooke MR, Burt DW, Cheshire JK, Woo P (1989) Identification of *cis*-acting sequences responsible for phorbol ester induction of serum amyloid A gene expression via a nuclear factor κB-like transcription factor. Moll Cell Biol 9:1908–1916

Evans E, Courtois GM, Kilian PL, Fuller GM, Crabtree GR (1987) Induction of fibrinogen and a subset of acute phase response genes involves a novel monokine which is mimicked by phorbol esters. J Biol Chem 262:10850–10854

Frisch SM, Ruley HE (1987) Transcription from the stromelysin promotor is induced by interleukin-1 and repressed by dexamethasone. J Biol Chem 262:16300–16304

Gauldie J, Richards C, Harnish D, Lansdorp P, Baumann H (1987) Interferon β2/B-cell stimulatory factor type 2 shares identity with monocyte-derived hepatocyte-stimulating factor and regulates the major acute phase protein response in liver cells. Proc Natl Acad Sci USA 84:7251–7255

Hardon EM, Frain M, Paonessa G, Cortese R (1988) Two distinct factors interact with the promoter regions of several liver-specific genes. EMBO J 7:1711–1719

Koj A, Kurdowska A, Magielska-Zero D, Rokita H, Sipe JD, Dayer JM, Demczuk S, Gauldie J (1987) Limited effects of recombinant human and murine interleukin I and tumour necrosis factor on production of acute phase proteins by cultured rat hepatocytes. Biochem Int 14:553–560

Lowell CA, Potter DA, Stearman RS, Morrow JF (1986a) Structure of the murine serum amyloid A gene family. J Biol Chem 261:8442–8452

Lowell CA, Stearman RS, Morrow JF (1986b) Transcriptional regulation of serum amyloid A gene expression. J Biol Chem 261:8453–8461

Oliviero S, Morrone G, Cortese R (1987) The human haptoglobin gene: transcriptional regulation during development and acute phase induction. EMBO J 6:1905–1912

Perlmutter DH, Goldberger G, Dinarello CA, Mizel SB, Colten HR (1986) Regulation of class III major histocompatibility complex gene products by interleukin-1. Science 232:850–852

Prowse KR, Baumann H (1988) Hepatocyte-stimulating factor, β2 interferon, and interleukin-1 enhance expression of the rat α1-acid glycoprotein gene via a distal upstream regulatory region. Mol Cell Biol 8:42–51

Ramadori G, Sipe JD, Dinarello CA, Mizel SB, Colten HR (1985) Pretranslational modulation of acute phase hepatic protein synthesis by murine recombinant interleukin 1 (IL-1) and purified human IL-1. J. Exp Med 162:930–942

Rienhoff H Jr, Groudine M (1988) Regulation of amyloid A gene expression in cultured cells. Mol Cell Biol 8:3710–3716

Rosoff PM, Savage N, Dinarello CA (1988) Interleukin-1 stimulates diacylglycerol production in T lymphocytes by a novel mechanism. Cell 54: 73–81

Sack GH Jr (1988) Serum amyloid A (SAA) gene variations in Familial Mediterranean Fever. Mol Biol Med 5:61–67

Sen R, Baltimore D (1986) Inducibility of κ immunoglobulin enhancer-binding protein NF-κB by a post-translational mechanism. Cell 47:921–928

Strunk RC, Sessions Cole F, Perlmutter DH, Colten HR (1985) γ-interferon increases expression of class III complement genes C2 and factor B in human monocytes and in murine fibroblasts transfected with human C2 and factor B genes. J Biol Chem 260:15280–15285

Williams JM, DeLoria D, Hansen JA, Dinarello CA, Loertscher R, Shapiro HM, Strom TB (1985) The events of primary T cell activation can be staged by use of sepharose-bound anti-T3 (64.1) monoclonal antibody and purified interleukin-1. J Immunol 135:2249–2255

Woo P, Sipe J, Dinarello CA, Colten HR (1987) Structure of a human serum amyloid A gene and modulation of its expression in transfected L cells. J Biol Chem 262:15790–15795

Chapter 3

Transcriptional Regulation of Acute Phase Response Genes with Emphasis on the Human C-reactive Protein Gene

G. Ciliberto

Introduction

Acute phase response (APR) is the name given to the complex changes in the concentration of the circulating plasma proteins ensuing from inflammations, infections, stress or traumas. The variation in the level of acute phase reactants is only part of an overall response to local or generalized injury, which includes (i) local events such as vasodilatation; (ii) secondary responses such as neutrophil chemotaxis; and (iii) systemic reactions such as pain, fever, decreased serum zinc and iron concentration, increased serum copper concentration, and other effects, all of which have recently been reviewed (Kushner, 1982; Pepys and Baltz 1983; Koj 1987; Fey and Fuller 1987; Fey and Gauldie 1989).

The acute phase was discovered by Avery's group in the early 1940s (MacLeod and Avery 1941). They demonstrated a constant and substantial increase of C-reactive protein (CRP) during inflammation. Plasma CRP concentration increases rapidly after the onset of inflammation, but also rapidly returns to basal level after the inflammatory stimulus has been removed. Thus the prototype acute phase reactant CRP serves as a clinical marker of inflammation. Indeed, a very large group of plasma proteins, called positive acute phase reactants, are enhanced during the inflammatory process. The increase ranges from 2- to 1000-fold, depending on the protein and on the species. Other plasma proteins, e.g. albumin and lipoproteins, decrease in response to inflammation and are therefore called negative acute phase reactants (Kushner 1982).

Studies on the molecular mechanisms underlying APR indicate that changes in the plasma concentrations of acute phase reactants are caused by changes in their rate of synthesis, and a correlation has been established between the level of protein in the blood and the quantity of the corresponding mRNA in hepatocytes. In the case of the positive reactants, this can be the consequence of increased stability of a pre-existing mRNA or of an increased rate of gene transcription. In most of the cases examined so far, this latter mechanism has been shown to be responsible for the modulation of acute phase gene expression. In this chapter I shall review current knowledge about the regulation of plasma protein expression during inflammation at the transcriptional level, with the emphasis on the human C-reactive protein gene.

The Physiological Role of the Acute Phase Response

The physiological relevance of APR is not yet completely understood. It is generally accepted that local or generalized injuries, whatever the aetiologic agent involved, are responsible for changes in homeostasis. The organism tends to respond to these changes by releasing substances, mainly proteins and hormones, into the plasma. This is how proteins, non-protein hormones and leukocytes are distributed to organs and tissues.

The proteins present in plasma are synthesized mainly in the liver (Putnam 1975). The hepatocytes, therefore, have acquired, during evolution, a key role in the response to inflammation.

The involvement of some of the acute phase reactants in defence reactions against tissue damage is easily explained. For example, proteinase inhibitors like α1-antitrypsin (α1-AT), α1-antichymotrypsin (ACH) α2-macroglobulin (α2-M), α1-macroglobulin (α1-M) and α1-inhibitor III (α1-I3) antagonize the activity of proteinases released from macrophages and granulocytes (Gordon 1976; Esnard and Gauthier 1980; Fey and Fuller 1987). Fibrinogen participates in blood clotting and wound healing (Furie and Furie 1988). Haptoglobin and haemopexin, which are involved in the removal of free globin and haem groups, respectively, are called into play because of the increased haemolysis that follows inflammations, infections and traumas (Bowmann and Kurowsky 1982; Muller-Eberhard and Liem 1974). However, a single function has not been attributed to other acute phase reactants and therefore their precise role in APR is still obscure. The difficulty in assigning a role to these proteins is compounded because they lack naturally occurring mutants. In addition, they are very often the major acute phase reactants, whose concentration increases several-fold: C-reactive protein (CRP), serum amyloid P (SAP), serum amyloid A (SAA), and α1-acid glycoprotein (α1-AGP) (Kushner 1982; Fey and Fuller 1987). Each of these proteins has several activities. For example, pentraxins (CRP and SAP) efficiently bind negatively charged molecules such as phosphorylcholine, a component of the plasma membranes (reviewed by Pepys and Baltz 1983). Thus CRP opsonizes bacteria, parasites and plasma membranes, thereby facilitating their clearance by phagocytic cells. It has been also postulated that these molecules play an important role in the complexation, followed by phagocytosis, to DNA (Robey et al. 1984) released from lysed cells at the site of inflammation. In addition, some of the major acute phase reactants are active as general modulators of the immune response (James 1980; Hudig et al. 1981; Ades et al.

1982; Pepys and Baltz 1983; Robey and Jones 1986). It appears that their function during APR is to exert a feedback control on the activation of monocytes and lymphocytes. In conclusion, APR can be considered a sort of unspecific first-aid defence of the organism against a broad array of disrupting events.

It is important to stress that the APR phenomenon is species-specific (Kushner 1982; Fey and Fuller 1987). The proteins involved differ from species to species. High responsiveness can be observed in one species and no responsiveness in another. For instance, α2-M is a major reactant in the rat (Gordon 1976), but does not participate in the APR in man (Lonberg-Holm et al. 1987). The behaviour of the two members of the pentraxin gene family, SAP and CRP, is more complex and intriguing. The former is highly induced in mouse but not in man, while the latter is the prototype positive reactant in man and rabbit, with levels increased 100- to 1000-fold in several diseases, but only slightly induced in mouse (reviewed by Pepys and Baltz 1983).

Recent work with transgenic mice, in which the human CRP gene has been introduced into the mouse genome, has demonstrated that the information for inducibility is carried by the gene itself and that the transcriptional machinery responsible for tissue-specific and inducible expression in the mouse species efficiently recognizes *cis*-acting regulatory sequences in the human gene (Ciliberto et al. 1987b). The lack of induction of the natural murine CRP therefore depends on the inefficiency of interaction of its promoter or enhancer sequences with otherwise potentially active transcription factors (Ciliberto et al. 1987b).

It is not clear why promoter sequences of the same gene have diverged from one species to another to give different degrees of gene inducibility. It is tempting to speculate that species-specificity in the APR and therefore species-specific differences in promoter sequences are the consequence of selective pressures based on the need for each individual species to present a certain pattern of serum protein production in order to counteract better the effects of stress, trauma or inflammation. Answers to the questions raised in this chapter will come only from genetic systems where it is possible to study the effect of mutations introduced into a specific APR gene, and to substitute a major reactant in one species with that from a different species.

Systems for the Study of the Acute Phase Response

The acute phase response was first studied in whole animals in which experimental inflammation was induced by intraperitoneal or subcutaneous injection of various irritants such as turpentine, bacterial lipopolysaccharide (LPS), croton oil, or by other types of injuries such as burns, ischaemic necrosis, tumour implantation etc. Although these studies gave a comprehensive view of the overall pattern of changes occurring in the plasma level of several proteins, they gave no indication as to the molecular mechanisms responsible. In an attempt to clarify these mechanisms, the effect of the injection of individual monokines has been explored. IL-1 injection has been shown to lead to increased SAA and fibrinogen synthesis (Sipe et al. 1982; Kampschmidt and Mesecher 1985). However, induction by IL-1 might be the result of an indirect effect: this monokine stimulates several cell types which in turn produce factors capable of modulating hepatocyte gene expression (reviewed by Fey and Gauldie 1988).

The advent of recombinant DNA techniques was a breakthrough in the study of the molecular biology of APR, and genes for several plasma proteins from various organisms have been cloned. The acquisition of nucleic acid probes for the detection of specific RNAs resulted in the determination of the steady-state RNA content in cells and tissues before and after the onset of inflammation. As a complementary analysis, nuclear run-on experiments have been performed aimed at determining the participation of transcriptional regulation in the activation or repression of acute phase reactants (Birch and Schreiber 1986; Lowell et al. 1986; Gehring et al. 1987). However, run-on studies can be strongly biased by spurious transcripts arising from upstream promoters and reading through the gene, and therefore run-on studies have recently given way to gene transfer techniques in the study of the transcriptional regulation of gene expression.

The promoter and/or the enhancer region of hepatic genes can be isolated, linked to a suitable test-gene and transferred back into cultured cells of both hepatic and non-hepatic origin (Ott et al. 1984; Ciliberto et al. 1985; D'Onofrio et al. 1985). Gene transfer studies became possible because of the availability of cultured cell lines derived from hepatomas, in particular the two human cell lines hep3B and hepG2 (Knowles et al. 1980) and the rat FAO (Deschatrette and Weiss, 1974; Deschatrette et al. 1979, 1980), in which foreign DNA can be easily introduced.

The other major breakthrough in the study of APR was the technique whereby the APR is reproduced in cultured hepatocytes.

Conditioned medium from monocytes (MoCM) modulates the production of several plasma proteins in primary cultures of mouse and rat hepatocytes (Woloski and Fuller 1985; Baumann et al. 1983a, b; Koj et al. 1984). Explanted hepatocytes are, however, more difficult to cultivate. Their life-span in culture is limited and experiments with them are more subject to variability caused in part by damage to the cells during their isolation and in part by contamination with other cell types. More recently, Darlington et al. (1986) showed that conditioned medium from monocytes stimulated for 48 hours with bacterial endotoxin (LPS) also determines in the human hepatoma cell line hep3B the whole spectrum of changes of plasma protein production typical of the APR. A similar response can be generated by conditioned medium from mouse macrophages (Fuller et al. 1987). MoCM therefore contains one or more factors which can be called hepatocyte stimulating factors (HSFs), which are the modulators of the APR. Similar molecules, at least functionally, are also produced and released from other cells, epidermal cells, fibroblasts and human transformed cell lines (Baumann et al. 1984, 1986, 1987a,b).

Monokines Responsible for the Modulation of Liver-specific Gene Expression

The whole spectrum of changes in liver gene expression observed during inflammation is not caused by the action of a single factor. Monocytes secrete a series of biologically active peptides and proteins that act in nanomolar amounts on a broad range of target cells (Nathan 1987). Among all monokines, those shown to be involved in the APR are interleukin-1 (IL-1), tumor necrosis factor (TNF), transforming growth factor-β (TGF-β) and interleukin-6 (IL-6). Together with these factors, glucocorticoids appear to act as important cofactors for full induction or repression of APR genes both in vivo and in vitro (Bauer et al. 1984; Gross et al. 1984; Baumann et al. 1984; Kulkarni et al. 1985; Northemann et al. 1988a,b).

The expression of negative acute phase reactants is modulated by TGF-β. Steady-state mRNA levels for albumin, apolipoproteins AI (Apo-AI) and AII (Apo-AII) are strongly decreased in hep3B cells after administration of purified porcine TGF-β, whereas the synthesis of haptoglobin and factor B remains unmodified (G. Morrone and R. Cortese, personal communication).

Recombinant IL-1 induces some acute phase genes such as SAA, human C3 and factor B and decreases the synthesis of albumin, but does not exert any effect on other acute phase reactants such as CRP or haptoglobin (Ramadori et al. 1985; Perlmutter et al. 1986a; Darlington et al. 1986). It is surprising that fibrinogen synthesis, which increases during the acute phase, is strongly repressed both by recombinant and purified IL-1 in human hep3B and hepG2 cells (Darlington et al. 1986; Baumann et al. 1987b). IL-1 can only, therefore, determine a small part of the APR in in vitro cultured hepatoma cells.

Tumor necrosis factor has an even more restricted spectrum of activity. This monokine has been shown only to induce complement C3 and factor B and to a certain extent ACH, while it has no effect on the synthesis of other acute phase reactants (Darlington et al. 1986; Perlmutter et al. 1986b; Andus et al. 1987). The effect exerted by TNF on albumin is not clear: in some cases a decreased synthesis has been observed (Perlmutter et al. 1986b), while in others albumin synthesis was unaffected (Darlington et al. 1986).

Only recently, interleukin-6 (IL-6) has been identified as the major modulator of the APR (Gauldie et al. 1987). IL-6 is a new name for interferon-$\beta 2$ (IFN-$\beta 2$) or BSF-2. It is a 26 kDa glycoprotein with pleiotropic functions (reviewed by Billiau 1987; Seghal et al. 1987; Kishimoto and Hirano 1988; Wong and Clark 1988) produced not only in monocytes but also in fibroblasts, in lymphocytes, in the EJ bladder carcinoma cell line T24 and in several other cell types (Fey and Gauldie 1989). IL-6 is a potent stimulator of B-cell differentiation and of immunoglobulin release from plasma cells. Recombinant IL-6 is a potent inducer of the expression of fibrinogen, ACH, haptoglobin, factor B, haemopexin, SAA, CRP and AGP (Gauldie et al. 1987; Baumann et al. 1987b; Castell et al. 1988; Morrone et al. 1988). A combination of IL-1, IL-6 and dexamethasone reproduces the whole response in hepG2 cells (Gauldie et al. 1987).

On the basis of their responsiveness to IL-1 and IL-6, acute phase reactants have been classified into two groups (Baumann et al. 1987a). Class I comprises complement C3, albumin and AGP (modulated mainly by IL-1); class II comprises ACH, fibrinogen, haemopexin and haptoglobin (modulated mainly by IL-6). However, the distinction into two groups is only indicative for two reasons. First, haptoglobin belongs to class II in human cells, but to class I in rat cells. Moreover, several genes like CRP, AGP, ACH and haptoglobin itself are induced by more than one monokine. The assignment to class I or class II is based on the cytokine that gives the higher level of induction.

The use of different systems has generated some conflicting results. For example, hepG2 and hep3B respond to IL-6 in a slightly different manner:

haptoglobin synthesis is more sensitive to IL-6 induction in hep3B than in hepG2. In hepG2, IL-1 stimulates haptoglobin RNA about 2-fold and IL-6 about 10-fold. The administration of both monokines gives additive effects with a final level of stimulation similar to that obtained with unfractionated MoCM (Baumann et al. 1987b). In contrast, in hep3B full stimulation of haptoglobin is achieved by IL-6 alone: the mRNA level is higher than that reached with MoCM (Morrone et al. 1988). In the case of CRP, low levels of mRNA for this gene can be detected after induction with MoCM only in hep3B but not in hepG2 (Darlington et al. 1986; Lei et al. 1985 and G. Ciliberto, unpublished observations). Finally, dexamethasone is required to obtain maximal response in hepG2. Dexamethasone alone is poorly active, but potentiates the activity of each individual monokine (Baumann et al. 1987b). On the other hand, full gene inducibility is obtained in hep3B cells without the addition of dexamethasone (Morrone et al. 1988). It is not possible to say which of the two cell types better resembles the normal hepatocyte. Based on several observations, hepG2 is a more differentiated cell line than hep3B (De Simone and Cortese 1988). However, further studies are needed to clarify this issue. In particular, the recent cloning of IL-1 and IL-6 receptors (Sims et al. 1988; Yamasaki et al. 1988) will allow a precise comparison of the number of receptors on the surface of normal hepatocytes and hepatoma cells in culture.

The signal transduction pathway for IL-6 is unknown. It has been proposed that IL-6 acts via activation of protein kinase C. In fact, fibrinogen production is enhanced by TPA (Evans et al. 1987). This is not the case for induction of haptoglobin and CRP genes in hep3B cells (G. Morrone, S. Oliviero and G. Ciliberto, 1988, unpublished work). It is possible therefore that this cytokine is responsible for the activation of several independent cellular pathways.

The effect of IL-6 is exerted at the transcriptional level. Fusions between the promoter of several liver-specific genes and the coding region of the bacterial chloroamphenicol acetyl transferase (CAT) gene were constructed and transfected into hep3B cells. The ability of each promoter to stimulate transcription was determined by CAT assays with cell extracts prepared 36 hours after gene transfection (Morrone et al. 1988). Each batch of cells transfected with a particular construct was split into three aliquots. One was incubated with normal medium; a second was treated with 10% MoCM; and the third was treated with recombinant IL-6. Several promoters are stimulated by IL-6, including those for haptoglobin, Factor B, AGP and CRP. The results obtained with the gene transfer experiments mirror those obtained by Northern analysis. Similar results were obtained by transfection of the rat AGP 5′ flanking region in hepG2 cells (Prowse and Baumann 1988).

The study of the effect of monokines on CRP induction has been difficult due to the low level of endogenous CRP expression in hepatoma cells. In contrast, the transfected CRP promoter fused to the CAT gene is highly inducible both in hep3B and hepG2, and in the former at least 10- to 20-fold better than in the latter (Arcone et al. 1988). An explanation for the different behaviour of endogenous and transfected CRP is that the endogenous gene is somehow inactivated, perhaps by altered chromatin configuration or methylation, and it is not accessible to transcription factors.

The high level of induction shown by the transfected CRP promoter fusions makes it relatively easy to explore inducibility by single monokines. A construction carrying 250 bp of the CRP 5′-flanking region is induced by both IL-1 and IL-6, but not by TNF in hep3B cells.

Several acute phase reactants are induced by more than one monokine and in most cases this effect is additive. An exception is AGP where IL-1 and IL-6 act in a co-operative manner (Baumann et al. 1987b). Also in the case of CRP, the same molecules show co-operative stimulation of gene transcription if present in sub-optimal quantities (U. Ganter, R. Arcone and G. Ciliberto, 1988, unpublished work).

Detailed studies of epithelial cells and on a human squamous carcinoma cell line COLO-16 showed that the conditioned medium of these cells also reproduces the APR in hepatoma cells or primary cultures of rat hepatocytes (Baumann et al. 1984, 1986). Fractionation studies demonstrated the existence of two HSFs, named HSF-I and HSF-II. These two factors have now been partially purified and their approximate molecular weight determined (Baumann 1987a). HSF-I is functionally equivalent to IL-1 and HSF-II to IL-6. Surprisingly, these two molecules do not react with antibodies raised against IL-1 and IL-6 and also have a different molecular weight.

In conclusion, the APR is mediated by several factors. The complete response is the result of their combined action (additive or cooperative). The major cell-type responsible for the production of HSFs is the monocyte-macrophage. In these cells the various monokines are regulated in a different manner depending on the irritating stimuli (Northoff et al. 1987; U. Ganter, J. Bauer and G. Ciliberto, 1988, unpublished work). Therefore, in different pathologic conditions the relative proportion of cytokines released from mononuclear phagocytes can vary considerably, thus leading to different patterns of activation of acute phase reactants. Furthermore, HSFs of identical or similar nature are also produced in other tissues and cell types and sometimes constitutively by tumour cells. This might explain the chronic alteration in the levels of some plasma proteins observed in cancer.

Transcriptional Regulation – the Mechanism

There is increasing evidence that sufficient information for induction of acute phase reactants is located in the 5′ flanking region of several genes and that regulation therefore takes place mainly at the transcriptional level.

Liver-specific Promoters and Enhancers

The acute phase reactants are a special group of liver-specific genes in which, superimposed on liver-specific expression, modulation of gene expression occurs. The structure and function of liver-specific promoters and enhancers have been reviewed recently (De Simone and Cortese 1988).

The activity of liver-specific promoters and enhancers is achieved through the interaction of these sequences with transcription factors which are predominantly or exclusively expressed in hepatocytes. The immediate question is therefore whether transcriptional regulation of acute phase genes depends on inducible liver-specific *trans*-acting factors or whether it is due to the combined action of general transcription factors whose concentration is modulated during inflammation with liver-specific factors constitutively present in hepatocytes.

After initial studies that identified the minimal sequences required for liver-specific expression of several genes (Ott et al. 1984; Ciliberto et al. 1985; D'Onofrio et al. 1985; Costa et al. 1986; Godbout et al. 1986; Heard et al. 1987), considerable effort has been invested in identifying the transcription factors capable of interacting with them (Cerenghini et al. 1987; Lichtsteiner et al. 1987; De Simone et al. 1987; Courtois et al. 1987; Costa et al. 1988; Grayson et al. 1988a,b; Monaci et al. 1988; Hardon et al. 1988). Liver-specific promoters and enhancers do not share an obvious sequence or structure, like that found in other classes of tissue-specific genes (reviewed by De Simone and Cortese 1988). Transcriptional control regions are, in general, DNA segments of length 50–100 bp, composed of modules of 10–20 base pairs, each capable of interacting with a specific nuclear factor (for a review, see Jones et al. 1988). In liver-specific promoters there is a combination of binding sites both for ubiquitous and liver-specific factors. This has been found in the mouse and rat albumin promoter (Lichtsteiner et al. 1987; Cerenghini et al. 1987), in the mouse and human α1-AT promoter and enhancer regions (Monaci et al. 1988; Grayson et al. 1988a), in the mouse transthyretin gene (Costa et al. 1988) and in the rat α-fibrinogen promoter (Courtois et al. 1988). Many studies have been directed at purifying and cloning the genes for these factors. The genes for AP1 (Bohmann et al. 1987) a general transcription factor, which has a binding site on the mouse α1-AT and TTR 5′-flanking regions (Costa et al. 1988) and for two other factors, named CBP and NF-1, which bind to the albumin promoter, have been cloned (Landschultz et al. 1988; Santoro et al. 1988; Paonessa et al. 1988). Of these, only CBP is liver-specific.

Regulation of liver-specific gene expression is rather complex. In rat liver, for example, two proteins have been found that have the property of binding in a mutually exclusive manner to the same sequence motif on the human α1-AT promoter. Comparison of binding and transcriptional studies on wild type and mutant templates has led to the idea that one of the two factors is a positive regulator and that the other might act as a repressor (Monaci et al. 1988). Negative control has been shown to be responsible for liver-specific expression of the retinol binding protein gene (Colantuoni et al. 1987). Also, in the case of albumin and fibrinogen genes, two mutually exclusive nuclear proteins interact with the same sequence, but these factors have been found in different rat cell-types. One (APF/HNF-1/LFB1) is found in cells where the gene is actively transcribed, the other (vAPF) in dedifferentiated hepatocytes where the gene is inactive (Cerenghini et al. 1988). In mixing experiments vAPF is dominant over APF. This interesting observation opens the possibility of explaining the phenomenon of extinction by which hybrids between hepatocytes and other cell types lose the ability to synthesize several plasma proteins, including albumin (Mével-Ninio and Weiss 1981). The phenomenon of extinction led to the postulation of the existence of a negative control. It is possible therefore that vAPF is somehow involved in this negative control.

Promoters Induced During the Acute Phase

Three promoters for acute phase induced genes have been studied in detail in the search of sequences responsible for the regulation of the expression of AGP, haptoglobin and CRP genes.

The rat and human AGP regulatory regions

Rat AGP is positively regulated by glucocorticoids, IL-1 and IL-6 (Baumann et al. 1983a, 1984; Baumann and Maquat 1986). These three different hormones act in a cooperative manner: all three are necessary to give full gene expression but each one individually gives only a limited degree of induction (Prowse and Baumann 1988). Two DNA segments are necessary for inducibility (Fig 3.1). A proximal promoter element is located between position −120 to −64 and is responsible for dexamethasone induction (Baumann and Maquat 1986). A 150-

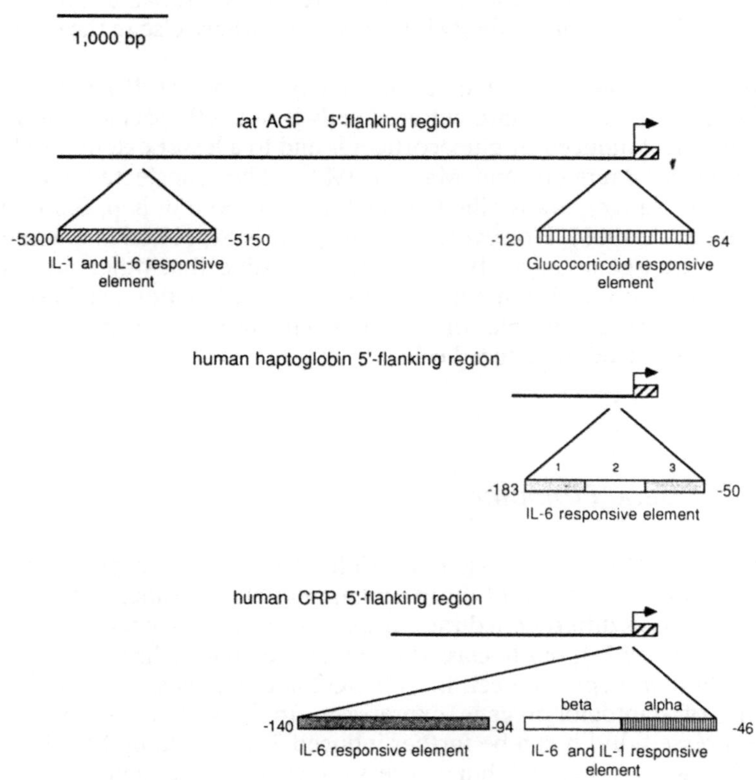

Fig. 3.1. Schematic representation of the sequences required for acute phase induction of the rat AGP gene and of the human haptoglobin and CRP genes. *Black bars* represent the 5′-flanking regions which have been analysed for each gene. *Arrows* indicate the start sites of transcription.

nucleotide-long segment from position −5300 to −5150 (distal element) is responsible both for IL-1 and IL-6 induction (Prowse and Baumann 1988). It can act as an enhancer because it can be moved closer to the proximal element without losing its function. The proximal element does not share any homology with a canonical glucocorticoid responsive element (GRE). On the other hand, induction by glucocorticoids of the AGP gene is atypical because, unlike other genes regulated by this hormone, such as the MMTV promoter or the metallothionein promoter, it requires ongoing protein synthesis (Klein et al. 1987).

Human and rat AGP control regions have a different structure. Transgenic mice carrying only one (the most active) of the three AGP genes present in the human genome, AGP-A, have been obtained (Dente et al. 1987, 1988). This 6.6 kb DNA segment carries only 1.2 kb of the 5′-flanking region, but in these transgenic mice AGP is highly induced upon experimental inflammation and its preferential liver-specific expression is also maintained. No further functional dissection of this segment has been done in vivo. Close inspection of the 5′-flanking region of the human gene does not reveal any obvious homology with the rat AGP −5300 to −5150 area defined by Prowse and Baumann (1988) (L. Dente, 1988, personal communication). In addition, in vitro experiments where the 5′-flanking region fused to the CAT gene was transfected into hep3B cells gave only a 3-fold stimulation of CAT expression in the presence of MoCM (L. Dente, 1988, personal communication). Therefore, it is probable that the information essential for gene inducibility is located somewhere else in the 6.6 kb segment.

Another unsolved problem is the tissue-specificity of the AGP gene. The promoter of this gene transfected into cultured cells is not cell-specific. The rat AGP promoter is in fact induced by glucocorticoids and to a lesser extent also by monokines in L-cells (Baumann and Maquat 1986). The transfected human AGP-A gene is constitutively transcribed at similar levels both in hep3B and in HeLa cells (Dente et al. 1988). However, the same gene is expressed in a tissue-specific manner in transgenic mice. It is therefore possible that other mechanisms, such as changes in the chromatin structure or in methylation established during development, are responsible for a liver-specific mode of transcription, which cannot be reproduced in cultured cells.

The Haptoglobin Gene Promoter

Haptoglobin is constitutively expressed in adult liver and its expression is increased approximately 3- to 5-fold during inflammation (Kushner 1982). Its induction by monokines is different in different species. It belongs in fact to class I in rat (Baumann et al. 1987a), but to class II in human cell lines (Baumann et al. 1987b). The two human hepatoma cell lines hepG2 and hep3B show a slightly different pattern of haptoglobin gene expression. In hepG2 haptoglobin is constitutively expressed, but it can be further induced 3- to 4-fold by MoCM or IL-6. In hep3B,however, haptoglobin expression is almost undetectable in unstimulated cells and its expression can be induced several-fold. The same behaviour is shown by fusions between the promoter and the reporter CAT gene (Oliviero et al. 1987). DNA transfections in hepG2 have allowed the definition of the minimal sequences necessary for tissue-specific expression. Hep3B cells have been used for the identification of sequences involved in induction. They both map within a DNA segment from position −183 to −30 with respect to the cap site (Oliviero et al. 1987) (Fig. 3.1).

A detailed mutagenesis of the −180 to −50 region has shown that three distinct segments (HP1, HP2 and HP3) are required for gene expression (S. Oliviero and R. Cortese, 1988, personal communication). Regions HP1 and HP3 bind the same nuclear proteins. Gel retardation experiments with oligonucleotides reproducing HP1 and HP3 give rise to a series of shifted bands with the same sequence specificity but different size. The most intriguing finding is the appearance of a new shifted band with the same DNA binding specificity, as revealed by DNA footprinting experiments, in induced hep3B cells. This extra shift is undetectable in uninduced cells, but becomes the major one after MoCM or IL-6 stimulation. It is obvious to speculate that it has something to do with induction of haptoglobin gene expression. However, the mechanisms responsible for the variation in the pattern of gel retardation given by regions HP1 and HP3 with extracts from induced and uninduced cells have not yet been identified. This can be due to: (i) structurally different factors with the same specificity of binding; (ii) modifications of the same protein such as glycosylation or phosphorylation; or (iii) the presence of an additional protein to form a heterodimeric structure. More work is needed to clarify this issue. Region HP2 interacts with a different protein. Binding to this region does not change either qualitatively or quantitatively in stimulated cells. The same factor interacts with the CRP promoter (see below).

C-reactive Protein Gene Expression

C-reactive protein is the prototype human acute phase reactant. It is historically the first identified acute phase reactant and also the one which, together with SAA, shows the highest degree of induction in several pathologic conditions (reviewed by Pepys and Baltz 1983). For these reasons CRP has always been considered a useful marker for monitoring the evolution of diseases such as rheumatic disease, heart attacks, bacterial meningitis, cancer and also the acute exacerbations of chronic inflammations. Although the CRP cDNA was cloned in 1983 (Whitehead et al. 1983), a detailed study of CRP expression became possible only after the gene was cloned and characterized (Woo et al. 1985; Lei et al. 1985). When the cDNA probe became available it was soon found that the change in CRP protein production during inflammation is accompanied by increased levels of steady-state mRNA in the liver (Whitehead et al. 1983). However, studies on CRP expression have been hampered both by the extreme species-specificity (the CRP gene is poorly expressed in mice) and by the low level of expression of the endogenous gene in the two commonly used human hepatoma cells hepG2 and hep3B.

Over the last two years our group has expended considerable effort in attempting to elucidate the steps and mechanisms responsible for human CRP expression. We have used both in vivo and in vitro systems and several experimental approaches to overcome the problems mentioned above. The first step in our study was to isolate a cosmid clone carrying the human CRP gene

(Ciliberto et al. 1987). Although the human gene had been previously isolated and characterized independently in two laboratories from lambda libraries (Woo et al. 1985; Lei et al. 1985), the availability of a cosmid clone provided us with a longer piece of contiguous DNA in which the CRP gene is located in the centre with more than 15 kb of the 5′-flanking and 15 kb of the 3′-flanking sequence which also includes a small pseudogene for CRP.

The in vivo studies have been mainly directed to the generation of families of transgenic mice in which the human CRP gene is stably integrated in the mouse genome and is transmitted as a Mendelian trait (Ciliberto et al. 1987b). A 30 kb ClaI-ClaI fragment, with the gene located approximately in the middle, was isolated and introduced into fertilized eggs by microinjection, and two transgenic mouse lines were derived (Ciliberto et al. 1987b). In both, the CRP gene was exclusively transcribed in the liver and expression was strictly dependent on experimental inflammation. The CRP mRNA found in transgenic mice starts at the same nucleotide as in humans. The kinetics of induction both for RNA and protein synthesis was very fast: RNA was first detected in the liver as early as 2 hours after intraperitoneal injection of LPS, the protein after 6 hours in the serum. Human CRP levels in the sera of transgenic mice were comparable to those observed in human diseases. Nuclear run-on experiments suggested that regulation takes place primarily at the transcriptional level.

More recently, other transgenic mouse families have been generated with a shorter DNA segment with only 2.5 kb of the 5′-flanking of the CRP gene linked to the SV40 large-T antigen or to the H-ras oncogene coding regions. In all the families generated with the 2.5 kb 5′ flanking region no expression of the linked genes could be detected (U. Ruther and G. Ciliberto, unpublished work). The conclusion, therefore, is that other sequences located more upstream or downstream to the coding region play a major role in CRP expression in the whole animal.

The activity of the CRP promoter has been studied with transfection experiments in various cell lines. A 2.5 kb segment of the CRP flanking region was fused to the bacterial CAT gene and transfected into the non-hepatic cell line HeLa and into the two hepatoma cell lines hepG2 and hep3B. The CRP-flanking region gives CAT expression only in hepatoma cells and only after stimulation (Arcone et al. 1988). Therefore, it contains enough information for cell-specific and inducible expression. This is in striking contrast with the experiments in vivo with transgenic mice. A similar finding has been obtained with the human albumin promoter (Pinkert et al. 1987). Also here, the 150 bp promoter region is sufficient for cell-specific expression in transfected cells, but, in the whole animal it requires the help of an enhancer element located approximately 10 kb upstream to the cap site.

It is worth analysing the reasons for the discrepancy sometimes observed between in vivo or in vitro data. For the human α1-AT gene, 700 bp of the 5′-flanking region are sufficient for liver-specific expression both in vitro (Ciliberto et al. 1985) and in vivo (Shen et al. 1988). A detailed dissection of this area shows the existence of two important segments (De Simone et al. 1987; Shen et al. 1987). The promoter element, a region from nucleotide −137 to the start of transcription, contains the binding sites for several liver-specific transcription factors (Monaci et al. 1988; Hardon et al. 1988; Li et al. 1988). Upstream to this, another region, between nucleotides −350 and −150 has the property of a liver-specific enhancer (De Simone et al. 1987). These two elements are therefore located in close proximity to each other. In the albumin gene there is a similar organization: two elements are required, the promoter and the enhancer. In this case, however, they are separated by a 10 kb DNA segment. On the basis of these considerations one can assume that the combination of these two sequences is probably responsible for developmental expression in transgenic mice. In other words, the enhancer is important to establish a certain pattern of tissue-specific expression during development, probably via the acquisition of an open chromatin configuration or binding to the nuclear matrix. The promoter region itself contains sequences sufficient for transcription in liver cells once the tissue-specific pattern has been established. Therefore, I postulate the existence of a liver-specific enhancer or what can be called a "programmer sequence" also in the case of the CRP gene, the location of which can be either 5′ or 3′ with respect to the promoter. In the case of gene transfer experiments in cultured cells, the "programmer" is not necessary because the promoter element itself is already in an "open" conformation and therefore accessible for the interaction with transcription factors.

Gene transfer studies in hep3B cells showed, by progressive deletions of the original 2.5 kb-flanking sequence, that a high level of inducibility (about 100-fold) is also retained when only 94 bp of the 5′-flanking region are present (Arcone et al. 1988). Inducibility is dependent on the cooperative action of several factors including IL-1 and IL-6 (U. Ganter, R. Arcone and G. Ciliberto, unpublished work). A further deletion down to position −46 is only inducible 3- to 4-fold as judged by CAT assays. A possible interpretation of these results is that sequences responsible for induction are located between −94 and −46. This hypothesis is favoured by the finding that a −41/−121 segment of the CRP promoter confers inducibility to MoCM to a heterologous SV40 promoter (Arcone et al. 1988). In addition, single point mutations in the −80 to −60 area abolish inducibility (R. Arcone, C. Toniatti and G. Ciliberto, unpublished work). The structure of the CRP promoter is further complicated by the existence of another important signal for induction between positions −91 and −140 (Arcone et al. 1988) which is highly responsive to IL-6 (U. Ganter, R. Arcone and G. Ciliberto, unpublished work) (Fig. 3.1). In our laboratory a systematic search for nuclear proteins capable of binding to the CRP promoter was recently started (B. Majello, G. Arpaia and G. Ciliberto, unpublished work). Two nuclear proteins have been identified in rat liver nuclear extracts. A first protein (common to the HP2 area of the haptoglobin promoter) binds to a distal area between nucleotides −87 and −61 (β); a second protein binds to an intermediate region between positions −59 and −40 (α). Binding to region α is increased dramatically after cell treatment with IL-6.

Conclusions and Perspectives

Important advances have been made in recent years towards the identification of the mechanisms responsible for the regulation of acute phase genes. This has been possible because of the use of gene transfer experiments in hepatoma cells.

Cis-acting sequences responsive to IL-1 and IL-6 have been localized in the 5′-flanking region of several genes. Their architecture is rather heterogeneous (as shown in Fig. 3.1). Parallel to this, preliminary studies of the transcription factors interacting with promoters of genes induced during inflammation have revealed the involvement of several distinct factors with different size and DNA binding specificity. A simplistic expectation could have been that genes induced by the same cytokine, namely IL-6, share a similar sequence and also the same inducible transcription factor. The analysis of the AGP, CRP and haptoglobin promoters shows that this is not the case. Therefore, how many mechanisms are there of gene induction or repression during inflammation? How many proteins are involved and why is there such a multiplicity of factors? A detailed examination of these problems will help to elucidate the meaning of the APR and its evolutionary significance.

The possibility of preparing liver nuclear extracts capable of supporting liver-specific gene transcription allows an easier access to the parallel study of the promoter sequences and of transcription factors capable of interacting with them (Gorski et al. 1986). Faithful in vitro liver-specific transcription systems have been used for the study of the albumin and of the α1-AT promoters (Lichsteiner et al. 1987; Monaci et al. 1988). They have contributed enormously to the understanding of the physical distribution of transcription factors in the promoter areas and of the way they interact with each other. A full understanding of the mechanism of promoter activation during acute phase awaits the establishment of a similar "acute phase response" in in vitro transcription extracts.

Acknowledgements. I thank L. Dente, V. Colantuoni and C. Toniatti for discussions and criticism and J. Gilder for critical reading of the manuscript. I am grateful to S. Oliviero, G. Morrone, R. Cortese, G. Fey, M. Yaniv and S. Woo for informing me of their unpublished work. Some of the work reported here was supported by a CEC grant BAP I-116.

References

Ades EW, Hinson A, Chapnis-Celliar C, Arnaud P (1982) Modulation of the immuno response by plasma protease inhibitor. Scand J Immunol 15:109–113

Andus T, Heinrich PC, Bauer J, et al. (1987) Discrimination of hepatocyte-stimulating activity from human recombinant tumor necrosis factor α. Eur J Immunol 17:1175–1197

Arcone R, Gualandi G, Ciliberto G (1988) Identification of sequences responsible for acute-phase induction of human C-reactive protein. Nucl Acids Res 16:3195–3207

Bauer J, Birmelin M, Northoff GH, et al. (1984) Induction of rat α2-macroglobulin in vivo and in hepatocyte culture: synergistic action of glucocorticoids and a Kupffer cell derived factor. FEBS Lett 177:89–94

Baumann H, Maquat LE (1986) Localization of DNA sequences involved in dexamethasone-dependent expression of the rat α1-acid glycoprotein gene. Mol Cell Biol 6:2551–2561

Baumann H, Firestone CL, Burgess TL, et al. (1983a) Dexamethasone regulation of α1-acid glycoprotein and other acute phase reactants in rat liver and hepatoma cells. J Biol Chem 258:563–570

Baumann H, Jahreis GP, Gaines KC (1983b) Synthesis and regulation of acute phase plasma proteins in primary cultures of mouse hepatocytes. J Cell Biol 97:866–876

Baumann H, Jahreis GP, Sauder DN, et al. (1984) Human keratinocytes and monocytes release factors which regulate the synthesis of major acute phase plasma proteins in hepatic cells from man, rat and mouse. J Biol Chem 259:7331–7342

Baumann H, Hill RE, Sauder DN, et al. (1986) Regulation of major acute phase plasma proteins by hepatocyte stimulating factors of human squamous carcinoma cells. J Cell Biol 102:370–383

Baumann H, Onorato V, Gauldie J, et al. (1987a) Distinct sets of acute phase plasma proteins are stimulated by separate human hepatocyte-stimulatory factors and monokines in rat hepatoma cells. J Biol Chem 262:9756–9768

Baumann H, Richards C, Gauldie J (1987b) Interaction among hepatocyte-stimulating factor, interleukin 1 and glucocorticoids for regulation of acute phase plasma proteins in human hepatoma (HepG2). J Immunol 139:4122–4128

Billiau A (1987) Interferon-β2 as a promoter of growth and differentiation of B cells. Immunol Today 8:84–87

Birch H, Schreiber G (1986) Transcriptional regulation of plasma protein synthesis during inflammation. J Biol Chem 261:8077–8080

Bohmann D, Bos TJ, Nishimura T, et al. (1987) Human proto-oncogene c-jun encodes a protein with antigenic and enhancer binding properties of transcription factor AP-1. Science 238:1386–1392

Bowmann BH, Kurowsky (1982) Haptoglobin: the evolutionary product of duplication, unequal crossing over and point mutation. Adv Hum Genet 12:189–261

Catell JV, Gomez-Lechon MJ, David M, et al. (1988) Recombinant human interleukin-6 (IL-6/BSF-2/HSF) regulates the synthesis of acute phase proteins in human hepatocytes. FEBS Lett 232:347–350

Cerenghini S, Raymondjean M, Carranca AG, et al. (1987) Factors involved in control of tissue-specific expression of albumin gene. Cell 50:627–638

Cerenghini S, Blumenfeld M, Yaniv M (1988) A liver specific factor essential for albumin transcription differs between differentiated and dedifferentiated rat hepatoma cells. Genes Devel 2:957–974

Ciliberto G, Dente L, Cortese R (1985) Cell-specific expression of a transfected human α1-antitrypsin gene. Cell 41:531–540

Ciliberto G, Pizza MG, Arcone R, et al. (1987a) DNA sequence of a pseudogene for human C-reactive protein. Nucl Acids Res 14:5895

Ciliberto G, Arcone R, Wagner EF, et al. (1987b) Inducible and tissue-specific expression of human C-reactive protein in transgenic mice. EMBO J 6:4017–4022

Colantuoni V, Pirozzi A, Blance C, et al. (1987) Negative control of liver-specific gene expression: cloned human retinol-binding protein gene is repressed in HeLa cells. EMBO J 6:631–636

Costa RH, Eseng L, Darnell JE Jr (1986) Transcriptional control of the mouse prealbumin (transthyretin) gene: both promoter sequences and a distinct enhancer are cell specific. Mol Cell Biol 6:4697–4707

Costa RH, Grayson DR, Xanthopulos KG, et al. (1988) A liver-specific DNA-binding protein recognizes multiple nucleotide sites in regulatory regions of transthyretin, α1-antitrypsin, albumin and simian virus 40 genes. Proc Natl Acad Sci USA 85:3840–3844

Courtois G, Morgan JG, Campbell LA, et al. (1987) Interaction of a liver-specific nuclear factor with the fibrinogen and α1-antitrypsin promoters. Science 238:688–692

Darlington GJ, Wilson DR, Lachmann LB (1986) Monocyte-conditioned medium, interleukin-1, and tumor necrosis factor stimulate the acute phase response in human hepatoma cells in vitro. J Cell Biol 103:787–793

Dente L, Pizza MG, Metzpalu A, et al. (1987) Structure and expression of the genes coding for human α1-acid glycoprotein. EMBO J 6:2289–2296

Dente L, Tripodi M, Rüther U, et al. (1988) Expression of cloned human AGP genes in cultured cell lines and in transgenic mice. Genes Dev 2:259–266

Deschatrette J, Weiss MC (1974) Characterization of differentiated and dedifferentiated clones from a rat hepatoma. Biochemie 56:1603–1611

Deschatrette J, Moore EE, Dubois M, et al. (1979) Dedifferentiated variants of a rat hepatoma: analysis by cell hybridization. Somatic Cell Genet 5:697–718

Deschatrette J, Moore EE, Dubois M, et al. (1980) Dedifferentiated variants of a rat hepatoma: reversion analysis. Cell 19:1043–1051

De Simone V, Cortese R (1988) The transcriptional regulation of liver specific gene expression. In: McLean N (ed) Oxford surveys on eukaryotic genes, vol 5. Oxford University Press, Oxford, pp 51–90

De Simone V, Ciliberto G, Hardon E, et al. (1987) *Cis*- and *trans*-acting elements responsible for the cell-specific expression of the human α1-antitrypsin gene. EMBO J 6:2759–2766

D'Onofrio C, Colantuoni V, Cortese R (1985) Structure and cell-specific expression of a cloned human retinol binding protein gene: the 5′-flanking region contains hepatoma specific transcriptional signals. EMBO J 4:1981–1989

Esnard F, Gauthier F (1980) Purification and physiochemical characterization of a new rat plasma proteinase inhibitor, α1-inhibitor III. Biochim Biophys Acta 614:553–563
Evans E, Courtois GM, Kilian PL, Fuller GM, Crabtree GR (1987) Induction of fibrinogen and a subset of acute phase response genes involves a novel monokine which is mimicked by phorbol esters. J Biol Chem 262:10850–10854
Fey GH, Fuller GM (1987) Regulation of acute phase gene expression by inflammatory mediators. Mol Biol Med 4:323–338
Fey GH, Gauldie J (1989) The acute phase of the liver in inflammation. In: Popper M, Schaffner F (eds) Progress in liver diseases, vol 9. WB Saunders, Philadelphia (in press)
Fuller GM, Bunzel RJ, Woloski BM, et al. (1987) Isolation of hepatocyte stimulating factor from human monocytes. Biophys Biochem Res Comm 144:1003–1009
Furie B, Furie BC (1988) The molecular basis of blood coagulation. Cell 53:505–518
Gauldie J, Richards C, Harnish D, et al. (1987) Interferon β2/BSF2 shares identity with monocyte derived hepatocyte stimulating factor (HSF) and regulates the major acute phase protein response in liver cells. Proc Natl Acad Sci USA 84:7251–7255
Gehring MR, Shiels BR, Northemann W, et al. (1987) Sequence of rat α2-macroglobulin and acute phase control of its messenger RNA. J Biol Chem 262:446–454
Godbout R, Ingram R, Thilghman SM (1986) Multiple regulatory elements in the intergenic region between the α-fetoprotein and albumin genes. Mol Cell Biol 6:477–487
Goldman ND, Liu T-Y (1987) Biosynthesis of human C-reactive protein in cultured hepatoma cells is induced by a monocyte factor(s) other than interleukin 1. J Biol Chem 262:2363–2368
Gordon AH (1976) The α-macroglobulins of rat serum. Biochem J 159:643–650
Gorski K, Carneiro M, Schibler U (1986) Tissue-specific in vitro transcription from the mouse albumin promoter. Cell 47:767–776
Grayson DR, Costa RH, Xanthopoulos KG, et al. (1988a) A cell-specific enhancer of the mouse α1-antitrypsin gene has multiple functional regions and corresponding protein binding sites. Mol Cell Biol 8:1055–1066
Grayson DR, Costa RH, Xanthopoulos KG, et al. (1988b) One factor recognizes the liver-specific enhancers in α1-antitrypsin and transthyretin genes. Science 239:786–788
Gross V, Andus T, Tran-Thi TA, et al. (1984) Induction of acute phase proteins by dexamethasone in primary rat hepatocyte cultures. Exp Cell Res 151:46–54
Hardon EM, Frain M, Paonessa G, et al. (1988) Two distinct factors interact with the promoter regions of several liver-specific genes. EMBO J 7:1711–1719
Heard JM, Herbomel P, Ott M-O, et al. (1987) Determinants of rat albumin promoter tissue specificity analyzed by an improved transient expression system. Mol Cell Biol 7:2425–2434
Hudig D, Haverty T, Fulcher C, et al. (1981) Inhibition of human natural cytoxicity by macromolecular antiproteases. J Immunol 126:1569–1574
James K (1980) α2-macroglobin and its possible importance in the immune system. Trends Biochem Sci 5:43–47
Jones NC, Rigby PW, Ziff EB (1988) *Trans*-acting protein factors and the regulation of eucaryotic transcription: lessons from studies on DNA tumor viruses. Genes Dev 2:267–281
Kampschmidt RF, Mesecher M (1985) Interleukin-1 from P388D: effects upon neutrophils, plasma iron, and fibrinogen in rats, mice and rabbits. Proc Soc Exp Biol Med 179:197–200
Kishimoto T, Hirano T (1988) Molecular regulation of lymphocyte response. Ann Rev Immunol 6:485–512
Klein ES, Reinke S, Feigelsohn P, et al. (1987) Glucocorticoid-regulated expression from the 5′ flanking region of the rat α1-acid glycoprotein gene. J Biol Chem 262:523–532
Knowles BB, Howe CC, Aden DP (1980) Human hepatocellular carcinoma cell lines secrete the major plasma proteins and hepatitis B surface antigen. Science 209:497–499
Koj A (1987) Biological functions of acute phase proteins and the cytokines involved in their induced synthesis. In: Reutter W (ed) Modulation of liver cell expression. MTP Press, Lancaster, pp 331–342 (Falk symposium no. 43)
Koj A, Gauldie J, Regoeczi E, et al. (1984) The acute phase response of cultured rat hepatocytes. Biochem J 22:550–514
Kulkarni AB, Reinke R, Feigelson P (1985) Acute phase mediators and glucorticoids elevate α1-acid glycoprotein gene transcription. J Biol Chem 260:15386–15389
Kushner I (1982) The phenomenon of the acute phase response. Ann NY Acad Sci 389:39–48
Landschultz WH, Johnson F, Adashi EY, et al. (1988) Isolation of a recombinant copy of the gene encoding C/EBP. Genes Dev 2:786–800
Lei K-J, Liu T, Zon G, et al. (1985) Genomic DNA sequence for human C-reactive protein. J Biol Chem 260:13377–13383

Li Y, Shen R-F, Tsai SY, et al. (1988) Multiple hepatic *trans*-acting factors are required for in vitro transcription of the human α1-antitrypsin gene. Mol Cell Biol 83:4362–4369

Lichtsteiner S, Wuarin J, Schibler U (1987) The interplay of DNA-binding proteins on the promoter of the mouse albumin gene. Cell 51:963–973

Lonbera-Holm K, Reed D, Roberts RC, et al. (1987) Three high molecular weight proteinase inhibitors of rat plasma. J Biol Chem 252:438–445

Lowell CA, Stearman RS, Morrow JF (1986) Transcriptional regulation of serum amyloid A gene expression. J Biol Chem 261:8453–8461

MacLeod CM, Avery OT (1941) The occurrence during acute infections of a protein not normally present in the blood. J Exp Med 73:191–200

Mével-Ninio M, Weiss MC (1981) Immunofluorescence analysis of the time-course of extinction, reexpression and activation of albumin production in rat hepatoma–mouse fibroblast heterokaryons and hybrids. J Cell Biol 10:339–350

Monaci P, Nicosia A, Cortese R (1988) Characterization of the *cis*- and *trans*-acting elements necessary for the liver-specific transcription in vitro of the human α1-antitrypsin promoter. EMBO J 7:2075–2087

Morrone G, Ciliberto G, Oliviero S, et al. (1988) Recombinant interleukin 6 regulates the transcriptional activation of a set of human acute phase genes. J Biol Chem 263:12554–12558

Muller-Eberhard V, Liem HH (1974) In: Allison AC (ed), Structure and function of plasma proteins, vol 1. Plenum Press, London, pp 35–53

Nathan CF (1987) Secretory products of macrophages. J Clin Invest 79:319–326

Northemann W, Ueberberg H, Fey GH (1988a) Cordinated regulation of the rat α2-macroglobulin gene by inflammatory signals and glucocorticoids. Mol Gen (Life Sci Adv) 7:7–10

Northemann W, Shiels BR, Braciak TA, et al. (1988b) Structure and negative transcriptional regulation by glucocorticoids of the acute phase rat α1-inhibitor III gene. Biochemistry (in press)

Northoff H, Andus T, Tran-Thi TA, et al. (1987) The inflammation mediators interleukin 1 and hepatocyte stimulating factor are differently regulated in human monocytes. 17:707–711

Oliviero S, Morrone G, Cortese R (1987) The human haptoglobin gene: transcriptional regulation during development and acute phase induction. EMBO J 6:1905–1912

Ott M-O, Sperling L, Herbomel P, et al. (1984) Tissue-specific expression is conferred by a sequence from the 5′ end of the rat albumin gene. EMBO J 3:2505–2510

Paonessa G, Gounari F, Frank R, et al. (1988) Purification of a NF1-like DNA binding protein from rat liver and cloning of the corresponding cDNA. EMBO J 10:3115–3123

Pepys MB, Baltz ML (1983) Acute phase proteins with special reference to C-reactive protein and related proteins (pentraxins) and serum amyloid A protein. Adv Immunol 34:141–212

Perlmutter DH, Goldberger G, Dinarello CA, et al. (1986a) Regulation of class III major histocompatibility complex gene products by interleukin-1. Science 232:850–652

Perlmutter DH, Dinarello CA, Punsai PI, et al. (1986b) Cachectin/tumour necrosis factor regulates hepatic acute-phase gene expression. J Clin Invest 78:1349–1353

Pinkert CA, Ornitz DM, Brinster RL, et al. (1987) An albumin enhancer located 10 kb upstream functions along with its promoter to direct efficient, liver-specific expression in transgenic mice. Genes Dev 1:268–276

Prowse KR, Baumann H (1988) Hepatocyte-stimulating factor, beta 2 interferon and interleukin 1 enhance expression of the rat α1-acid glycoprotein gene via a distal upstream regulatory region. Mol Cell Biol 8:42–51

Putnam FW (ed) (1975). The plasma proteins: structure, function and genetic control. Academic Press, New York

Ramadori G, Sipe JD, Mizel SB, et al. (1985) Pretranslation modulation of acute-phase hepatic protein synthesis by murine recombinant interleukin 1 (IL-1) and purified IL-1. J Exp Med 162:930–942

Robey FA, Jones KD (1986) C-reactive protein, chromatin, immunomodulation of the acute phase response. In: Peeters H (ed), Protides of the biological fluids, vol 34. Pergamon Press, Oxford, pp 267–270

Robey FA, Jones KD, Tanaka T, et al. (1984) Binding of C-reactive protein to chromatin and nucleosome core particles. J Biol Chem 259:7311–7316

Rüther U, Tripodi M, Cortese R, et al. (1987) The human α1-antitrypsin gene is efficiently expressed from two tissue-specific promoters in transgenic mice. Nucl Acids Res 15:7519–7529

Santoro C, Mermod N, Andrews PC, et al. (1988) A family of human CCAAT-box-binding proteins active in transcription and DNA replication and expression of multiple cDNAs. Nature 334:218–224

Sehgal PB, May LT, Tamm I, et al. (1987) Human β2 interferon and B cell differentiation factor BSF-2 are identical. Science 235:731–732
Shen R-F, Clift SM, de Mayo JL, et al. (1987) Tissue-specific expression of the human α1-antitrypsin gene is controlled by multiple *cis*-regulatory sequences. Nucl Acids Res 15:8399–8415
Shen R-F, Clift SM, De Mayo JL, et al. (1988) Tissue-specific regulation of human α1-antitrypsin gene expression in transgenic mice. DNA (in press)
Sifers RN, Carlson JA, Clift SM, et al. (1987) Tissue-specific expression of the human α1-antitrypsin gene in transgenic mice. Nucl Acids Res 15:1459–1475
Sims JE, March CE, Losman D, et al. (1988) cDNA expression cloning of the IL1 receptor, a member of the immunoglobulin superfamily. Science 241:585–589
Sipe JD, Vogel SN, Sztein MB, et al. (1982) Interleukin 1 in acute phase. Serum amyloid A (SAA) and serum amyloid P (SAP) biosynthesis. Ann NY Acad Sci 389:137–150
Yamasaki K, Taga T, Hirata Y, et al. (1988) Cloning and expression of the human interleukin 6 (BSF/IFNβ2) receptor. Science 241:825–828
Whitehead AS, Bruns GAP, Markham AF, et al. (1983) Isolation of human C-reactive protein complementary DNA and localization of the gene to chromosome 1. Science 221:69–71
Woloski BM, Fuller GM (1985) Identification and partial characterization of hepatocyte stimulating factor leukemia cell lines: comparison with interleukin-1. Proc Natl Acad Sci USA 82:1443–1447
Wong GG, Clark SC (1988). Multiple actions of interleukin 6 within a cytokine network. Immunol Today 9:137–139
Woo P, Korenberg JR, Whitehead AS (1985) Characterization of genomic and complementary DNA sequence of human C-reactive protein, and comparison with the complementary DNA sequence of serum amyloid P component. J Biol Chem 260:13384–13388
Wu LC, Morley BJ, Campbell RD (1987) Cell-specific expression of the human complement factor B gene: evidence for the role of two distinct 5′ flanking elements. Cell 48:331–342

Chapter 4

Organization, Structure and Expression of Pentraxin Genes

A. S. Whitehead

Introduction

The acute phase response is a systemic reaction to an inflammatory stimulus (such as burns, surgery, infection or tissue trauma) and is characterized by a spectrum of physiological changes. Alterations in hormone and electrolyte levels, nitrogen balance and the concentrations of a wide range of serum proteins are readily observed. Quantitatively, the most dramatic change is the increase by up to a thousand-fold in the serum concentration of a limited number of acute phase proteins. In man, these are C-reactive protein (CRP) and serum amyloid A component (SAA). The CRP analogue, serum amyloid P component (SAP), is not an acute phase protein in man, though in the mouse it is.

CRP was originally named for its ability to bind to the C-polysaccharide of *Streptococcus pneumoniae* (Tillett and Francis 1930). Although its precise role in the acute phase response is unclear, human CRP has long been known to promote agglutination, bacterial capsular swelling, phagocytosis, and complement fixation through its calcium-dependent binding to phosphorylcholine (reviewed in Kushner et al. 1982). More recently, CRP has been shown to bind to chromatin with high affinity (Robey et al. 1984) and to be able to enhance macrophage tumoricidal activity (Zahedi and Mortensen 1986). The related serum protein, SAP, is a precursor of amyloid P component which is found in association with amyloid deposits in all forms of amyloidosis (reviewed by Pepys and Baltz 1983). Immunohistochemical evidence (Inoue et al. 1986) suggests that amyloid P component is a normal structural constituent of the microfibrils of connective tissue; amyloidosis may therefore be the result of aberrant amyloid P organization and/or deposition within tissue matrices. SAP, like CRP, has recently been shown to exhibit high affinity binding to chromatin (Pepys and Butler 1987).

CRP and SAP are pentraxins, proteins which have a discoid arrangement of five non-covalently bound subunits. Gross structural similarities and several other characteristics indicate that CRP and SAP are products of an ancestral gene duplication event. The existence of homologues of both proteins in a variety of phylogenetically diverse species (e.g. horseshoe crab [*Limulus*], plaice, mouse and man), suggests that the putative gene duplication event occurred early in evolution.

This chapter will explore three topics: (i) the structure, sequence and possible function of pentraxin molecules; (ii) the intrinsic pentraxin gene elements that may be relevant to gene activation and their possible role in the induction of acute phase protein synthesis; and (iii) the locations of the pentraxin genes in the human and mouse genomes and the possible existence of a complex of functionally related genes, including those encoding pentraxins.

The Structure of Pentraxin Proteins

The pentraxin family of proteins is so called due to the basic pentameric structures of its members. Mammalian pentraxins in their native configuration are composed of five monomeric non-covalently bound subunits; electron microscopy indicates that the monomers are arranged as a planar disc (Osmand et al. 1977). *Limulus* CRP differs in that it is arranged as a hexameric disc of six subunits (Fernandez-Moran et al. 1968).

The complete amino acid sequences of *Limulus* CRP (Nguyen et al. 1986a, b), rabbit CRP (Hu et al. 1986), human CRP (Oliviera et al. 1979; Lei et al. 1985; Woo et al. 1985), human SAP (Mantzouranis et al. 1985; Ohnishi et al. 1986), mouse SAP (Ishikawa et al. 1987) and Syrian hamster "female protein" (Dowton et al. 1985) have been determined, in some cases by direct protein sequencing in addition to derivation via nucleotide sequencing of complementary and genomic DNA clones.

The mammalian pentraxins consist of between 204 and 206 amino acid residues in the native monomer. These are preceded in the pre-molecules by typical hydrophobic leader sequences of 18 or 19 residues. Considering only amino acid identity, rabbit and human CRP are over 70% homologous, while pairwise comparisons within human, mouse and Syrian hamster (female protein) SAP exhibit levels of homology only slightly lower. Comparison of CRP and SAP sequences within a species, e.g. man, reveals approximately 50% amino acid identity, attesting to the remarkable level of conservation among pentraxins. The above levels of homology are even more striking when the calculations upon which they are based allow neutral amino acid substitutions.

Figure 4.1 depicts a comparison of the sequences of the mammalian pentraxins computed using the Genalign program. The similarity between the five proteins is apparent from examination of the consensus sequence obtained. There are 70 residues, approximately 35% of the total, that are conserved between all five mammalian pentraxins. Of these, exactly half are hydrophobic whereas the proportion of hydrophobic residues in the individual pentraxins ranges from 37.4% to 39.8%. This enriched conservation of hydrophobic residues most

```
       F.P.    1   TDLtgKVFVFPRqSETDyVKLIPrLdKPLQNFTvCFRaYSDLS  RphSLFSY
  Mouse SAP    1  qTDLkrKVFVFPRESETDHVKLIPhLEKPLQNFTLCFRtYSDLS  RsqSLFSY
  Human SAP    1  hTDLSgKVFVFPRESvTDHVnLItPLEKPLQNFTLCFRaYSDLS  RaYSLFSY
  Human CRP    1  QTDMSrKAFVFPKESDTSYVSLkAPLtKPLKAFTVCLhFYTeLSsTRGYSIFSY
 Rabbit CRP    1  QagMhkKAFVFPKESDnSYVSLnAqLkKPLKAFTVCLyFYTdLSmTRGYSIFSY

        con       qtdls-KvFVFPreS-tdyV-Li-pLeKPLqnFTvCfr-YsdLS-tRgySlFSY

       F.P.   52  nteygeNELLIYKEriGEYeLYIGnqgtkVhGvEEfaSPVHfCTsWESSSGIaE
  Mouse SAP   53  svkGRDNELLIYKEkVGEYSLYIGqsKVTVrGmEEylSPVHICTtWESSSGIvE
  Human SAP   53  nTqGRDNELLvYKErVGEYSLYIGrhKVTpkViEktpAPVHICvSWESSSGIaE
  Human CRP   55  ATkRQDNEILiFWSKDIGYSFtVGGsEIlFeVpEvtVAPVHICtSWESaSGIvE
 Rabbit CRP   55  ATrRQfNEILlFWSKDIGYSFsVGGdEIiFkVsdipVdPtHlCaSWESstGIaE

        con       -tk--dNElLiykek-geYslyiGgs--t-kv-eefv-PvH-CtsWESssGIaE

       F.P.  106  FWVNGKPWVKKgLQkgYTVKnkPSIiLgQEQDNYGGGFdnyQSFVGEigDLnMW
  Mouse SAP  107  FWVNGKPWVKKsLQreYTVKApPSIVLrQEQDNYGGGFqRSQSFVGEfsDLYMW
  Human SAP  107  FWiNGtPlVKKgLrqGYfVeAqPkIVLGQEQDSYGGkFdRSQSFVGEIGDLYMW
  Human CRP  109  FWVDGKPrVRKSLKKGYtVGAEASIILGQEQDSFGGnFEgSQSLVGDIGNVNMW
 Rabbit CRP  109  lWVDGKPmVRKSLKKGYilGpEASIILGQdQDSFGGsFEkqQSLVGDIGNVNMW

        con       fWvnGkPwVkKsL-kgYtv-aepsIiLgQeQDsyGGgF-rsQSfVGeigdlnMW

       F.P.  160  DsVLTPeeIksVYqgvPlePNILdWQALNYEmNGYaVIRPRcvalssynkis
  Mouse SAP  161  DyVLTPqdILfVYrdsPvnPNILnWQALNYEINGYVVIRPRVWd
  Human SAP  161  DsVLpPenILsaYqGtPlpaNILdWQALNYEIrGYViIKPlVWv
  Human CRP  163  DfVLSPdEINTIYlGGPFSPNVLnWRALkYEVqGEVftKPQLWP
 Rabbit CRP  163  DyaLSPeEINTIYaGGtFSPNVLdWReLtYqVrGEVhvKPQLWP

        con       D-vL-PeeI---Yqggp-spNiLdWqaLnYe--GyvvikP--wp--------
```

Fig. 4.1. Alignment of the amino acid sequences of the mammalian pentraxins, Syrian hamster female protein, mouse SAP, human SAP, human CRP and rabbit CRP using the "Genalign" program (Intelligenetics, Mountain View, CA). Listed below is a consensus sequence based on identity: *upper case letters* indicate residues conserved at the given position in all five proteins; *lower case letters* indicate residues present in a majority of the five proteins; *dashes* indicate residues for which no consensus choice could be made.

probably highlights regions important to monomer folding and subunit interactive faces in the native pentamer. That hydrophobic and hydrophilic periodicities for the human CRP and SAP molecules are extremely similar, especially over the N-terminal 100 amino acids, is apparent from the hydrophilicity plots depicted in Figure 4.2.

Based on similar comparative studies of rabbit, human and *Limulus* CRP, Liu et al. (1987) have identified two regions of probable functional importance. The first is between residues 51 and 66 of human CRP and possibly confers upon the CRPs and female protein the capacity to bind phosphorylcholine (PC). The second is between residues 133 and 147 of human CRP and may confer upon the CRPs the capacity to bind calcium ions. Searches of the Swiss protein database using the putative PC binding region of human CRP identified limited homologies with histone H4 from a range of species; searches using the putative calcium binding region identified limited homologies with phospholipases and phospholipase inhibitors.

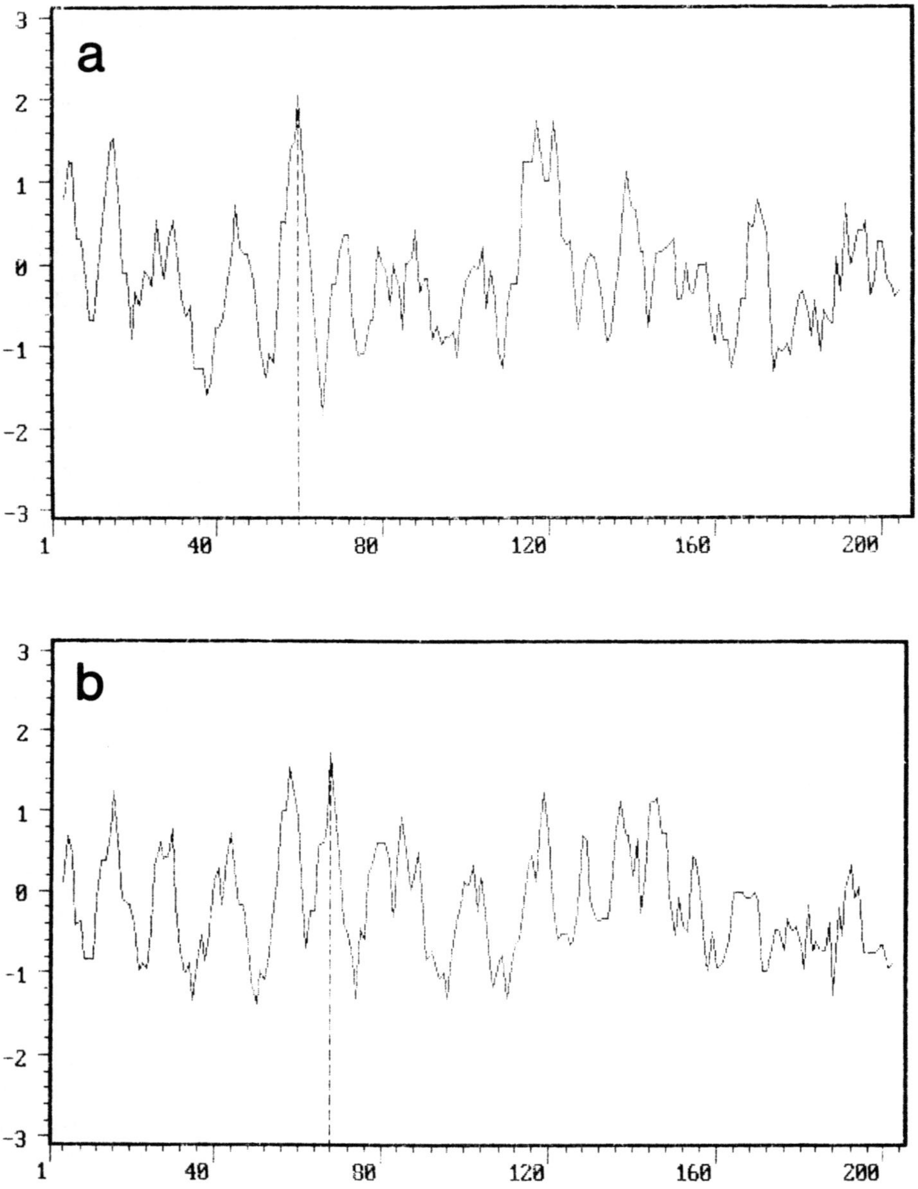

Fig. 4.2. Hydrophilicity plots of (a) human CRP and (b) human SAP using the "Antigen" program (Intelligenetics, Mountain View, CA). An average window of six amino acids was used. The *X-axis* indicates linear position from amino to carboxy terminus within the molecule; the *Y-axis* is an arbitrary scale from −3 (hydrophobic) to +3 (hydrophilic).

In addition, searches using the whole CRP amino acid sequence yielded a list of proteins with some regional sequence similarities including some that interact with lipoproteins and some that interact with nucleic acids, e.g. viral and bacterial DNA-directed DNA and RNA polymerases. The C terminal portion of human CRP has previously been reported (Woo et al. 1985) to have moderate homology with the C terminal region of *Drosophila* heat shock protein 26. Although such restricted stretches of homology are difficult to interpret, the possibility that pentraxins have the capacity to bind DNA in a specific manner and may be related to heat shock proteins is provocative.

The Structure of Pentraxin Genes

The structures of a number of pentraxin genes have been determined. Three functional CRP genes in *Limulus* (Nguyen et al. 1986b), a single functional CRP gene in rabbit (Hu et al. 1986), single functional CRP (Woo et al. 1985; Lei et al. 1985) and SAP genes in man (Ohnishi et al. 1986) and a single human CRP pseudogene (Goldman et al. 1987) have been characterized. The *Limulus* CRP genes are structurally simple. They have CAAT and TATAA promoter elements in the appropriate position in the 5′ region, a short 5′ untranslated region, an intron-less region encoding a pre-CRP with a leader sequence, and a longer than average length 3′ untranslated region of just under 300 base pairs. The predicted size of the mRNA is 1.3 kb. In *Limulus*, CRP is expressed constitutively at high levels. In contrast, the human and rabbit CRP genes have a more complex structure and contain elements that may be of importance to induction during the acute phase of inflammation. They both have elements in the 5′ region similar to heat shock consensus sequences; both have introns of almost 300 base pairs that essentially separate the sequence encoding the leader peptide from that encoding the bulk of the mature protein; the introns of each contain a purine–pyrimidine repeat sequence capable, in theory, of adopting the Z-DNA conformation; and both have an unusually long 3′ untranslated region of approximately 1.2 kb. The predicted and experimentally observed size of the mRNA is approximately 2.2 kb. The gene elements listed above are described below in a discussion of the human CRP gene and are shown in Fig. 4.3. It is interesting to note that the human SAP gene, which is not induced dramatically during acute inflammation and does not have heat shock consensus elements, has a similarly located intron that nevertheless is much smaller and lacks the purine–pyrimidine repeats, and

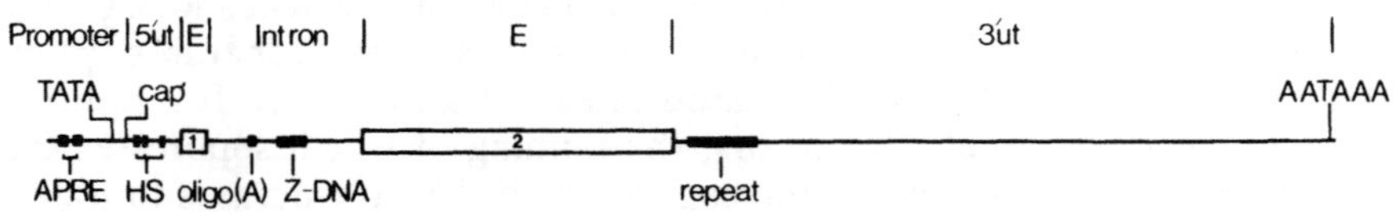

Fig. 4.3. Human CRP gene structure. Scale representation of the 2.4 kilobases of the human CRP gene. Regional features are listed above; *5′ut*, 5′ untranslated region; *3′ut*, 3′ untranslated region; *E*, regions corresponding to sequence of the first exon encoding the CRP leader sequence (*1*) and sequence of the second exon encoding the bulk of the mature CRP molecule (*2*). Individual elements are indicated on the linear representation of the gene; *TATA*, TATA box; *cap*, mRNA cap site; *AATAAA*, polyadenylation signal sequence; *APRE*, acute phase response elements; *HS*, heat shock consensus sequences; *oligo (A)*, stretch of adenines; *Z-DNA*, region of alternating purines and pyrimidines; *repeat*, intron-like tandem repeat.

specifies a normal length (i.e. 155 bases) 3′ untranslated region. The predicted and experimentally observed size of the human SAP mRNA is 1.1 kb.

Heat Shock Elements

The 5′ portion of the human CRP gene contains three elements similar to the heat shock consensus sequence CTnGAAnnTTCnAG first recognized in the control regions of *Drosophila* heat shock protein genes (Pelham 1982) and subsequently shown to be present in the control regions of a variety of heat shock genes in a wide range of species. In the human gene these regions lie between the TATA box and the codon specifying the pre-CRP N-terminal methionine, i.e. immediately downstream from the mRNA cap site (defined by Lei et al. 1985) between nucleotides 13 and 25 (CTagGACTTCtAG), between nucleotides 29 and 40 (CTGAActTTCAG), and between nucleotides 64 and 73 (gTGAATTCAG) of the mRNA. The rabbit CRP gene contains heat shock consensus sequences in similar positions. In *Drosophila*, the presence of one or more heat shock consensus sequences in the 5′ regions of the heat shock genes is necessary for heat-induced chromosome puffing and the coincident increase in transcription of heat shock protein RNAs (Simon et al. 1985). The heat shock consensus sequence analogues listed above may have a similar physiological role in enhancement of CRP synthesis in the fever conditions usually associated with the acute phase response.

NMR spectral analysis (Ott and Eckstein 1985) and X-ray crystallography (Dickerson and Drew 1981) of the dodecamer d(CGCGAATTCGCG), whose central portion is similar to that of the heat shock consensus sequence, supports the contention that the importance of these elements in temperature induced gene expression may be related to their intrinsic thermal instability. In these studies, a structural discontinuity, manifested as anomalous twist angles, was observed immediately before and after the central hexamer; and temperature dependent conformational changes were generated by the presence of the central A–T base pairs. In addition, transcription factors involved in induction of gene expression have been shown to bind to the heat shock consensus sequences of heat shock protein genes (Parker and Topol 1984). It is possible that thermally induced conformational changes or binding of transcription factors, or a combination of both, effect alterations in the expression of the CRP gene through one or more of the heat shock consensus sequences. Alternatively, the position of the heat shock consensus sequences in the 5′ untranslated region of the CRP mRNA, rather than upstream of the TATA box as is usual in authentic heat shock genes, allows speculation that their role may not be simply as targets for transcription factors that mediate induction via increased rates of transcription. It is entirely possible that they are targets for factors, the binding of which suppresses the initiation of transcription or blocks the capacity of RNA polymerases to utilize the CRP gene as a template for synthesis of an intact CRP pre-RNA, thereby facilitating the de-induction of the CRP gene during the later stages of the acute phase response. If such factors exist and are identical to those responsible for increased heat shock protein gene expression, it should be possible to demonstrate their capacity to co-ordinate the induction of heat shock protein synthesis and de-induction of CRP synthesis in the in vivo and in vitro models of the acute phase response that are currently available.

The Purine–Pyrimidine Repeat Region and Oligo-A Stretch in the CRP Intron

The human CRP gene contains a single intron of 278 base pairs beginning after nucleotide 165 of the mRNA and following the codons specifying the first two amino acids of the mature CRP. Its position is similar to that observed for introns in some other genes; however, it contains two unusual repetitive elements: a run of 16 adenines from nucleotide 243 to nucleotide 258 and a block of $(GT)_{15}G(GT)_3$ from nucleotide 294 to nucleotide 330 of the pre-RNA. The latter purine–pyrimidine repeat region is also present in the rabbit CRP intron and is similar to those found in several other regions of the genome (reviewed by Rogers 1983). Such $(GT)_n$ sequences can, in theory, adopt the left-handed Z–DNA form and may play a role in gene regulation. Already, such elements have been shown to be targets for intracellular messengers such as cyclic-AMP receptor protein (Fried et al. 1983) and the Z-DNA binding proteins that activate specific genes through recognition of control regions in the SV40 genome (Azorin and Rich. 1985). The stretch of adenines (though not present in the rabbit CRP gene intron) may be important to post-transcriptional regulatory mechanisms in man. It has been demonstrated by isoelectric focusing and tryptic mapping (Schönfelder et al. 1985) that poly(A)-binding protein, a 70 kDa molecule that binds polyadenylated RNA and is associated with mRNA–protein complexes (messenger ribonucleoproteins or mRNPs), is identical to the 70 kDa mammalian heat shock protein (Hsp70). It is probable that Hsp70 is induced during the acute phase response to inflammation coincident with the onset of fever; it may play a role in facilitating CRP induction, perhaps via rapid processing and/or transport of newly synthesized CRP transcripts in addition to its role in mRNPs. This speculation is consistent with the observation that CRP synthesis and secretion rates change during the course of the acute phase response to inflammation (Macintyre et al. 1985).

The 3′ Untranslated Region

The 3′ untranslated region of the CRP mRNA is remarkable for its great length. It is reminiscent of the relatively long 3′ untranslated sequence necessary for the efficient activation and transcription of *Drosophila* heat shock genes (Simon et al. 1985). The CRP 3′ untranslated region may fulfill a similar function or contribute to postranscriptional RNA processing and transport or to mRNA stability. Additionally, there are 5′ and 3′ consensus splice junctions beginning at nucleotides 1079 (AGGT) and 1214 (P_{y7}CAG) respectively. Within this intron-like region there is an almost perfect extensive repeat element, comprising nucleotides 1083–1130 and 1131–1178. Although there is no experimental evidence for two different CRP mRNA transcripts in human acute phase liver (Tucci et al. 1983), it is possible that there is a poly-RNA for CRP and that different mRNA processing occurs in tissue other than liver, and/or in nonacute phase liver. Together with the homology of portions of the CRP amino acid sequence with various heat shock proteins (Woo et al. 1985; Whitehead, unpublished work) and the presence of heat shock consensus sequences in the mammalian CRP gene, the unusual length of the 3′ untranslated region of the mRNA suggests that CRP may be a mammalian heat shock protein.

Elements Responding to Cytokines

The acute phase proteins are induced, at least in part, by cytokines such as interleukin-1 (IL-1) and interleukin-6 (IL-6). Using a chloramphenicol acetyl transferase (CAT) assay system in which a range of 5′ fragments of the CRP gene were coupled to a promoterless CAT gene, Arcone et al. (1988) have shown that monocyte conditioned medium can act on two genetic elements. Relative to the TATA box, these are between nucleotides −94 to −50 and upstream of nucleotide −106. The dramatic induction of CAT activity by monocyte conditioned medium indicated that the above regions contain acute phase responsive elements (APREs). Using purified recombinant IL-6 a convincing but much smaller induction is achieved (Morrone et al. 1988), indicating that IL-6 is responsible for some, but not all, of the increase in APRE-directed CRP gene expression previously reported to be mediated by monocyte conditioned medium (Darlington et al. 1986; Goldman and Liu 1987).

Genetics

The genes encoding human CRP and SAP were originally mapped to chromosome 1 using cDNA probes in the Southern blot analysis of DNA samples from human–rodent somatic cell hybrids containing a limited number and varied composition of human chromosomes (Whitehead et al. 1983; Mantzouranis et al. 1985). Subsequently, *in situ* hybridization experiments allowed localization of the human CRP and SAP genes to band q2.1 (Floyd-Smith et al. 1986). A human CRP pseudogene has also been defined and mapped to within 7 kb of the functional CRP gene (Goldman et al. 1987). Thus, in man, there are at least three closely linked pentraxin loci. The *in situ* hybridization studies showed no significant accumulation of specific hybridization signals at any position on human chromosome 1 other than band q2.1 or on any other chromosome. Therefore, if other closely related pentraxin genes or pseudogenes remain to be discovered, they must be clustered rather than dispersed.

In mouse, a locus controlling endogenous serum SAP levels, which differ between strains, has been mapped via the assessment in recombinant inbred animals of progenitor-strain-specific serum SAP levels and correlation with progenitor-strain-specific allele profiles (Mortensen et al. 1985). A subsequent study (Whitehead et al. 1988) mapped the mouse SAP structural gene via correlation of the presence of progenitor-strain-specific SAP restriction fragment length polymorphisms with progenitor-strain-specific allele profiles. These analyses established that the mouse SAP gene maps to the distal region of mouse chromosome 1; in the latter study a recombination frequency between the SAP and Ly-9 loci of less than 0.7 units was calculated, thereby indicating extremely close linkage. Very close linkage to the α-spectrin gene (Seldin et al. 1987) is also implied by the above.

The mapping of the human pentraxin genes and the mouse SAP gene is to regions that are extensively homologous between the two species (Lalley et al.

1988). The genes for apolipoprotein A–II, renin, SAP, α-spectrin, peptidase C, alkaline phosphatase, complement factor H, the CR2 receptor and antithrombin 3 have all been mapped to the q arm of human chromosome 1 and to the distal portion of mouse chromosome 1. This region of mouse chromosome 1 also contains several other loci with immune functions, i.e. the lymphocyte cell surface associated molecules Ly-5, Ly-17, Ly-22 and Ly-33. The Ly-17 locus has recently been shown to be indistinguishable from the Fcγ2b/γ1R locus that is itself closely linked to another Fc receptor gene, FcεRIα and to the α-spectrin gene (Huppi et al. 1988).

In addition to the pentraxin genes, band q2.1 of human chromosome 1 carries the H4F2 histone gene cluster (Green et al. 1984) and the genes for two inflammation-associated macrophage products (MRP-8 and MRP-14) each of which is probably secreted in addition to being expressed on macrophage cell surfaces (Odink et al. 1987). MRP-14 and MRP-8 are both produced by infiltrate macrophages in chronically inflamed tissue whereas only MRP-14, which is identical to the cystic fibrosis associated antigen (Dorin et al. 1987), is expressed by infiltrate macrophages in acutely inflamed tissues.

It seems likely that Fc receptors and lymphocyte cell surface molecules will be found close to the pentraxin genes in the human genome and that genes for CRP, the MRP macrophage inflammatory proteins and histones will be found close to the SAP locus in the mouse genome. The sublocalization of the above genes suggests the existence of a complex of closely linked genes with roles related to immunity and inflammation. Given the limited homology of CRP and SAP to histones and their high affinity for chromatin, an additional theme of such a complex might be the clustering of genes whose products interact with DNA.

Acknowledgements. This work was supported in part by a grant from the Council for Tobacco Research, USA–Inc., and by National Science Foundation grant DCB–8615767. I am a Pew Scholar in the Biomedical Sciences. I thank Shirley Lerner for excellent secretarial assistance.

References

Arcone R, Gualandi G, Ciliberto G (1988) Identification of sequences responsible for acute-phase induction of human C-reactive protein. Nucl Acids Res 16:3195–3207

Azorin F, Rich A (1985) Isolation of Z–DNA binding protein from SV40 minichromosomes: evidence for binding to the viral control region. Cell 41:365–374

Darlington GJ, Wilson DR, Lachman LB (1986) Monocyte-conditioned medium, interleukin 1, and tumour necrosis factor stimulate the acute phase response in human hepatoma cells in vitro. J Cell Biol 103:787–793

Dickerson RE, Drew HR (1981) Structure of a B DNA dodecamer. II. Influence of base sequence on helix structure. J Mol Biol 149:761–786

Dorin JR, Novak M, Hill RE, Brock DJH, Secher DS, van Heyningen V (1987) A clue to the basic defect in cystic fibrosis from cloning the CF antigen. Nature 326:614–617

Dowton SB, Woods DE, Mantzouranis EC, Colten HR (1985) Syrian hamster female protein: analysis of female protein primary structure and gene expression. Science 228:1206–1208

Fernandez-Moran H, Marchalonis J, Edelman GM (1968) Electron microscopy of a hemaglutinin from *Limulus polyphemus*. J Mol Biol 32:467–469

Floyd-Smith G, Whitehead AS, Colten HR, Francke U (1986) The human C-reactive protein gene (CRP) and serum amyloid P component gene (APCS) are located on the proximal long arm of chromosome 1. Immunogenetics 24:171–176
Fried MG, Wu H-M, Crothers DM (1983) CAP binding to B and Z forms of DNA. Nucl Acids Res 11:2479–2494
Goldman ND, Liu T-Y (1987) Biosynthesis of human C-reactive protein in cultured hepatoma cells is induced by a monocyte factor(s) other than interleukin 1. J Biol Chem 262:2363–2368
Goldman ND, Liu T, Lei K-J (1987) Structural analysis of the locus containing the human C-reactive protein gene and its related pseudogene. J Biol Chem 262:7001–7005
Green L, van Antwerpen R, Stein J, Stein G, Tripputi P, Emanuel B, Seldin J, Croce C (1984) A major human histone gene cluster on the long arm of chromosome 1. Science 226:838–840
Hu S-I, Miller SM, Samols D (1986) Cloning and characterization of the gene for rabbit C-reactive protein. Biochemistry 25:7834–7839
Huppi K, Mock BA, Hilgers J, Kochan J, Kinot J-P (1988) Receptors for Fc and Fc are linked on mouse chromosome 1. J Immunol 141:2807–2810
Inoue S, Leblond CP, Grant DS, Rico P (1986) The microfibrils of connective tissue: II. Immunohistochemical detection of the amyloid P component. Amer J Anat 176:139–152
Ishikawa N, Shigemoto K, Masuyama N (1987) The complete nucleotide and deduced amino acid sequence of mouse serum amyloid P component. Nucl Acids Res 15:7186
Kushner I, Volanakis JE, Gewurz H (Eds) (1982) C-reactive protein and the plasma protein response to tissue injury. Ann NY Acad Sci 389
Lalley PA, O'Brien SJ, Creau-Goldberg N, Davisson MT, Roderick TH, Echard G, Womack JE, Graves JM, Doolittle DP, Guidi JN (1988) Report of the committee on comparative mapping. Cytogenet Cell Genet 47:368–389
Lei K-J, Liu T, Zon G, Soravia E, Liu T-Y, Goldman ND (1985) Genomic DNA sequence for human C-reactive protein. J Biol Chem 160:13377–13383
Liu T-Y, Syin C, Nguyen NY, Suzuki A, Boykins RA, Lei K-J, Goldman N (1987) Comparison of protein structure and genomic structure of human, rabbit, and *Limulus* C-reactive proteins: possible implications for function and evolution. J Prot Chem 6:263–271
Macintyre SS, Kushner I, Samols D (1985) Secretion of C-reactive protein becomes more efficient during the course of the acute phase response. J Biol Chem 260:4169–4173
Mantzouranis EC, Dowton SB, Whitehead AS, Edge MD, Bruns GAP, Colten HR (1985) Human serum amyloid P component: cDNA isolation, complete sequence of pre-serum amyloid P component and localization of the gene to chromosome 1. J Biol Chem 260:7752–7756
Morrone G, Ciliberto G, Oliviero S, Arcone R, Dente L, Content J, Cortese R (1988) Recombinant interleukin 6 regulates the transcriptional activation of a set of human acute phase protein genes. J Biol Chem 263:12554–12558
Mortensen RF, Le PT, Taylor BA (1985) Mouse serum amyloid P-component (SAP) levels controlled by a locus on chromosome 1. Immunogenet 22:367–375
Nguyen NY, Suzuki A, Boykins RA, Liu T-Y (1986a) The amino acid sequence of *Limulus* C-reactive protein. J Biol Chem 261:10456–10465
Nguyen NY, Suzuki A, Cheng S-M, Zon G, Liu T-Y (1986b) Isolation and characterization of *Limulus* C-reactive protein genes. J Biol Chem 261:10450–10455
Odink K, Cerletti N, Bruggen J, Clerc RG, Tarcsay L, Zwadlo G, Gerhards G, Schlegel R, Sorg C (1987) Two calcium binding proteins in infiltrate macrophages of rheumatoid arthritis. Nature 330:80–82
Ohnishi S, Maeda S, Shimada K, Arao T (1986) Isolation and characterization of the complete complementary and genomic DNA sequences of human serum amyloid P component. J Biochem 100:849–858
Oliviera EB, Gotschlich, EC, Liu T-Y (1979) Primary structure of human C-reactive protein. J Biol Chem 254:489–502
Osmand AP, Friedenson B, Gewurz H, Painter RH, Hofmann T, Shelton E (1977) Characterization of C-reactive protein and the complement subcomponent C1t as homologous proteins displaying cyclic pentameric symmetry (pentraxins). Proc Natl Acad Sci USA 74:739–743
Ott J, Eckstein F (1985) ^{31}P NMR spectral analysis of the dodecamer d(CGCGAATTCGCG). Biochemistry 24:2530–2535
Parker CS, Topol J (1984) A *Drosophila* RNA polymerase II transcription factor binds to the regulatory site of an hsp70 gene. Cell 37:273–283
Pelham HRB (1982) A regulatory upstream promoter element in the Drosophila Hsp70 heat-shock gene. Cell 30:517–528

Pepys MB and Baltz ML (1983) Acute phase proteins with special reference to C-reactive protein and related proteins (pentraxin) and serum amyloid A protein. Adv Immunol 34:141–212

Pepys MB, Butler PJG (1987) Serum amyloid P component is the major calcium-dependent specific DNA binding protein of the serum. Biochem Biophys Res Commun 148:308–313

Robey FA, Jones KD, Tanaka T, Liu T-Y (1984) Binding of C-reactive protein to chromatin and nucleosome core particles. A possible physiological role of C-reactive protein. J Biol Chem 259:7311–7316

Schönfelder M, Horsch A, Schmid HP (1985) Heat shock increases the synthesis of the poly (A)-binding protein in HeLa cells. Proc Natl Acad Sci USA. 82:6884–6888

Seldin MF, Morse HC, D'Hoostelaere L, Britten JL, Steinberg AD (1987) Mapping of alpha-spectrin on distal mouse chromosome 1. Cytogenet Cell Genet 45:52–54

Simon JA, Sutton CA, Lobell RB, Glaser RL, Lis JT (1985) Determinants of heat shock-induced chromosome puffing. Cell 40:805–817

Tillett WS, Francis T Jr (1930) Serological reactions in pneumonia with non-protein somatic fraction of pneumococcus. J Exp Med 52:561–571

Tucci A, Goldberger G, Whitehead AS, Kay RM, Woods DE, Colten HR (1983) Biosynthesis and postsynthetic processing of human C-reactive protein. J Immunol 131:2416–2419

Whitehead AS, Bruns GAP, Markham AF, Colten HR, Woods DE (1983) Isolation of human C-reactive protein complementary DNA and localization of the gene to chromosome 1. Science 221:69–71

Whitehead AS, Rits M, Michaelson J (1988) Molecular genetics of mouse serum amyloid P component (SAP): cloning and gene mapping. Immunogen 28:388–390

Woo P, Korenberg JR, Whitehead AS (1985) Characterization of genomic and complementary DNA sequence of human C-reactive protein, and comparison with the complementary DNA sequence of serum amyloid P component. J Biol Chem 260:13384–13388

Zahedi K, Mortensen RF (1986) Macrophage tumoricidal activity induced by human C-reactive protein. Cancer Res 46:5077–5083

Chapter 5

ApoSAA: Structure, Tissue Expression and Possible Functions

E. P. Benditt, R. L. Meek and N. Eriksen

Introduction

Background

Amyloidosis is a classical problem in pathology (Virchow 1860). Investigations in our laboratory on amyloid substance, and its relation to the acute phase reaction, began more than 25 years ago and we wish to recount some history and then present a résumé of recent work dealing with the questions that are pertinent to the problems of serum amyloid A (SAA) as an acute phase reactant.

Investigations of amyloid substance have centered about the following questions: (i) of what is amyloid substance composed? (ii) what is the structure of the major components of the amyloid substance? (iii) where do they come from? (iv) how do the components get there? (v) what factors determine their deposition in the amyloid substance? Modern molecular biology of proteins and genetic constituents has enlarged our horizons enormously, presented new questions about, and provided new means for dissecting a pathological process and its related physiology.

Our studies of amyloidosis began at the University of Washington in 1961 with the aim of investigating the nature of amyloid substance using tools and techniques that were then becoming available, including ultrastructural analysis and modern biochemical methods. The first case from which we obtained tissues for our studies was a case with chronic inflammation of the bowel and systemic amyloidosis (Benditt et al. 1962). The amyloid-rich tissues from this case yielded a polypeptide of about 8.5 kDa not seen in tissues from nonamyloidotic persons.

Chronic inflammation, particularly of the suppurative variety, has long been known to be associated in some persons with the appearance of amyloid deposits (Bell 1930); since the early part of the twentieth century it was believed that this association related to the involvement of antibodies in the deposition of the amyloid substance. This idea seemed to be further strengthened by the fact that horses used to produce antitoxins frequently succumbed to amyloidosis. This point of view prevailed when we began our studies. Under these circumstance it is not surprising that our first reports, that the amyloidosis found in association with chronic inflammation has as its major constituent a peptide, amyloid A protein, unrelated by chemical or immunological criteria to any known immunoglobulin, made little impact (Benditt and Eriksen 1964, 1966; Benditt et al. 1962, 1971) Our case was helped by the investigations of Husby et al. (1972) in Oslo and Levin et al. (1972) in New York who reported finding a similar entity. The development of antibodies made against the degraded amyloid fibrils from this form of amyloidosis revealed the presence in blood of a related protein (Levin and Franklin 1973). An interesting feature observed early was that the circulating concentration of this immunologically related entity was highly variable (Levin et al. 1972). It was quickly ascertained that normal persons had very low circulating levels, and that markedly elevated serum concentrations were found following surgery or in association with an acute infection (Rosenthal and Franklin 1975). This behaviour is similar to that of other "acute phase reactants" in plasma; however, the range of variation in concentration of SAA is much greater than that of most other acute phase reactants.

SAA is an Apolipoprotein

Recognition of the circulating relative of the amyloid fibril protein (SAA) led to characterization of the precursor protein. By molecular sizing chromatography, there appeared to be two main sizes of the serum immunoreactive material, one of approximately 180 kDa and another of about 12 kDa. A significant feature of the structure of the amyloid protein was revealed when Segrest and co-workers at Baylor in Houston, a group interested in lipoproteins, found when examining protein sequence data that amyloid protein A shared with apolipoprotein AI a distribution of charged and uncharged amino acids with the capacity to form amphipathic helices (Segrest et al. 1976). This suggested the possibility that the large serum protein reacting with antibodies to amyloid A protein was not an aggregate or polymer of the smaller 12 kDa protein as had been postulated by others, but that the latter might be part of an apolipoprotein complex. A simple test of this hypothesis showed that the 12 kDa element was in fact an apolipoprotein, which we have named apoSAA (Benditt and Eriksen 1978; Parmelee et al. 1982). ApoSAA appears in several lipoproteins but is found mainly in the HDL3 fraction of the plasma, the fraction with an apparent size of about 180 kDa. SAA now appears to be a family of SAA lipoproteins found in all mammals and birds studied thus far (Benditt and Eriksen 1973; Husebekk et al. 1988).

Once the nature of the amyloid precursor protein and its amino acid sequence was established, it became possible to clone the genes and this was done for the SAAs of mouse by two laboratories, one in Baltimore and one in Japan (Lowell et al. 1986; Yamamoto and Migita 1985). This information has been used by us and by others to develop nucleotide probes specific for the individual members of the murine family of genes. These reagents provided the means for exploring the cell source and regulation of the proteins of the SAA gene family in the mouse. Using these probes, and antibodies that react with the mouse SAA_1 and SAA_2 proteins we have examined the expression of mRNA for the individual genes in normal mice and in mice during the acute phase response as well as during the development of experimentally induced amyloidosis. The following is a résumé of recent results followed by some speculations about the possible role(s) of the apoSAA family of genes in normal physiology and the response to injury.

Where are SAAs Made and Where do They go?

We have used studies of cellular interactions and metabolism of apoSAA to search for clues to the function of SAA. Our first examination of tissue distribution of intravenously administered radio-labelled apoSAA in mice showed a widespread distribution of SAA among tissues and rapid clearance rate from plasma, half life <1 hour (Hoffman and Benditt 1983; Kushner 1982; Benson et al. 1977). In recent studies combining investigations of protein distribution by immunocytochemistry and mRNA expression specific for the three SAAs, we have been able to identify individual cell types which take up apoSAA during the acute inflammation induced by casein injection: it has been possible to observe the appearance of apoSAA proteins in tissue locations and cells and to relate this to specific SAA mRNA expression (Meek et al. 1989). Such observations permit inferences with reference to the question of whether there is local synthesis of the proteins or whether they are taken up from the circulation. Using RNA blot hybridization on extracted RNA from different tissues of mice after casein injection, we detected SAA_1 and SAA_2 mRNA in liver, but not in kidney, testis, spleen and adrenal gland; however, immunohistochemical staining showed the presence of the protein in these extrahepatic locations. Specifically, SAA was found within epithelial cells of kidney proximal convoluted tubules and its accumulation followed a time course that paralleled SAA levels in the circulation. Similar immunocytochemical findings were reported for kidney by Miura et al. (1985). Although SAA is an apoprotein of HDL, a particle too large to be filtered by glomerular epithelium, the SAA_1 and SAA_2 are reversibly associated with HDL and a small amount of apoSAA may well circulate unbound to the lipid particle carrier (Hoffman and Benditt 1982a,b). If so, this free apoSAA could be filtered and resorbed in the kidney as has been postulated for rat apo A-I found in kidney proximal convoluted tubules (Glass et al. 1983).

In Leydig cells of the testis, SAA uptake was delayed by approximately 12 hours compared with the appearance of SAA in liver, but thereafter rapidly reached high levels (Meek et al. 1989). This order of events may reflect a regulated uptake mechanism that is delayed compared to the kinetics of SAA plasma elevation. Similarly, 20 hours after casein injection, when circulating SAA was at the highest levels, adrenal cortical cells displayed high levels of SAA. This is of interest because the two cell types that displayed this pattern of SAA uptake are both involved in steroid synthesis, suggesting perhaps that one function of SAA is related to steroid hormone formation and/or lipid metabolism performed in these sites.

In the spleen, SAA was found localized to parafollicular regions. It was first detected by 16 hours after casein injection when circulating SAA levels had reached their peak, and plateaued by 24 hours. Not all cells exhibited SAA immunoreactivity, and a more thorough investigation is required to establish the precise identity of cells that take up SAA. Localization of SAA to the splenic parafollicular zones is of special interest since this is a major site of amyloid deposition. What special task SAA may be performing there is not yet clear.

Murine amyloid protein A is derived from circulating apoSAA_2 by a process that involves selective removal of only apoSAA_2 from the circulating pool of HDL containing both apoSAA_1 and apoSAA_2 (Hoffman et al. 1984; Meek et al. 1986) and deposition of the 8.5 kDa NH_2-terminal AA derivative in spleen. The monotypic character of murine AA is not due to local producation of SAA_2 as indicated by evidence below. Hence there may be selective binding sites for SAA_2 leading to fibril deposition (Meek et al. 1986).

Patterns of Cell and Tissue Expression of SAA mRNAs in Mice

Liver is the major site of SAA synthesis. Injection of LPS elevates the steady state levels of SAA mRNA for each of the three SAA genes several hundred to 2000-fold (Meek and Benditt 1986; Lowell et al. 1986). After casein injection a similar induction of SAA_1 and SAA_2 but not SAA_3 mRNAs is seen (Meek and Benditt 1986). Hepatocytes are considered to be the cells responsible for SAA synthesis, since in primary culture they synthesize and secrete SAA_1 and SAA_2 (Kushner 1982; Hoffman and Benditt 1982; Benson and Kleiner 1980). By *in situ* hybridization of liver, we observed that hepatocytes expressed high levels of SAA_1 and SAA_2 mRNA whereas Kupffer and other liver cells did not (Meek et al. 1989).

ApoSAA genes are expressed in extrahepatic tissues in response to lipopolysaccharide (LPS) injection; this expression is almost exclusively limited to SAA_3 (Meek and Benditt 1986). However, the kidney and intestine are exceptions. After injection of LPS into mice, the kidneys express high levels of SAA_1, SAA_2 and SAA_3 mRNA (Meek and Benditt 1986). *In situ* hybridization revealed that SAA_1/SAA_2 mRNA is expressed in epithelial cells of proximal and distal convoluted tubules of kidney and that ileum and large intestine express SAA_1 mRNA in epithelial cells lining the mucous membranes (Meek et al. 1989). In regard to the function of these SAAs expressed in kidney and intestine, two possibilities come to mind: (i) apoSAAs at these sites may simply be secreted into the circulation to become associated with HDL and perform the same function as the SAAs secreted from liver; (ii) alternatively, there may be some local function performed by the SAAs.

We have found by blot analyses of extracted RNA that SAA_3 mRNA is expressed in all tissues examined after mice were treated with endotoxin (LPS) (Meek et al. 1986). Estimates showed the highest level of SAA_3 mRNA in non-hepatic cells to be in peritoneal macrophages and the signal strength was equivalent to that found in liver RNA of LPS-injected mice. Such macrophage-related cells are resident within many tissues and monocyte/macrophages accumulate in tissues that are sites of injury (Rosenstreich 1981). Because LPS-activated macrophages express high levels of SAA_3 mRNA, it seemed reasonable to ask whether or not a single dispersed cell system, such as the reticuloendothelial system, might be the source of extrahepatic SAA_3 mRNA. An SAA_3 specific cRNA probe was employed for *in situ* hybridization to histological sections of tissues obtained after injection of mice with LPS (Benditt and Meek 1989). The observations with *in situ* hybridization showed extrahepatic SAA_3 expression largely by adipocytes. In testis and spleen certain other non-adipocytes also yielded signals with the SAA_3 cRNA probe.

In liver sections from LPS-injected mice all hepatocytes expressed SAA_3 mRNA. However, some cells other than hepatocytes also expressed the SAA_3 mRNA, including cells lining sinusoids and central veins and cells adjacent to bile duct epithelium. Some of these SAA_3 mRNA-containing cells in liver appear to be Kupffer cells. This finding in liver differs from the expression of SAA_1 and SAA_2 mRNA in liver which is restricted to hepatocytes (Meek et al. 1989).

The finding of extrahepatic expression of SAA_3 mRNA by adipocytes associated with organs such as aorta, lung, ileum, large intestine and adrenal gland was unexpected and revealed that more than one dispersed system of cells is responsible for the SAA_3 mRNA expression found in the extrahepatic sites. In addition other cells expressing SAA_3 mRNA include Leydig cells of testis and other as yet undefined cells in the spleen. Since monocyte/macrophages express high levels of SAA_3 mRNA; SAA_3 expression in the spleen could be due to resident macrophages, free and/or fixed, which populate the parafollicular zones.

It is clear from our experiments that the murine family of SAA genes is differentially and widely expressed in various tissues in response to inflammatory stimuli and that the pattern of expression depends upon the inducers of the acute phase response. As we have indicated above, SAA expression in kidney is distinct from that in liver; SAA_1 and SAA_2 mRNA are both elevated to high levels after casein injection in liver but not in kidney. However, after LPS injection, all three SAA mRNAs are expressed in liver and kidney (Meek and Benditt 1986). Inducers of inflammation are known to stimulate production of various cytokines such as interleukin-1 (IL-1), interleukin-6 (IL-6) and tumour necrosis factor-α (TNF-α) (Selinger et al. 1980; Ramadori et al. 1985; Dinarello et al. 1986; Gauldie et al. 1987; Gittan et al. 1987). Individually or synergistically these molecules in turn mediate production of various acute phase proteins (Gauldie et al. 1987; Gittan et al. 1987). Il-1 has been shown to have SAA-inducing activity when injected into mice (Ramadori et al. 1985) and in cultured primary mouse hepatocytes (Kushner 1982; Selinger et al. 1980; Ramadori et al. 1985). Since the patterns of SAA mRNA expression in liver and kidney following LPS treatment are different from that seen after casein challenge, we suggest that induction of SAA mRNA may not be mediated by the same cytokines or combination of cytokines in liver and kidney.

Rats Have SAA-related Genes

Among all mammals studied the apoSAAs appear to be highly conserved. Rats have been puzzling. Rats do not develop amyloidosis and SAA has not been found as an apoprotein of rat HDL (Baltz et al. 1987). We have looked for SAA gene expression in rats in the following way (Meek et al. 1989): RNA extracted from various tissues of normal and injured rats was examined by RNA blot hybridization with SAA cDNA and cRNA probes for the three murine SAA genes. An SAA mRNA of approximately 400 nucleotides related to mouse SAA_1 and SAA_2 mRNAs was observed to reach a high level in rat liver 24 hours after injection of bacterial lipopolysaccharide (LPS). No extrahepatic tissues were found to express the SAA_1/SAA_2-related mRNA. Turpentine induced two hepatic SAA_1/SAA_2-related mRNAs of 400 and 500 nucleotides in length. Cell free translation of liver SAA_1/SAA_2-related mRNA, hybrid selected from both LPS- and turpentine-injected rats, produced a single protein with an estimated molecular weight of 8 kDa. This rat liver SAA-related mRNA appears to lack a highly conserved coding region for portions of the two amphipathic helical domains and the joining sequence found in other mammals. An mRNA related to mouse SAA_3 was found expressed at a high level in lung after LPS but not following turpentine injection. This mRNA was also expressed at high levels in ileum and large intestine of control rats and was not found in the liver of control or challenged rats. Thus the rat also has representation of the SAA gene family and the mRNA is expressed in rats under situations of injury response similar to those found in mice; these data lend further weight to the importance of the SAA gene family as part of an injury-responding mechanism of vertebrates.

Summary

We have presented a résumé of some history and some recent findings from our laboratory that bear upon the the structure and function of the acute phase reactant, the SAA(s). In the mouse there are three genes in the SAA family. Two of the gene products circulate, mainly associated with the HDL fraction of the plasma lipoproteins. These apoproteins show very rapid disappearance time compared with the other apoproteins, such as apo AI. Although both apoSAA_1 and apoSAA_2 circulate, only the SAA_2 is a precursor of the amyloid fibrils that appear in the perifollicular regions of the spleen during the early phase of amyloid deposition.

The apoSAAs, although differing in amino acid sequence, share the amphipathic helical structure with HDL apoproteins, such as apo AI. In the mouse all three genes are transcribed in the liver of animals challenged with certain inflammation-inducing agents, e.g. lipopolysaccharide. On the other hand, casein induces only SAA_1 and SAA_2. This clearly indicates different responses to different mediators, i.e. IL-1 versus IL-6 versus TNF. Extrahepatic expression of SAA genes can be found for all three genes in the mouse. However, the SAA_3 gene is transcribed most regularly in extrahepatic sites and most strikingly in the fat. Cells that have been found thus far to transcribe the SAA genes include: macrophages, hepatocytes, fat cells, epithelial cells of the convoluted tubules of the kidney and some epithelial cells of the intestine, and interstitial cells of the testis.

The SAA gene family is highly conserved among the vertebrates. The rat is not an exception, and has at least two genes that respond to injurious stimuli in a fashion similar to that found in other vertebrates.

From these facts we can draw some tentative conclusions. Plasma apolipoproteins play a central role in lipid metabolism (Mahley et al. 1984). They are known to function as signals in cell receptor-mediated uptake processes, as co-factors in enzymatic processing of lipoprotein lipids and as determinants of lipoprotein particle structure. However, very little is known in regard to the function of the apoSAAs. It seems possible that their role in the host response to injury is analogous to functions known for other apolipoproteins. The SAAs may play a role in lipid metabolism by influencing: (i) the rate of cholesterol transport between cells and HDL particles, (ii) intracellular transport of lipids (steroids) or (iii) activities of enzymes involved in steroid biosynthesis. We have identified several cell types that take up and others that may synthesize different apoSAAs and it remains to elucidate the interaction of apoSAA with these cells.

The fact that these are genes that respond as part of an acute phase response indicates that they are part of an early responding network of defensive molecules. A likely possibility is that they are conveyors of necessary lipids, including cholesterol, to needed locations like the adrenal and the Leydig cells, in addition to aiding in removal of lipid debris resulting from injured cells.

The behaviour of SAA_3 and its apparent lack of expression in the serum, in spite of its manifest transcription, raises the interesting possibility that these molecules could function either inside the cell or in some way bound to the cell. Something similar is seen in rats where no SAA is found associated with lipoproteins, but at least two forms of SAA genes are expressed in response to tissue injury.

Acknowledgements. We wish to acknowledge the excellent assistance of Ron Hanson and Susan Reed, and the patient help in manuscript preparation of Betty BeCraft and Virginia Wejak. This work was supported in part by USPHS grants HL-03174 and HL-40079.

References

Baltz ML, Rowe IF, Caspi D, Turnell WG, Pepys MB (1987) Acute-phase high-density lipoprotein in the rat does not contain serum amyloid A protein. Biochem J 242:301–303

Bell ET (1930) Pathology. Lea and Febiger, Philadelphia, pp 77–78

Benditt EP, Eriksen N (1964) Amyloid. II. Starch gel electrophoretic analysis of some protein extracted from amyloid. Arch Pathol 78:325–330

Benditt EP, Eriksen N (1966) Amyloid. III. A protein related to the subunit structure of human amyloid fibrils. Proc Natl Acad Sci USA 55:308–316

Benditt EP, Eriksen N (1973) The chemical classes of amyloid substance in men, monkeys and ducks. In: Peeters H (ed) Protides of the biological fluids (Proceedings of the twentieth colloquium, Brugge, Belgium 1972). Pergamon Press, Oxford, pp 81–85

Benditt EP, Eriksen N (1978) Amyloid protein SAA is associated with high density lipoprotein from human serum. Proc Natl Acad Sci USA 74:4025–4028

Benditt EP, Meek RL (1989) Expression of the third member of the serum amyloid A gene family in mouse adipocytes. J Exp Med 169:1841–1846

Benditt EP, Lagunoff D, Eriksen N, Iseri OA (1962) Amyloid. Extraction and preliminary characterization of some proteins. Arch Pathol 74:323–330

Benditt EP, Eriksen N, Hermodson MA, Ericsson LH (1971) The major proteins of human and monkey amyloid substance: common properties including N-terminal amino acid sequences. FEBS Lett 19:169–173

Benson MD, Kleiner L (1980) Synthesis and secretion of serum amyloid protein A (SAA) by hepatocytes in mice treated with casein. J Immunol 124:495–499

Benson MD, Scheinberg MA, Shirahama T, Cathcart ES, Skinner M (1977) Kinetics of serum amyloid protein A in casein-induced murine amyloidosis. J Clin Invest 59:412–417

Dinarello CA, Cannon JG, Wolff SM, Bernheim HA, Beutler B, Cerami A, Figari IS, Palladino MA, O'Connor JV (1986) Tumor necrosis factor (cachectin) is an endogenous pyrogen and induces production of interleukin 1. J Exp Med 163:1433–1450

Gauldie J, Richards C, Harnish D, Lansdorp P, Baumann H (1987) Interferon B2/B-cell stimulating factor type 2 shares identity with monocyte-derived hepatocyte-stimulating factor and regulates the major acute phase protein response in liver cells. Proc Natl Acad Sci USA 84:7251–7255

Gittan JD, Colten HR (1987) Molecular biology of the acute phase plasma proteins. Lymphokines 14:123–153

Glass CK, Pittman RC, Keller GA, Steinberg D (1983) Tissue sites of degradation of apoprotein A-I in the rat. J Biol Chem 258:7161–7167

Hoffman JS, Benditt EP (1982a) Changes in high density lipoprotein content following endotoxin administration in the mouse: formation of serum amyloid protein-rich subfractions. J Biol Chem 257:10510–10517

Hoffman JS, Benditt EP (1982b) Secretion of serum amyloid protein and assembly of serum amyloid protein-rich high density lipoprotein in primary mouse hepatocyte culture. J Biol Chem 257:10518–10522

Hoffman JS, Benditt EP (1983) Plasma clearance kinetics of the amyloid-related high-density lipoprotein apoprotein, serum amyloid protein (apoSAA) in the mouse. Evidence for a rapid clearance. J Clin Invest 71:926–934

Hoffman JS, Ericsson LH, Eriksen N, Walsh KA, Benditt EP (1984) Murine tissue amyloid protein AA. NH_2-terminal sequence identity with only one of two serum amyloid proteins (apoSAA) gene products. J Exp Med 159:641–646

Husby G, Sletten K, Michaelsen TE, Natvig JR (1972) Alternative, non-immunoglobulin origin of amyloid fibrils. Nature New Biol 238:187

Husebekk A, Husby G, Sletten K, Skogen B, Nordstoga M (1988) Characterization of bovine amyloid proteins SAA and AA. Scand J Immunol 27:739–743

Kushner I (1982) The phenomenon of the acute phase response. Ann NY Acad Sci 389:39–48

Levin M, Franklin EC (1973) Immunologic studies of the major nonimmunoglobulin protein of amyloid. I. Identification and partial characterization of a related serum component. J Exp Med 138:373–380

Levin M, Franklin EC, Frangione B, Pras M (1972) The amino acid sequence of a major nonimmunoglobulin component of some amyloid fibrils. J Clin Invest 51:2773–2776

Lowell CA, Potter DA, Stearman RS, Morrow JF (1986a) Structure of the murine serum amyloid A gene family. J Biol Chem 261:8442–8452

Lowell CA, Stearman RS, Morrow JF (1986b) Transcriptional regulation of serum amyloid A gene expression. J Biol Chem 261:8453–8461

Mahley RW, Innerarity TL, Rall SC Jr, Weisgraber KH (1984) Plasma lipoproteins: apolipoprotein structure and function. J Lip Res 25:1277–1294

Meek RL, Benditt EP (1986) Amyloid A gene family expressions in different mouse tissues. J Exp Med 164:2006–2017

Meek RL, Benditt EP (1989) Rat tissues express serum amyloid A protein-related mRNAs. Proc Natl Acad Sci USA 86:1890–1894

Meek RL, Hoffman JS, Benditt EP (1986) Amyloidogenesis. One serum amyloid A isotype is selectively removed from the circulation. J Exp Med 163:499–510

Meek RL, Eriksen N, Benditt EP (1989) Serum amyloid A one and two in the mouse: sites of uptake and mRNA expression. Amer J Pathol 135:411–419

Miura K, Takahashi Y, Shirasawa H (1985) Immunohistochemical detection of serum amyloid A protein in the liver and the kidney after casein injection. Lab Invest 53:453–463

Parmelee DC, Titani K, Ericsson LH, Eriksen N, Benditt EP, Walsh KA (1982) Amino acid sequence of amyloid-related apoprotein (apoSAA1) from human high density lipoprotein. Biochemistry 21:3298–3303

Ramadori G, Sipe JD, Dinarello CA, Mizel SB, Colten HR (1985) Pretranslational modulation of acute phase hepatic protein synthesis by murine recombinant interleukin-1 (IL-1) and purified human IL-1. J Exp Med 162:930–942

Rosenstreich DL (1981) The macrophage. In: Oppenheim JJ (ed) Cellular function in immunity and inflammation. Elsevier/North-Holland, New York, pp 127–159
Rosenthal CJ, Franklin EC (1975) Variation with age and disease of an amyloid A protein-related serum component. J Clin Invest 55:746–753
Segrest JP, Pownall HJ, Jackson RL, Glenner GG, Pollock PS (1976) Amyloid A; amphipathic helixes and lipid binding. Biochemistry 15:3187–3191
Selinger MJ, McAdam KPWJ, Kaplan MM, Sipe JD, Vogel SN, Rosenstreich DL (1980) Monokine-induced synthesis of serum amyloid A protein by hepatocytes. Nature 285:498–500
Virchow R (1860) Cellular pathology. John Churchill, London, pp 366–384
Yamamoto K, Migita S (1985) Complete primary structures of two major murine amyloid A proteins deduced from cDNA sequences. Proc Natl Acad Sci USA 82:2915–2919

Chapter 6

Regulation of Biosynthesis and Secretion of Human C-reactive Protein and Serum Amyloid A

I. Kushner, M. K. Ganapathi and S. S. Macintyre

Introduction

It was the discovery of C-reactive protein (CRP) in 1930 that first focused attention on the phenomenon that is now known as the acute phase response (Tillett and Francis 1930). The term acute phase response has both a broad and a narrow definition. In the broadest sense it refers to the abrogation, following tissue injury or infection, of many of the homeostatic processes that maintain stability in vertebrates in the face of an inconstant external environment (Kushner 1982). The most familiar example of homeostasis is the maintenance of a relatively constant body temperature; the most familiar example of the acute phase response, in the broad sense, is fever. Apparently the *milieu interieur* which is optimal in normal circumstaces is not appropriate for dealing with the threat to body integrity represented by a broad variety of inflammatory stimuli; a different homeostatic state is temporarily instituted.

In the narrower, and more frequently employed sense, the acute phase response refers to the re-orchestration of the pattern of gene expression of plasma proteins in hepatocytes following inflammatory stimuli. Over the years, CRP has been the best studied acute phase protein. Since the original discovery of CRP, it has become apparent that it is normally present in plasma in only trace amounts, but that there may be dramatic increases in the rate of synthesis, and consequent plasma concentrations, following stimulus. Levels which are 1000-fold or more greater than normal may be seen in severely infected individuals. The only other human plasma protein known to display a change of comparable magnitude is serum amyloid A (SAA); these two proteins are the major acute phase proteins in man (Kushner and Mackiewicz 1987). Progressively increasing numbers of hepatocytes are recruited to synthesis of CRP and SAA over the course of the first several days following stimulus. (Macintyre et al. 1982; Shirahama and Cohen 1985).

Human C-reactive protein is a polymeric protein composed of five identical non-glycosylated subunits held together by non-covalent bonds. The subunits are arranged in cyclic pentameric symmetry within a single plane. The physicochemical characteristics of CRP are given in Table 6.1 (Macintyre 1988). CRP is an ancient molecule; a protein which meets our current definition of C-reactive protein can be detected as a constitutive element in the haemolymph of the horseshoe crab, *Limulus polyphemus* (Robey and Liu 1981). Over the course of evolution, CRP has undergone two remarkable sets of changes (Kushner 1986). Firstly, there is tremendous variability among vertebrates in both the resting levels of CRP [ranging from less than 0.02 $\mu g\ ml^{-1}$ in the mouse (Siboo and Kulisek 1978) to values over 500 $\mu g\ ml^{-1}$ in the rat (de Beer et al. 1982)] and the magnitude of the response to inflammatory stimuli [ranging from none in the cow (Maudsley et al. 1987) to 1000-fold or more in humans and rabbits]. Secondly, human CRP displays a number of functional capabilities involving interactions with elements of the inflammatory and immune responses, such as complement (Kaplan and Volanakis 1974), phagocytic cells (Buchta et al. 1987; Ballou et al. 1989), platelets (Bout et al. 1986) and lymphocytes (Vetter et al. 1986), which do not exist in the ancestral species, the horseshoe crab.

Table 6.1. Physicochemical characteristics of human CRP (Macintyre 1988)

Molecular weight	
Polymeric	120 000
Subunit	24 000
Sedimentation coefficient, $S^{0}_{w,20}$(S)	6.5
Partial specific volume, $\bar{v}$ (ml g^{-1})	0.735
Extinction coefficient, $E_{1\%}^{1 cm}$ at 280 nm	19.5
Isoelectric point, pI	7.9
Binding constant (K) for phosphorylcholine (M)	1.6×10^{-5}

There are four sub-areas of investigation which are being pursued in studies of the regulation of synthesis and secretion of CRP and other acute phase proteins. The greatest effort has been committed to delineating the role of extracellular signals in enhancing or diminishing gene expression in hepatocytes. It is now apparent that several cytokines play a major role in this process. Accordingly, studies of the effect of these regulatory peptides will form the bulk of this chapter. The effect of cofactors, such as corticosteroids, acting in conjunction with cytokines on acute phase changes is gradually being elucidated. The second area of investigation, about which data are beginning to appear, is study of the intracellular signal transducing mechanisms by which extracellular signals alter expression of acute phase protein genes. The third area is the molecular biology of the acute phase response—study of the nuclear elements involved in regulating the expression of these genes. Data relating to this subject are discussed by Ciliberto in Chapter 3. Finally, several studies exploring changes in the secretion of CRP during the acute phase response, as distinct from its synthesis, have appeared recently.

Regulation of Biosynthesis

Role of Cytokines

Model Systems and Their Interpretation

Before discussing the experimental data which have a bearing on induction of CRP or other acute phase proteins, it will be useful to review the model systems that are employed in such studies and to put into perspective our current understanding of how cytokines work. Four model systems have been employed to study regulation of acute phase protein biosynthesis. These are: (i) primary hepatocyte cultures; (ii) hepatoma derived, transformed cell lines; (iii) cells transfected with the gene of interest; and (iv) cells transfected with the 5′-flanking region of genes of interest fused to a reporter gene. In no model can one be confident that findings truly reflect in vivo events. It is extremely difficult to exclude the presence of contaminating non-parenchymal cells in primary hepatocyte cultures; such cells may themselves serve as sources of biologically active cytokines. Transformed cell lines, while largely free of this concern, are intrinsically abnormal and may differ from normal cells in some of the pathways by which extracellular signals are transmitted to genetic regulatory elements, or indeed, may differ in the regulatory elements themselves. Similar concerns pertain to studies in which genes or putative promoter regions have been transected. In addition, the number and location of transfected genes in transformed cells may affect responses. Finally, in those instances in which fibroblasts are employed as host cells, these cells may themselves produce active cytokines.

Furthermore, one might expect conflicting findings from different laboratories employing what is nominally the same cell line. A number of such examples are now appearing. For example, compare the effect of interleukin-1 (IL-1) on α1-antichymotrypsin (α1-ACH) synthesis in Hep 3B cells in Helfgott et al. (1989) and in Ganapathi et al. (1988). In addition to intentional or unintentional differences in experimental conditions (Schimmel 1988), genetic differences may exist. It is well recognized that cell lines, even after cloning, manifest genetic drift, and contain a constantly changing range of genotypes (Freshney 1983). It is not surprising that phenotypic variation would arise as a result.

The role of cytokines in acute phase protein induction must be interpreted in the context of our current understanding of the factors influencing cellular responses to multifunctional cytokines. There is increasing awareness that informational content does not reside in individual cytokines; rather, the effect of cytokines depends on the pattern or set of regulatory peptide molecules to which a cell is exposed (Sporn and Roberts 1988). There are clearly a number of intracellular mechanisms, as yet poorly understood, which integrate the variety of signals that are received by specific receptors and transmitted to the interior of the cell. For example, a cytokine may alter the distribution and affinity of receptors for other cytokines (e.g. Bonin and Singh 1988). In addition, other factors such as the corticosteroids have been shown to be capable of conditioning the response of cells to cytokines (e.g. Baumann et al. 1987a).

Review of Data

Pre IL-6. Approximately eight years ago, attention focused on interleukin-1 (IL-1) as an acute phase protein inducer, following reports of in vivo and in vitro induction of SAA by this cytokine (McAdam et al. 1982). Over the next few years, it was proposed that IL-1 was a primary mediator of the acute phase response (Dinarello 1984). It soon became widely accepted that the entire acute phase response could be attributed to this cytokine. However, a number of investigators were able to identify acute phase protein inducing activities which they could not attribute to IL-1, and employed the term hepatocyte stimulating factor(s) (HSF) to refer to such activities (reviewed by Fey and Fuller 1987). In the autumn of 1986, the ability of another cytokine, tumor necrosis factor α (TNF-α), to induce synthesis of several acute phase complement components was reported (Perlmutter et al. 1986).

At approximately the same time, induction of CRP in a cell system was reported for the first time, employing conditioned medium from lipopolysaccharide (LPS)-stimulated monocytes (Darlington et al. 1986); and not long afterwards, similar findings were reported from another laboratory (Goldman and Liu 1987). Neither group was able to induce CRP with IL-1 or TNF preparations. Further support for the view that CRP induction required cytokines other than IL-1 or TNF was found in our own studies in Hep 3B cells (Ganapathi et al. 1988). In addition, to our surprise, our studies also indicated that the same statement could be made for SAA induction, an iconoclastic view at the time. Incubation of Hep 3B cells with conditioned medium from LPS-stimulated monocytes led to substantial induction of both SAA and CRP (Fig. 6.1). In the case of CRP, there was a dose-dependent increase of up to 10- to 20-fold in synthesis. The degree of SAA induction could not be estimated quantitatively since no SAA was detected in control cultures. Increased synthesis of both acute phase proteins was accompanied by parallel increases in the levels of the specific mRNAs.

When we evaluated the effect of the defined cytokines IL-1β and TNFα (kindly provided by Drs C. A. Dinarello and B. A. Beutler, respectively), we found that human recombinant (r)IL-1β and human rTNFα, either alone or in combination (Fig. 6.1) did not cause induction of CRP or SAA in Hep 3B cells, although they were effective in altering synthesis of other acute phase proteins (see below). Medium from the human keratinocyte cell line, COLO-16 (kindly provided by Dr H. Baumann), which exhibits HSF activity, was also incapable of inducing SAA or CRP.

A different pattern of response to cytokines was found in studies of α1-protease inhibitor. Synthesis of this acute phase protein was increased by both conditioned medium and medium from COLO-16 but not by IL-1β, TNFα, or IL-1β plus TNFα (Fig. 6.1). Parallel accumulation of the specific mRNA was found. Yet a third pattern of response was noted in Hep 3B cells for albumin, ceruloplasmin and α1-antichymotrypsin (α1-ACH), all of which responded to both conditioned medium and IL-1β; these preparations led to decreased synthesis of albumin and to increased accumulation of ceruloplasmin and α1-ACH. We studied the effect of TNFα on only albumin and ceruloplasmin; in each case it had the same effect as IL-1.

Previous studies of SAA production had yielded contradictory findings. IL-1 alone had been shown to induce SAA in both mouse and (in our own experiments) rabbit primary hepatocyte cultures and in mouse fibroblasts

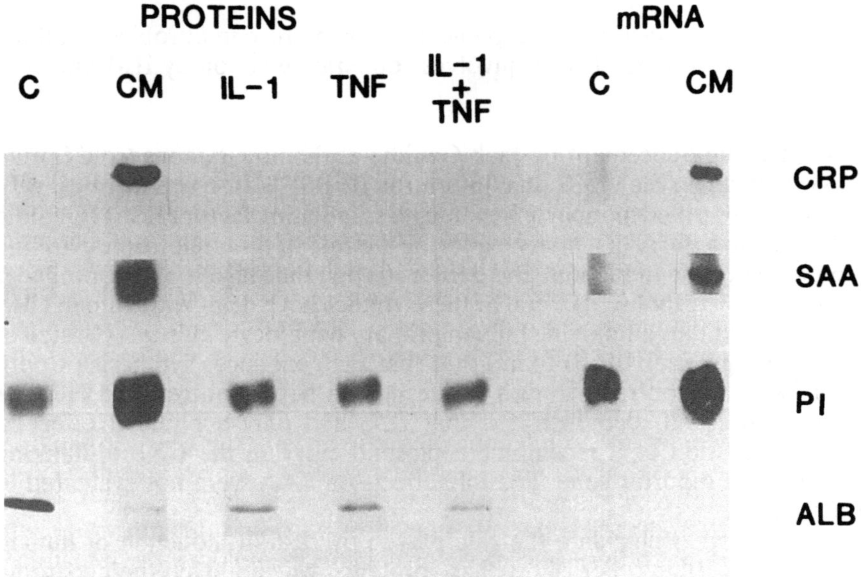

Fig. 6.1. Effect of cytokines on synthesis and mRNA levels of acute phase proteins in Hep 3B cells. Hep 3B cells were incubated for 18–24 hours in the absence (*C*) or presence (*CM*) of monocyte conditioned medium, IL-1β, TNFα and IL-1β plus TNFα. The medium was removed, cells were washed with Hanks Balanced Salt Solution and radiolabelled in methionine-free medium to which 0.1 mCi ml^{-1} L-[^{35}S] methione was added. CRP, SAA, α1-protease inhibitor (*PI*) and albumin (*ALB*) present in the culture medium were analysed by immunoprecipitation followed by fractionation on sodium dodecyl sulphate polyacrylamide gel electrophoresis. The two lanes to the *right* show mRNA accumulation in control and CM-treated cells. CRP and SAA mRNAs were determined by S1 nuclease analysis and PI mRNA was determined by Northern blot analysis (from Kushner I, Ganapathi M, Schultz D. The acute phase response is mediated by heterogeneous mechanisms. Ann NYAS 557: 19–30, 1989. By permission of the New York Academy of Science).

transfected with the human SAA gene (Ramadori et al. 1985; Mackiewicz et al. 1988; Woo et al. 1987). However, consistent with our findings, IL-1 alone had not induced SAA in Hep 3B cells (Perlmutter et al. 1986). Although several possible explanations for this discrepancy came to mind, these findings raised the possibility that IL-1 did not itself induce SAA, but rather that it had stimulated production of an authentic SAA-inducing cytokine by fibroblasts or contaminating non-parenchymal cells.

To test the hypothesis that IL-1 was capable of inducing production of an SAA-inducing factor in fibroblasts we evaluated the ability of preparations of unstimulated and IL-1-stimulated conditioned medium from fibroblasts (kindly provided by Dr Paul E. DiCorleto and Ms Carol de la Motte) to induce SAA in Hep 3B cells. Induction of SAA was not observed with IL-1 alone, with unstimulated fibroblast conditioned medium or with a combination of IL-1 plus unstimulated fibroblast conditioned medium. However, incubation of Hep 3B cells with conditioned medium from fibroblasts stimulated with IL-1 resulted in an increase in SAA production. These results suggested that the induction of the SAA gene transfected into fibroblasts could be due to the action of another cytokine (or cytokines) induced in fibroblasts by IL-1 and capable of acting alone or in combination with IL-1. In the light of subsequent studies it is possible that this cytokine is IL-6, since IL-1 is capable of inducing IL-6 in fibroblasts and the induction of SAA in Hep 3B cells involves cooperative action by IL-1 and IL-6 (*vide infra*).

Post IL-6. The announcement by Jack Gauldie at the now-famous Cold Spring Harbor symposium in May 1987, that interferon β2/BSF 2/ IL-6 was identical with the previously described monocyte hepatocyte stimulating factor (HSF) (Gauldie et al. 1987) led to a flurry of activity exploring the role of this pleiotropic cytokine in acute phase protein induction, and demonstrating that IL-6 is a major inducer of the acute phase response. Several of these studies have dealt with human CRP and SAA. Two studies employing human primary hepatocyte cultures (Castell et al. 1988; Moshage et al. 1988) found that that IL-6 enhances synthesis of both CRP and SAA. In contrast, studies in the human hepatoma cell line Hep 3B (Morrone et al. 1988) demonstrated that IL-6 had only a minimal effect in enhancing expression of a recombinant plasmid carrying the CRP 5′-flanking region fused to a reporter gene. Possible effects on SAA were not evaluated in the latter study.

In order to further elucidate the role played by IL-6 in induction of human CRP, SAA and certain other acute phase proteins, we employed the hepatoma cell lines Hep 3B and NPLC/PRF/5 as model systems. When we examined the ability of antibodies directed against human recombinant derived IL-6 to inhibit enhancement of CRP and SAA synthesis by conditioned medium in these two cell lines, we found that induction of these two acute phase proteins was completely inhibited by antibodies to rIL-6 (obtained from Dr T. Kishimoto and from Dr P. B. Sehgal) in both cell lines (Table 6.2). The ability of conditioned medium to cause induction of both α1-protease inhibitor and fibrinogen and to decrease synthesis of albumin was also completely inhibited by anti-IL-6 in Hep 3B cells. Both partially purified IL-6 (from Drs L. T. May and P. B. Sehgal) and rIL-6 (from Dr Sehgal and from Dr Kishimoto) by themselves increased synthesis of CRP and SAA in the NPLC/PRF/5 cell line (Table 6.2), in a concentration-dependent manner. In contrast, in the Hep 3B cell line, none of the preparations of IL-6 by themselves led to an increase in synthesis of either CRP or SAA (Table 6.2). Recombinant IL-6 alone was, however, capable of inducing other acute phase protein changes in Hep 3B cells; this cytokine induced synthesis of α1-protease inhibitor and fibrinogen (Table 6.3) and decreased albumin synthesis in these cells. Studies of the effect of IL-1 in combination with IL-6 on induction of CRP and SAA revealed that addition of either IL-1α or IL-1β to IL-6 did not further increase the response of CRP or SAA in the NPLC/PRF/5 cell line above that observed with IL-6 alone. In Hep 3B cells, however, the combination of IL-1α or IL-1β with IL-6 led to substantial induction of both CRP and SAA synthesis (Table 6.2).

These results demonstrate that IL-6 plays an important role in regulating synthesis of CRP and SAA in two human hepatoma cell lines. The different findings in the two hepatoma lines we studied probably reflect differences between these transformed cell lines in the intracellular signal transduction mechanisms by which cytokines control expression of the CRP and SAA genes. However, our studies do not permit a firm conclusion as to whether IL-6 alone or IL-6 plus IL-1 is required for CRP and SAA induction in vivo, or indeed, whether other cytokines might not be required as well, for a maximal response. In addition, these studies demonstrate that the intracellular mechanisms through

Table 6.2. Induction of CRP and SAA in two human cell lines

	NPLC		Hep 3B	
	CRP	SAA	CRP	SAA
CM	+	+	+	+
CM + a-IL-6	−	−	−	−
IL-6	+	+	−	−
IL-6 + IL-1	+	+	+	+

+: induction.
−: no induction.
a-IL-6: antibodies to IL-6.

Table 6.3. Effect of cytokines and induction of four acute phase proteins in Hep 3B cells

Treatment	CRP	SAA	α1-Protease Inhibitor	Fibrinogen
Control	0	0	+	0
CM (20%)	+++	+++	+++	+++
CM (20%) + Anti-IL-6	0	0	+	0
IL-6 (5 ng ml^{-1})	0	0	+++	++++
IL-1α ORβ (20 u ml^{-1})	0	0	+	ND
IL-6 (5 ng ml^{-1}) + IL-1α OR β	++	++	ND	ND

0, +, ++, +++, ++++: indicate magnitude of acute phase protein production.
ND: not done.

which IL-6 enhances CRP and SAA expression can be dissociated from those regulating gene expression of α1-protease inhibitor, fibrinogen and albumin in Hep 3B cells.

We observed no effect of IL-6 alone on CRP synthesis and secretion by Hep 3B cells, while Morrone et al. (1988) demonstrated a modest effect of IL-6 on expression of a transfected reporter gene fused to the CRP 5′-flanking region. The results of the latter authors also differed from ours in that they found that IL-6 failed to influence α1-protease inhibitor or albumin synthesis in their stock of Hep 3B cells. It may be that the difference in CRP findings is more apparent than real, since the two assay systems employed are vastly different. On the other hand, as previously indicated, there are a number of instances in which different laboratories report conflicting findings in what is presumably the same cell line.

It is of interest that the apparent Mr of SAA and CRP-inducing activity in our conditioned medium, found to be about 18 kDa by gel permeation chromatography, differed from the approximately 30 kDa reported by Goldman and Liu (1987). This difference raises the possibility that some of the forms of IL-6 (May et al. 1988) derived from different cell types are more potent than others in induction of CRP or SAA in different model systems. The data currently available do not permit an answer to this question.

Role of Cofactors: Signal Transduction

It is becoming increasingly apparent that interaction between cytokines and other factors may be required for optimal induction of several acute phase proteins. Corticosteroids are the best studied cofactors; their importance in different species and for different proteins may vary (Baumann et al. 1987b; Otto et al. 1987). Studies carried out in rat primary hepatocyte cultures and rat hepatoma cell lines indicated that dexamethasone is capable of enhancing induction of several acute phase proteins by HSFs, and that dexamethasone by itself increases synthesis of acute phase proteins in rat primary hepatocyte cultures. Dexamethasone has also been shown to synergize the effect of HSFs on induction of several, but not all, human acute phase proteins in the human hepatoma cell line Hep G2. However, by itself it was not capable of affecting synthesis of these proteins in this cell line. Studies in human primary hepatocyte cultures (Castell et al. 1988) revealed that induction of SAA by IL-6 required the presence of dexamethasone, whereas induction of CRP by IL-6 was not dependent on this glucocorticoid. This finding suggests that there may be differences in the intracellular signal pathways responsible for induction of some acute phase proteins even when extracellular signals required for induction appear to be similar.

Our studies of the effect of the methylxanthine, caffeine, also suggest that there may be differences in the intracellular mechanisms regulating synthesis of CRP and SAA. Caffeine potentiated the induction of CRP by monocyte-conditioned medium by 3- to 4-fold in both Hep 3B and NPLC/PRF/5 cells. The effect of caffeine was dose-dependent; maximal response was observed at a concentration of 2 mM. A similar potentiating effect of caffeine was observed for the induction of CRP by IL-6 and IL-1 in Hep 3B and by IL-6 alone in NPLC/PRF/5 cells. However, caffeine did not significantly potentiate induction of SAA by conditioned medium in Hep 3B cells. No significant effect of caffeine was observed on induction of two other acute phase proteins α1-protease inhibitor and α1-ACH, or on decrease in synthesis of albumin by monocyte-conditioned medium.

The mechanism by which caffeine potentiates induction of CRP is not yet known, but seems to be independent of increase in either intracellular cAMP or Ca^{2+}. This conclusion is based on several lines of experimental evidence: (i) caffeine, either alone or in the presence of conditioned medium, did not lead to an increase in intracellular levels of cAMP in Hep 3B cells; (ii) two other methylxanthines, theophylline and 3-isobutyl 1-methylxanthine, which are more potent inhibitors of cAMP phosphodiesterase, did not mimic the potentiating effect of caffeine on the induction of synthesis of CRP by conditioned medium; (iii) Forskolin, which is capable of increasing cAMP levels in Hep 3B cells, and 8-bromo-cAMP also did not potentiate CRP induction by conditioned medium; and (iv) the Ca^{2+} ionophores A23187 and ionomycin, although capable of increasing intracellular levels of Ca^{2+} in Hep 3B cells, did not mimic the effects of caffeine.

During the past few years, significant progress has been made on defining extracellular signals that regulate synthesis of acute phase proteins. However, the mechanisms by which these signals transduce their message are poorly understood. A recent report (Evans et al. 1987) demonstrated that protein kinase C might serve as an intermediate in the activation of the fibrinogen genes as well as some other acute phase protein genes. They found that the phorbol ester, 12-O-tetradecanoyl-phorbol-13-acetate (TPA), and the synthetic diacylglycerol analogue, 1-oleoyl-2-acetylglycerol (OAG) were capable of mimicking the effect of HSF on increasing mRNA levels of fibrinogen in the rat hepatoma cell line, FAZA. TPA was also capable of inducing mRNA levels of α2-macroglobulin. The effect of TPA on induction of γ-fibrinogen mRNA was also observed in a human hepatoma cell line, Hep G2.

Studies carried out in our laboratory on the effect of phorbol esters on synthesis of CRP, SAA and α1-protease inhibitor in Hep 3B cells revealed that both TPA and phorbol 12,13-didecanoate (PDD) inhibited induction of CRP and SAA by conditioned medium and by IL-1 plus IL-6 in Hep 3B cells. The inhibitory effect of phorbol esters was not observed for synthesis of total proteins secreted into culture medium nor for induction of α1-protease inhibitor. Thus, these results suggest that synthesis of CRP and SAA in hepatocytes is subject to bidirectional regulatory mechanisms, the negative control of which may perhaps involve protein kinase C.

Heterogeneity in the Acute Phase Response

Taken together, these findings underline the heterogeneous nature of the intra- and extra-cellular mechanisms responsible for acute phase protein induction. The concentrations of different acute phase proteins increase at variable rates, to variable extents. In addition to the defined cytokines described above, interferon γ has been shown capable of acute phase protein induction (Miura et al. 1987); multiple cytokines are thus capable of inducing acute phase proteins and multiple cell types may serve as sources of these cytokines. Different cytokines or combinations of cytokines are capable of inducing different acute phase proteins. Differences between cell lines in their patterns of response to cytokines and differences within a cell line of the pattern of plasma protein response to a single cytokine indicate that a number of intracellular signal transducing pathways are involved in acute phase protein induction. Similarly, differences in the effects of caffeine and dexamethasone suggest differences in intracellular mechanisms. Clearly, studies carried out in vitro have greatly enhanced our understanding of the mechanisms involved in the acute phase response. However, these experiments, which employ a limited number of defined components, may not necessarily represent the entire in vivo scenario, which is certainly highly complex.

Regulation of CRP Secretion

While it is likely that increased transcriptional activity is the major mechanism responsible for the elevations in plasma concentration of the acute phase proteins, an additional regulatory mechanism, occurring at a post-translational level, has recently been demonstrated in the case of CRP. In studies employing rabbit primary hepatocyte cultures, secretion of newly synthesized CRP was found to become progressively more efficient during the course of the acute phase response. Whether this phenomenon has a significant impact on plasma CRP levels is not clear at present, but CRP provides a particularly useful tool in investigation of the mechanisms which regulate the sorting and secretion of plasma proteins.

Dynamics of CRP Secretion by Rabbit Hepatocytes

Initial studies of the synthesis and secretion of CRP by rabbit hepatocytes employing isolated liver perfusion (Kushner et al. 1980) and primary hepatocyte cultures (Macintyre et al. 1983) showed that rates of CRP secretion were linear during the periods of perfusion (8 hours) or culture (18–24 hours). Rates of secretion were found to correlate with the degree of inflammatory response attained following in vivo stimulus, as judged by serum CRP level at time of sacrifice. Insights into the possibility of a post-translational regulatory step first became apparent when both intracellular CRP content and rate of CRP secretion were determined in a series of cultures prepared from rabbits manifesting varying degrees of the acute phase response (Macintyre et al. 1985). While rate of secretion increased by 100-fold, intracellular content increased by only 17-fold, resulting in a progressive decrease in the ratio of intracellular content to rate of secretion (Table 6.4). This ratio, termed "residence time", which can be viewed as representing the time required to secrete the amount of CRP contained within the intracellular pool, was found to decrease from about 18 hours in cultures from unstimulated animals to about 2.5 hours in cells from a maximally responsive rabbit, suggesting more efficient secretion during the course of the acute phase response. In contrast, residence times for albumin did not change. In pulse-chase labelling experiments, the time required for secretion of 50% of pulse-labelled CRP varied from well over 6 hours in cells from a minimally responsive animal to about 75 min in cells from a highly responsive animal, again indicating changes in efficiency of CRP secretion. In contrast, both the half-time for albumin secretion and the residence time for albumin were constant (about 45 min) despite the expected 3- to 4-fold decrease in albumin synthesis in cells from responsive animals.

Table 6.4. Relationship of intracellular protein content to rate of protein secretion

CRP				Albumin		
Serum [CRP] ($\mu g\ ml^{-1}$)	Secretion rate, A ($ng\ h^{-1}$ per 10^6 cells)	Intracellular content, B (ng per 10^6 cells)	Ratio B/A (h)	Secretion rate, A ($ng\ h^{-1}$ per 10^6 cells)	Intracellular content, B (ng per 10^6 cells)	Ratio B/A (h)
2	1.2	22	18		ND	
2	2.3	32	14	1900	2900	1.5
15	8.3	84	10		ND	
46	8.3	92	11	1100	900	0.8
38	23	180	7.7	330	690	2.1
55	25	150	5.8		ND	
90	56	140	2.5	770	780	1.0
118	62	160	2.6	470	430	0.9
140	65	340	5.2	630	680	1.1
180	120	380	3.2	1300	1000	0.8

At about the time this work was performed, three other studies (Fitting and Kabat 1982; Lodish et al. 1983; Fries et al. 1984) indicated that different secretory proteins and integral plasma membrane proteins traversed the secretory pathway at differing rates. A prevalent theory explained these differences by postulating that such proteins interacted with a hypothetical transport receptor having different affinities for each protein. However, our findings that an individual protein, CRP, displayed widely differing kinetics under different circumstances argued against this explanation, since it is unlikely that such an affinity for CRP would change during the acute phase response. Instead, we postulated that the CRP secretion is regulated either by (i) a specific receptor which itself is regulated during the acute phase response, or by (ii) concentration-dependent diffusion, whereby the relative concentration of individual proteins within the intracellular pool determined their rate of exit from this pool.

Subcellular Localization of the Intracellular Pool of CRP

To clarify this point, we isolated subcellular fractions from cultured hepatocytes which had been incubated with [^{35}S]methionine under pulse-chase conditions (Macintyre et al. 1986). Results indicated that the rate-limiting step in the secretion of CRP was transfer from the endoplasmic reticulum (ER) to Golgi, in agreement with reports for several other secretory proteins (Fitting and Kabat 1982; Lodish et al. 1983; Fries et al. 1984). In cells from a relatively uninflamed rabbit, a 75 min chase resulted in exit from the ER of 71% of pulse-labelled albumin, but only 16% of labelled CRP (Table 6.5). When similar studies were done on cells from a more responsive animal, the rate of transfer of labelled CRP increased from 16% to 44%, while albumin transport was unchanged, indicating a selective increase in the efficiency of CRP transport.

Table 6.5. Transport of radiolabelled proteins within subcellular fractions

Experiment	A	B
CRP secretion (ng h^{-1} per 10^6 cells)	7.1	22
CRP/Alb secretion	1.1%	3.0%
% CPM chased from RER		
Alb	71%	67%
CRP	16%	44%
ng CRP/ng Alb		
RER	19%	10%
SER	20%	9.0%
Golgi	1.2%	3.5%

Alb:albumin.

The ratio of CRP to albumin in the ER fractions (rough ER and smooth ER) was notably higher than in the Golgi in both experiments (Table 6.5, A versus B), in agreement with the premise that the ER represents the location of the intracellular CRP pool. Of particular interest was the finding that the ratio of CRP to albumin within the ER was 9%–10% in the cells with a more rapid transport of CRP (Experiment B) whereas the relative ratio within the ER was actually higher (19%–20%) in the culture in which CRP was secreted more slowly (Experiment A). Thus, it appeared unlikely that the relative concentration of CRP within the ER is the driving force determining its rate of exit from this compartment. We postulated that the observed differences in half-times of CRP secretion were due to a specific mechanism which serves either to retain CRP within the ER of unstimulated hepatocytes or to facilitate exit of CRP from the ER in stimulated hepatocytes.

CRP is Specifically Retained Within the Endoplasmic Reticulum

Studies to distinguish between these two mechanisms employed the technique of gene transfer. A fusion gene containing the mouse metallothionein (MMT) I promoter linked to the protein coding region of the rabbit CRP gene was introduced into the human hepatoma cell line, NPLC/PRF/5, and the nonliver cell line, HeLa (Hu et al. 1988). By varying the extracellular zinc concentration, the MMT promoter could be induced to drive the synthesis of rabbit CRP, producing a 10-fold range of rates of secretion which were equivalent to those obtained in primary hepatocyte cultures from moderately responsive rabbits.

Residence times and half-times for secretion of rabbit CRP were determined in both transfected cell lines incubated with a range of zinc concentrations. In contrast to our findings in primary hepatocyte cultures, we found that the residence times and half-times for secretion were constant in both cell lines, regardless of the overall rates of rabbit CRP secretion. Furthermore, the observed half-times were very rapid (75 min), even in transfected cells secreting rabbit CRP at low rates. We reasoned that it is unlikely that NPLC/PRF/5 cells, much less HeLa cells, are better able than rabbit hepatocytes to express a facilitative transport receptor which recognizes rabbit CRP. Rather we concluded that a specific mechanism serves to retain CRP within the ER of rabbit hepatocytes and that the transfected lines were unable to retain rabbit CRP, either because these transformed cells are not expressing the retention element, or because the putative equivalent human factor does not recognize rabbit CRP. Our hypothesis that CRP secretion is regulated by preferential retention rather than facilitative exit from the ER was supported by a recent report that the rate of bulk flow of lumenal contents from the ER is very rapid (half-time of about 10 min) and that delays in transport of lumenal proteins may be the result of specific interactions within the ER (Wieland et al. 1987).

In recent studies (Macintyre et al. 1988), we have investigated the interaction between rabbit CRP and isolated hepatic microsomes which have been permeabilized with low concentrations of detergents. In competitive binding studies employing ^{125}I-labelled CRP we have tentatively identified the presence of specific trypsin-sensitive binding to permeabilized rough, but not smooth, microsomes. Whether such a binding site is related to the observed changes in retention of CRP within the ER during the acute phase response is not clear at present.

Although the functional significance of variable retention of CRP within the ER is unknown, at least two intriguing possibilities can be envisioned. The retention phenomenon may serve to create an intracellular reservoir of CRP which can be released, by an unknown mechanism, during the acute phase response in order to provide the host with a very rapid bolus of circulating CRP. Alternatively, it is possible that CRP serves an unknown function within the ER, and during the acute phase response this function is over-ridden by the need for increased secretion of CRP into the circulation. Clarification of the possible functions of the intracellular retention of CRP will require considerably more investigation.

Conclusions

1. IL-6 plays a major role in induction of both CRP and SAA in two human hepatoma cell lines and in primary hepatocyte cultures.

2. There is a strong cell line-dependent influence on the ability of IL-6 to enhance gene expression of CRP and SAA.

3. The intracellular mechanisms through which IL-6 enhances CRP and SAA gene expression can be dissociated from those regulating gene expression of α1-protease inhibitor, fibrinogen and albumin in Hep 3B cells.

4. There may be differences between CRP and SAA in some of the mechanisms regulating their biosynthesis, since caffeine potentiates CRP, but not SAA induction, and since their induction may differ in the need for glucocorticoids.

5. These findings lend further support to the view that heterogeneous extra- and intra-cellular mechanisms are involved in induction of acute phase proteins.

6. There is an increase in the efficiency with which CRP is secreted by rabbit hepatocytes over the course of the acute phase response. The mechanisms which regulate this change are unknown.

Acknowledgements. We wish to thank Ms Debra Schultz and Mrs Nancy Kessler for their outstanding editoral and secretarial assistance. This work was supported by National Institutes of Health grants AG–02467 and AR–34313, the Irma Bender Research Fund of the Cuyhoga County Hospital Foundation, Cuyhoga County Hospital Grant IMED–203 and Cancer Center Grant P30 CA43703.

References

Ballou S, Buniel J, Macintyre S (1989) Specific binding of human C-reactive protein to human monocytes in vitro. J. Immunol 142:2708–2713

Baumann H, Richards C, Gauldie J (1987a) Interaction among hepatocyte-stimulating factors, interleukin 1, and glucocorticoids for regulation of acute phase plasma protein in human hepatoma (HepG2) cells. J Immunol 139:4122–4128

Baumann H, Onorato V, Gauldie J, Jahreis GP (1987b) Distinct sets of acute phase plasma proteins are stimulated by separate human hepatocyte-stimulating factors and monokines in rat hepatoma cells. J Biol Chem 262:9756–9765

Bonin PD, Singh JP (1988) Modulation of interleukin-1 receptor expression and interleukin-1 response in fibroblasts by platelet-derived growth factor. J Biol Chem 263:11052–11055

Bout D, Joseph M, Pontet M, Vorng H, Deslee D, Capron A (1986) Rat resistance to schistosomiasis: platelet-mediated cytotoxicity induced by C-reactive protein. Science 231:153–156

Buchta R, Pontet M, Fidkin M (1987) Binding of C-reactive protein to human neutrophils. FEBS Lett 211:165–168

Castell JV, Gomez-Lechon MJ, David M, Hirano T, Kishimoto T, Heinrich PC (1988) Recombinant human interleukin-6 (IL-6/BSF-2/HSF) regulates the synthesis of acute phase proteins in human hepatocytes. FEBS Lett 232:347–350

Darlington GJ, Wilson DR, Lachman LB (1986) Monocyte-conditioned medium, interleukin-1, and tumor necrosis factor stimulate the acute phase response in human hepatoma cells in vitro. J Cell Biol 103:787–793

de Beer FC, Baltz ML, Munn EA et al. (1982) Isolation and characterization of C-reactive protein and serum amyloid P component in the rat. Immunol 45:55–70

Dinarello CA (1984) Induction of acute phase reactants by interleukin-1. In: Weissmann G (ed) Advances in inflammation research, vol 8. Raven Press, New York, pp 203–225
Evans E, Courtois GM, Kilian PL, Fuller GM, Crabtree GR (1987) Induction of fibrinogen and a subset of acute phase response genes involves a novel monokine which is mimicked by phorbol esters. J Biol Chem 262:10850–10854
Fey GH, Fuller GM (1987) Regulation of acute phase gene expression by inflammatory mediators. Mol Biol Med 4:323–338
Fitting T, Kabat D (1982) Evidence for a glycoprotein "signal" involved in transport between subcellular organelles. J Biol Chem 257:14011–14017
Freshney RI (1983) Instability, variation and preservation In: Culture of animal cells. Alan R. Liss, New York, pp 189–190
Fries E, Gustafsson L, Peterson PA (1984) Four secretory proteins synthesized by hepatocytes are transported from endoplasmic reticulum to Golgi complex at different rates. EMBO J 3:147–152
Ganapathi MK, Schultz D, Mackiewicz A, et al. (1988) Heterogeneous nature of the acute phase response. Differential regulation of human serum amyloid A, C-reactive protein and other acute phase proteins by cytokines in Hep 3B cells. J Immunol 141:564–569
Gauldie J, Richards C, Harnish D, Lansdorf P, Baumann H (1987) Interferon β2/B-cell stimulatory factor type 2 shares identity with monocyte-derived hepatocyte-stimulating factor and regulates the major acute phase protein response in liver cells. Proc Natl Acad Sci USA 84:7251–7255
Goldman ND, Liu T-Y (1987) Biosynthesis of human C-reactive protein in cultured heptoma cells is induced by a monocyte factor(s) other than interleukin-1. J Biol Chem 262:2363–2368
Helfgott DC, Tatter SB, Santhanam U, Clarick RH, Bhardwaj LT, Sehgal PB (1989) Multiple forms of IFN-β2/IL-6 in serum and body fluids during acute bacterial infection. J Immunol 142:948–953
Hu S-I, Macintyre SM, Schultz D, Kushner I, Samols D (1988) Secretion of rabbit C-reactive protein by transfected human cell lines is more rapid than by cultured rabbit hepatocytes. J Biol Chem 263:1500–1504
Kaplan MH, Volanakis JE (1974) Interaction of C-reactive protein complexes with the complement system. J Immunol 112:2135–2147
Kushner I (1982) The phenomenon of the acute phase response. Ann NY Acad Sci 389:39–48
Kushner I (1986) How C-reactive protein has evolved. In: Peeters H (ed) Protides of the biological fluids, vol 34. Pergamon Press, Oxford, pp 259–261
Kushner I, Mackiewicz A (1987) Acute phase proteins as disease markers. Disease Markers 5:1–11
Kushner I, Ribich WN, Blair JB (1980) Control of the acute phase response. J Lab Clin Med 96:1037–1045
Lodish HF, Kong N, Snider M, Strous GJAM (1983) Hepatoma secretory proteins migrate from rough endoplasmic reticulum to Golgi at characteristic rates. Nature 304:80–84
Macintyre SS (1988) C-reactive protein. Meth Enzym 163:383–399
Macintyre SS, Schultz D, Kushner I (1982) Synthesis and secretion of C-reactive protein by rabbit primary hepatocyte cultures. Biochem J 210:707–715
Macintyre SS, Schultz D, Kushner I (1983) Biosynthesis of C-reactive protein. Ann NY Acad Sci 389:76–87
Macintyre SS, Kushner I, Samols D (1985) Secretion of C-reactive protein becomes more efficient during the course of the acute phase response. J Biol Chem 260:4169–4173
Macintyre SS, Samols D, Kushner I (1986) Regulation of C-reactive protein secretion by cultured rabbit hepatocytes. In: Peeters H (ed) Protides of the biological fluids, vol 34. Pergamon Press, Oxford, pp 251–254
Macintyre S, Samols D, Kushner I, Ballou S (1988) Retention of C-reactive protein within the endoplasmic reticulum is regulated during the acute phase response. FASEB J 2:A548
Mackiewicz, Ganapathi MK, Schultz D, Samols D, Reese J, Kushner I (1988) Regulation of rabbit acute phase protein biosynthesis by monokines. Biochem J 253:851–857
Maudsley S, Rowe IF, De Beer FC et al. (1987) Identification and isolation of two pentraxins from bovine serum. Clin Exp Immunol 67:662–673
May LT, Ghrayeb J, Santhanum U et al. (1988) Synthesis and secretion of multiple forms of β2-interferon/B-cell differentiation factor 2/hepatocyte-stimulating factor by human fibroblasts and monocytes. J Biol Chem 263:7760–7766
McAdam KPWJ, Li J, Knowles J et al. (1982) The biology of SAA: identification of the inducer, in vitro synthesis, and heterogeneity demonstrated with monoclonal antibodies. Ann NY Acad Sci 389:126–136
Morrone G, Ciliberto G, Oliviero S et al. (1988) Recombinant interleukin 6 regulates the transcriptional activation of a set of human acute phase genes. J Biol Chem 263:12554–12558

Miura N, Prentice HL, Schneider PM, Perlmutter DH (1987) Synthesis and regulation of the two human complement C4 genes in stable transfected mouse fibroblasts. J Biol Chem 262:7298–7305

Moshage HJ, Roelofs HMJ, vanPelt JF et al. (1988) The effect of interleukin-1, interleukin-6 and its interrelationship on the synthesis of serum amyloid A and C-reactive protein in primary cultures of adult human hepatocytes. Biochem Biophys Res Commun 155:112–117

Otto JM, Grenett HE, Fuller GM (1987) The coordinated regulation of fibrinogen gene transcription by hepatocyte-stimulating factor and dexamethasone. J Cell Biol 105:1067–1072

Perlmutter DH, Dinarello CA, Punsal PI, Colten HR (1986) Cachetin/tumor necrosis factor regulates hepatic acute phase gene expression. J Clin Invest 78:1349–1354

Ramadori G, Sipe JD, Dinarello CA, Mizel SB, Colten HR (1985) Pretranslational modulation of acute phase hepatic protein synthesis by murine recombinant interleukin 1 (IL-1) and purified human IL-1. J Exp Med 162:930–942

Robey FA, Liu T-Y (1981) Limulin: A C-reactive protein from *Limulus polyphemus*. J Biol Chem 256:969–975

Rodriguez Boulan E, Kreibich G, Sabatini DD (1978) Spatial orientation of glycoproteins in membranes of rat liver rough microsomes. J Cell Biol 78:874–893

Schimmel P (1988) Who will watch the watchdogs? Cell 54:596

Shirahama T, Cohen A (1985) Immunocytochemical study of hepatocyte synthesis of amyloid AA. Amer J Pathol 118:108–115

Siboo R, Kulisek E (1978) A fluorescent immunoassay for the quantification of C-reactive protein. J Immunol Methods 23:59–67

Sporn MB, Roberts AB (1988) Peptide growth factors are multifunctional. Nature 332:217–219

Tillet WS, Francis T Jr (1930) Serological reactions in pneumonia with a non-protein somatic fraction of pneumococcus. J Exp Med 52:561–585

Vetter ML, Gewurz H, Baum LL (1986) The effects of C-reactive protein on human cell-mediated cytotoxicity. J Leukocyte Biol 39:12–25

Wieland FT, Gleason ML, Serafine TA, Rothman JE (1987) The rate of bulk flow from the endoplasmic reticulum to the cell surface. Cell 50:289–300

Woo P, Sipe J, Dinarello CR, Colten HR (1987) Structure of a human serum amyloid A gene and modulation of its expression in transfected L cells. J Biol Chem 262:15790–15795

Chapter 7

Molecular Regulation of the Acute Phase Complement Proteins

H. R. Colten and J. D. Gitlin

Introduction

The complement genes and many other acute phase protein genes are expressed in liver and in extrahepatic tissues (Alper et al. 1980; Kushner and Feldmann 1978; Kushner 1982). The hepatocyte is a major source of the plasma proteins encoded by these genes. The original definition of an acute phase reponse included changes in plasma protein concentration but recent data suggest that changes in expression of these genes occur at the local site of tissue injury or inflammation as well as in liver (reviewed in Colten and Dowton 1986). These observations raise important questions about the mechanisms accounting for tissue specific control of acute phase protein synthesis following tissue injury or inflammation.

In this chapter we will focus on control of synthesis of the complement proteins C2 and factor B because these highly homologous proteins are rate-limiting components in the alternative and classical complement activation pathways (Muller-Eberhard 1975). Moreover, study of expression of the C2 and Bf genes provides a specific example for the general principle that a complex, overlapping, but not coincident, set of cytokines regulates the species and tissue specific expression of several of the acute phase genes. Recent evidence also suggests that full expression of the local response of these genes is developmentally regulated (St John Sutton et al. 1986).

The complement proteins have been grouped in several supergene families based on similarities in structure, function and chromosomal localization (Fig. 7.1). Among these, the genes encoding C2, factor B and C4 constitute the class III genes of the major histocompatibility complex (MHC) on human chromosome 6. The 3′ terminus of the C2 gene is upstream of, and close to, the 5′ terminus of the factor B gene (~400 base pairs) (Carroll et al. 1984). Two C4 genes (C4A and C4B) lie approximately 30 kb downstream from the C2 and

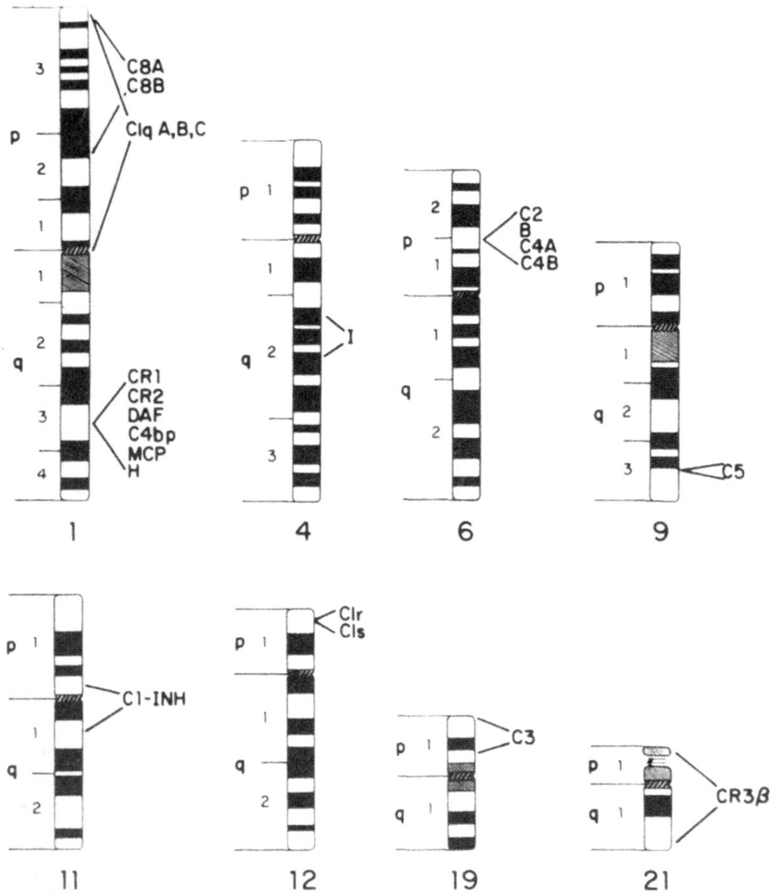

Fig. 7.1. Chromosomal localization of the human complement genes, from Wetsel and Colten (1988).

factor B genes. The C4 loci are separated by 10 kb, and each has a cytochrome P-450 steroid 21-hydroxylase (21-OH) gene within 1.5 kb of the 3′ terminus (Carroll et al. 1985a,b; White et al. 1985). The order of the genes in the direction of transcription is C2-BF-C4A-21OHA-C4B-21OHB, and they have been mapped within a 0.7 centimorgan region between LHA-B and HLA-DR (Olaisen et al. 1983; Whitehead et al. 1985). A map of the human MHC gene complex is shown in Fig. 7.2.

The organization of the human class III MHC genes and orientation of the complex with respect to the MHC class I and class II genes are similar to the corresponding murine genes on chromosome 17 (Chaplin et al. 1985; Perlmutter et al. 1985; Muller et al. 1987; Carroll et al. 1987). In both species the genes for tumor necrosis factor (TNF) α and β also are within the MHC. Factor B and C2 are similar in structural and functional features, which suggests that they were derived from a common ancestral gene (Bentley 1986). Although many of the genes in this region show extensive polymorphism, both fine structure (Sackstein et al. 1983) and overall organization indicate a high degree of evolutionary conservation. We have taken advantage of this fact in our initial efforts to define

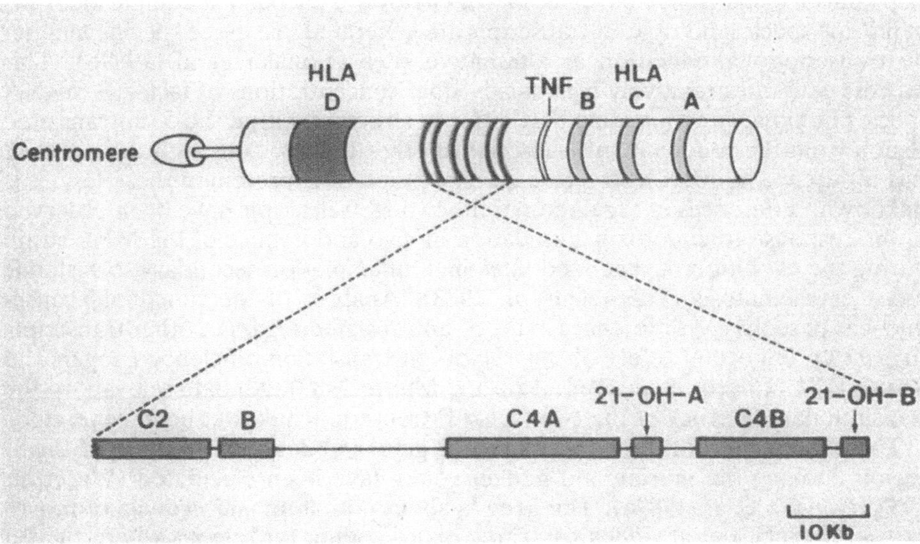

Fig. 7.2. Map of the human major histocompatibility complex, from Perlmutter and Colten (1987). The class III genes are shown oriented in the direction of transcription left to right. Pulse field electrophoretic analysis of the region shows that the C2 gene is nearest to TNF and the 21-OH gene nearest the HLA-D complex (Carroll et al. 1987; Muller et al. 1987).

the structural domains critical for specific cytokine and tissue specific regulation of complement gene expression (see below).

Factor B

Early evidence that factor B is an acute phase reactant prompted studies of the regulation of this gene by bacterial products and by well-characterized cytokines in primary cell culture and cell lines in order to ascertain the mechanism controlling net factor B production. This is important, particularly at extrahepatic sites of tissue inflammation, since factor B is a constituent of the alternative complement activation pathway, an antibody-independent, first-line, primitive host defence mechanism. That is, a cleavage product of factor B (Bb), the carboxy terminal fragment bearing the active site serine protease domain, is part of an unstable enzyme complex (C3bBb) that is rate limiting in the cleavage of complement protein C3, the lynch pin of the complement cascade.

Factor B is a single chain ~95 kilodalton (kDa) glycoprotein that is synthesized in hepatocytes, mononuclear phagocytes and in other cells in culture including fibroblasts and enterocytes (reviewed in Colten 1986). Factor B messenger RNA (~2.5 kb) programmes synthesis of an 83 kDa precursor polypeptide (Woods et al. 1982) to which is added core sugars via a dolichol phosphate-dependent pathway. Post-synthetic cleavage of signal peptide, modification of the carbohydrate and secretion of the factor B protein is accomplished in 30–60 min

(Perlmutter et al. 1984; Cole et al. 1985). Factor B is encoded at a single locus but in murine species multiple Bf transcripts are generated in a tissue specific manner by transcriptional initiation at alternative sites (Nonaka et al. 1989b). This process generates relatively high steady state concentrations of factor B mRNA species in kidney and intestine that differ in structure within the 5′ untranslated region from the predominant transcript in other tissues. Although the effect of this difference in mRNA structure on expression of B protein in these tissues is unknown, differences in regulation of the two Bf transcripts have been observed in the response to endotoxin stimulation in vivo and in mice of the MRL strain during the evolution of renal and intestinal inflammation secondary to systemic lupus erythematosis (Passwell et al. 1988). Analysis of the functional consequences of sequence differences in the 5′ untranslated region of other transcripts suggest an important effect of this region on translation efficiency (Logan and Shenk 1984; Thireos et al. 1984; Aziz and Munro 1987). No data relevant to the translational efficiency of the two factor B transcripts have yet been generated.

The organization of the human factor B gene and detailed structure of the 5′ region flanking the murine and human genes have been elucidated (Wu et al. 1987; Nonaka et al. 1989a). The gene is about 6 kb long and is divided into 18 exons (Campbell et al. 1984a,b). Three exons coding for regions within the Ba fragment show significant sequence homology with each other; i.e. they may be products of tandem duplication. These exons encode repeating homology units of 60 amino acids which are also found in C2, C1r, C4BP, CR1, factor H, IL-2 receptor, β_2-glycoprotein-1 and clotting factor XIII (Reid et al. 1986). With the exception of a single exon, the organization of the factor B gene in the region encoding the carboxy-terminus of the Bb fragment is similar to that observed in chymotrypsin, elastase, kallikrein, trypsin and C2. That is, the amino-terminus of the Bb fragment is generated from exons that do not share homology with other serine proteases. This is consistent with the observation that both factor B and C2 are unique among the serine proteases in that the catalytic chains are approximately twice the size of those in other serine proteases (Campbell et al. 1984a).

There is also a high degree of conservation of factor B primary structure between mice and humans (Sackstein et al. 1983). For the 477 amino acids determined in the Bb fragment of murine factor B, 83% of the residues are identical with the corresponding region of the human protein. There is also 59% sequence identity in the 3′ untranslated (UT) region, and an unusual polyadenylation signal is found in both species. The functional significance of homology in noncoding sequences of the 3′ flanking region is unknown, but studies of other transcripts suggest the importance of 3′ UT structure in determining mRNA catabolism; i.e. the stability of specific mRNA species.

About 400 base pairs separate the end of the C2 gene and the most 3′ of the two murine factor B transcriptional initiation sites. It is this region and the 3′ untranslated region of C2 that determine the independent constitutive and regulated expression of factor B transcription. Nucleotide sequence analysis of the murine C2/Bf intergenic region reveals a high degree of identity with the corresponding human sequence (Nonaka et al. 1989a). For this region, 65% identity was observed between the murine and human sequences in the 426 positions compared. This is greater than the extent of nucleotide identity between these two species in the 3′ untranslated region of either the Bf or the C2 gene or in the 5′ untranslated region of the Bf gene (~59%). The homology of the intergenic segment is comparable to the coding region of the first exon of the Bf gene (66%, 70 positions). Some possible regulatory DNA elements are also particularly conserved both in position and nucleotide sequence in this DNA segment. For instance, an interferon response element is found in the mouse sequence in a position identical to the human counterpart, and 22 out of the 28 nucleotides in the element are identical. Three heat shock consensus sequences, two in the 3′ end of the C2 gene and one in the intergenic region, reported for the human sequence (Wu et al. 1987) are also found in the mouse sequence. These results suggest a selective pressure to conserve not only transcribed sequences but also the intergenic segment between the C2 and Bf genes. The functional studies of this region (cited below) support this concept.

The Second Component, C2

The synthesis and secretion of C2 are, with one important exception, similar to that of factor B. That is, the secreted C2 glycoprotein is derived from a pre-C2 which is translated from a 2.8 kb polyadenylated mRNA, subjected to the action of a signal peptidase, glycosylated, and secreted with kinetics similar to that for factor B (Cole et al. 1980, 1985). Extracellular cleavage generates the active enzymatic form of C2. In addition, two C2 primary translation products of lower molecular mass (79 and 70 kDa) which remain cell-associated have been identified (Perlmutter et al. 1984). The function of these two forms of C2 has not been determined. Recent studies of murine C2 structure have revealed at least two forms of C2 mRNA (N. Ishikawa et al. unpublished work). These differ by the presence or absence of a 21 base pair segment within a region encoding the charge relay residues of the active serine protease site. The two species of C2 mRNA are generated by a mechanism involving alternative splicing. The shorter C2 mRNA is a candidate template for one of the cell-associated C2 proteins, but additional studies are needed to establish this conclusively.

Constitutive Expression of Factor B and C2

Several classical studies suggested that constitutive hepatic expression of the MHC class III genes is under genetic control. For instance, factor B plasma levels differ by about 10-fold between B10.WR and B10.SM mouse strains (Paolucci and Shreffler 1983) and C2 levels differ by about 3-fold in B10.K as compared to B10.PL mice (Gorman et al. 1980). We (Falus et al. 1987) examined the mechanism of genetic and tissue specific differences in constitutive expresssion of C2 and factor B genes by analysis of hepatic and macrophage mRNA from mice with a B10 background but differing in H-2 type (H2-k, H-2u, H-2d, and H-2f).

For C2, hepatic mRNA corresponds in amount to the relative plasma C2 concentrations (H-2K > H-2d > H-2f > H-2u) and for factor B there is a similar correspondence between plasma levels and mRNA, but the rank order differs (H-2u > H-2k = H-2d > H-2f). Control of C2 and factor B expression also differs at extrahepatic sites; i.e. it is tissue specific. For instance, the peritoneal macrophage C2 mRNA content is identical in all four strains, whereas macrophage factor B biosynthesis and mRNA are greater in the H-2k and H-2d than in the H-2u or H-2f strains. Similar observations established tissue-specific genetically determined expression of the other MHC class III genes, C4 and the sex-limited protein (Hansen and Shreffler 1976; Rosa and Shreffler 1982; Newell et al. 1982; Newell and Atkinson 1983). Taken together, these data suggest that sequence differences in control regions that govern constitutive expression of the MHC class III genes are affected differentially by *trans*-acting tissue-specific factors and that the latter are also genetically determined. These conclusions may be extended by observations of MHC class III gene expression in other species. For instance, in several species C4 is expressed in substantial amounts in extrahepatic unstimulated macrophages (mice, guinea pig) but only minimally in human mononuclear phagocytes. Moreover, expression of C4 is stable during tissue culture of guinea pig peritoneal macrophages but rapidly decreases when murine macrophages or human monocytes are maintained in culture (Newell and Atkinson 1983; Sackstein and Colten 1984; Kulich et al., unpublished work).

Regulated Gene Expression

In human mononuclear phagocytes, a small but significant increase in C2 and factor B mRNA is induced by the lipid A component of endotoxin (LPS) (Strunk et al. 1985b). A corresponding increase in factor B protein synthesis (2- to 3-fold) is noted, but net C2 protein synthesis is not affected by lipopolysaccharide. These findings suggest that an increase in C2 translation rate may require a second signal, or that increased catabolism of C2 protein accounts for zero net change in C2 synthesis in the presence of LPS. In fibroblasts, a more substantial increase in factor B is induced by endotoxin; again, net C2 synthesis is affected only to a limited extent by LPS (Katz et al. 1988). Endotoxin injection into an LPS responsive (C3H/HeN) strain, but not the LPS-unresponsive strain (C3H/HeJ), induces a dose-dependent increase in factor B gene expression in liver, lung, kidney, intestine, spleen, heart and peritoneal macrophages (Ramadori et al. 1985b). The responsive cell in the heart that produces complement in response to LPS has not been identified. This observation is of interest, however, since complement activation apparently plays a role in the evolution of a myocardial infarct (Mollnes et al. 1988).

Recently, the purification of several cytokines to homogeneity and isolation of corresponding cDNA clones has allowed generation of well-defined reagents for examination of the effect of these polypeptides on complement gene expression. For these studies, attention has been directed to the C2 and factor B genes, but the importance of these mediators in the control of other genes has been recognized.

Interleukin-1 and Interferon-γ

Interleukin-1 (IL-1) regulates factor B expression in murine hepatocytes and in human hepatoma cell lines (Perlmutter et al. 1986b; Ramadori et al. 1985a). In addition, when recombinant-generated purified mouse IL-1 (rIL-1) is injected into mice (C3H/HeJ), increased factor B mRNA is observed in both hepatic and extrahepatic sites. Whether extrahepatic factor B gene expression (especially in peritoneal macrophages) is regulated directly by IL-1 (autocrine or paracrine regulation) or via other cytokines is uncertain because IL-1 can induce expression of monokines (e.g. IL-6) which affect Bf expression in these cells. However, in hepatoma cells independent pathways for IL-1 and IL-6 regulation of Bf expression have been identified (see below).

In primary murine hepatocyte cultures, recombinant-generated murine and human IL-1 induce a dose- and time-dependent, reversible increase in expression of the factor B gene and a decrease in the albumin gene expression, but have no effect on C2 expression (Ramadori et al. 1985a). This regulation is pretranslational, since the kinetics and direction of change in specific mRNA for factor B and albumin correspond to the change in synthesis of the respective proteins. The small size of the factor B 5′ flanking region (limited by the 3′ terminus of the IL-1 unresponsive C2 gene) provides a well-defined region to probe the structural basis for the differential effect of IL-1 on factor B and the closely related gene, C2. A cosmid bearing both genes on an ~30 kb segment of human DNA was introduced into mouse LK cells where each is constitutively expressed (Perlmutter et al. 1985). These MHC class III-transfected cells synthesize and secrete C2 and factor B proteins. In the transfectant, human factor B, but not C2, gene expression is increased when the cells are incubated with murine or human IL-1. The fidelity of transcription, translation and post-synthetic modification of B and C2 and their response to cytokine regulation indicates the utility of this system for elucidating the operative mechanisms.

Interferon-γ (IFN-γ) regulates expression of the C2 and factor B genes and the C4 genes in primary cell culture and in the MHC class-III transfectants (Strunk et al. 1985a; Miura et al. 1987). IFN-γ induces a dose-dependent increase in synthesis of C2 and factor B protein. Regulation is pretranslational, but the kinetics and magnitude of the response to IFN-γ for the two proteins differ. That is, in both primary cultures of human monocytes and in the class-III transfectants, the IFN-γ-induced increase in factor B occurs at an earlier time point and exceeds in magnitude the increase in C2 gene expression.

To examine the functional details of Bf expression, a series of mouse and human Bf promoter – chloramphenicol acetyl transferase (CAT) chimeric gene constructs were transfected into mouse L or human HepG2 cells. Analysis of expression of these fusion gene constructs (Fig. 7.3) revealed that (i) *cis*-acting DNA sequences identified, at least in part, in the 3′ untranslated region of the C2 gene (within the 400 bp upstream of the Bf cap site) mediate responsiveness to IL-1 and IFN-γ, (ii) the responsiveness to each mediator appears to be conferred by separate upstream regions similar in position and homologous in man and mouse, and (iii) the IL-1 responsive region in both species appears to have the characteristics of an enhancer element, i.e. the effect is independent of position and orientation of the relevant segment (Nonaka et al. 1989a).

Similar structural and functional studies of the region 5′ of the C2 gene are underway but the structural requirements for regulated C2 expression are not yet understood.

As mentioned above, synthesis of human C4 is also regulated by IFN-γ. A dose- and time-dependent increase in synthesis of C4 in HepG2 cells and mouse fibroblasts transfected with cloned C4A and C4B DNA has been demonstrated

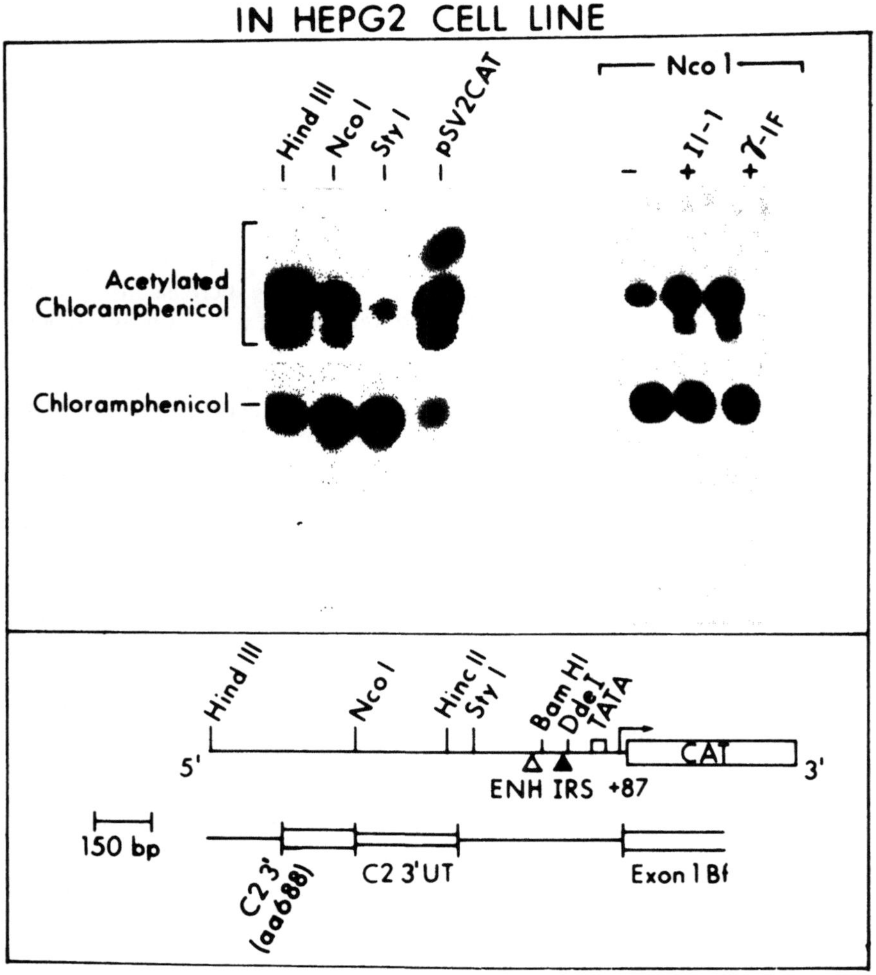

Fig. 7.3. Diagram of intergenic region between C2 and factor B, a chimeric consisting of this region and chloramphenicol acetyl transferase (CAT) gene and functional assay of deleted constructs (*left panel*). IL-1 and interferon-γ stimulated expression of the NcoI construct is shown in the *right panel*.

(Miura et al. 1987). Interferon induces a five-fold increase in steady state levels of C4 mRNA and in the rate of C4 synthesis in C4A transfectants and in HepG2 cells. In contrast, there is only a 1.5- to 2.0-fold increase in C4 expression in the C4B transfectants. Structural, functional and regulatory variants of murine C4 and Slp genes have been demonstrated (Shreffler et al. 1984). Such genetic variation includes a 20-fold difference among mouse strains in the circulating plasma levels of C4. Different plasma C4 concentrations in C4-high and C4-low mice are due to different steady state concentrations of hepatic C4 mRNA (Sackstein and Colten 1984). A difference in the rate of transcription is a possible explanation for this finding but a change in C4 mRNA stability cannot be excluded. Tissue-specificity of expression complicates analysis since macrophages from C4-high and C4-low strains synthesize similar amounts of C4 protein

(Sackstein and Colten 1984; Newell et al. 1982). Genetic variation in Slp expression appears to be even more complex. There are strains of mice which do not express Slp, strains in which Slp expression is androgen-related and strains in which Slp expression is constitutive. Control of androgen-independent constitutive Slp expression is associated with genes within the outside of the H-2 loci. Some of the former instances are in strains with duplications of the C4/Slp gene complex (Brown and Shreffler 1980; Hansen and Shreffler 1976; Levi-Strauss et al. 1985). Current studies by Nonaka, Takahashi and their colleagues (Nakayama et al. 1987) should reveal the molecular basis of these observations.

Tumor Necrosis Factor and Interleukin-6

Factor B is also regulated by the cytokines, tumor necrosis factor (TNF) (Perlmutter et al. 1986a) and interleukin-6 (IL-6, also known as interferon-$\beta 2$) (Perlmutter et al. 1988), each a structurally distinct polypeptide interacting with unique receptors. Since expression of several of the cytokines is induced by others, some investigators have suggested that regulation of the acute phase genes can be explained by all pathways converging on a proximate mediator such as IL-6 (Gauldie et al. 1987). The recent identification of a subclone of the human hepatoma HepG2 cell line unresponsive to IL-1 but responsive to TNF, IL-6 and IFN-γ, coupled with the use of cytokine-specific antibodies, indicates independent signal transduction mechanisms for IL-1 and IL-6 (Perlmutter et al. 1988). The mutant HepG2 cell expresses IL-1 receptors in comparable number and within an affinity similar to the parent cell line. In the mutant cells, IL-6, TNF and IFN-γ, but not IL-1, induce a dose-dependent upregulation of factor B. The IL-6 and TNF also decrease expression of albumin in the mutant cells. In the parent HepG2 cells, antibody to IL-1 or IL-6 specifically blocks the corresponding, but not the other cytokine-induced regulation of Bf expression. Finally, preliminary observations by Strunk and his colleagues suggest that cytokine-induced regulation of the complement genes is selectively modulated by other mediators.

Summary

Taken together, the data summarized in this review support the concept that for some genes, such as factor B and albumin, regulated expression can be accomplished by several cytokines (IL-1, TNF, IL-6) in a qualitatively and quantitatively similar manner, although via separate receptors and signal transduction pathways. Other acute phase genes are regulated by a more restricted set of cytokines. Examples of this more restricted control include C2 and C4 regulation by IFN-γ or fibrinogen and α1-antitrypsin by IL-6. In these, selectivity can be as restricted as to yield quantitative and kinetic differences in IFN regulation of the C4A and C4B genes, molecules that differ little in structure (four amino acid differences between these 200 kDa gene products) (Yu et al. 1986).

Thus, the acute phase protein response should be understood in the context of a programmed induction of multiple mediators of gene expression with variable selectivity in the genes and tissue specificity for each. Presumably, the redundancy of some of these mechanisms results from the multiplicity of stimuli that can elicit an acute phase response and the fundamental importance of this highly conserved mechanism for survival.

References

Alper CA, Raum D, Awdeh Z, Petersen BH, Taylor PD, Starzl TE (1980) Studies of hepatic synthesis in vivo of plasma proteins including orosomucoid, transferrin, alpha-1-antitrypsin, C8 and factor B. Clin Immunol Immunopathol 16:84–91

Aziz N, Munro HN (1987) Iron regulates ferritin mRNA translation through a segment of its 5′ untranslated region. Proc Natl Acad Sci USA 84:8478–8482

Bentley DR (1986) Primary structure of human complement component C2. Homology to two unrelated protein families. Biochem J 239:339–345

Brown LJ, Shreffler DC (1980) Female expression of the H-2 linked sex-limited protein (Slp) due to non-H-2 genes. Immunogenetics 10:19–29

Campbell RD, Bentley DR, Morley BF (1984a) The factor B and C2 genes. Phil Trans R Soc Lond (Biol) 306:367–378

Campbell RD, Law SKA, Reid KBM, Sim RB (1984b) Structure, organization and regulation of the complement genes. Ann Rev Immunol 6:161–195

Carroll MC, Campbell RD, Bentley DR, Porter RR (1984) a molecular map of the human major histocompatibility complex class III region linking complement genes C4, C2 and factor B. Nature 307:237–241

Carroll MC, Campbell RD, Porter RR (1985a) Mapping of steroid 21-hydroxylase genes adjacent to complement component C4 genes in HLA, the major histocompatibility complex in man. Proc Natl Acad Sci USA 82:521–525

Carroll MC, Palsdottir A, Belt KT, Porter RR (1985b) Deletion of complement C4 and steroid 21-hydroxylase genes in the HLA class III regin. EMBO J 4:2547–2552

Carroll MC, Katzman P, Alicot EM, Koller BH, Geraghty DE, Orr HT, Strominger JL, Spies T (1987) Linkage map of the human major histocompatibility complex including the tumor necrosis factor genes. Proc Natl Acad Sci USA 84:8535–8539

Chaplin DD, Woods DE, Whitehead AS, Goldberger G, Colten HR, Seidman JG (1985) Molecular map of the murine S region. Proc Natl Acad Sci USA 80:6947–6951

Cole FS, Auerbach HS, Goldberger G, Colten HR (1985) Tissue-specific pre-translational regulation of complement production in human mononuclear phagocytes. J Immunol 134:2610–2616

Cole FS, Matthews WJ, Marino JT, Gash DJ, Colten HR (1980) Control of complement synthesis and secretion in bronchoalveolar and peritoneal macrophages. J Immunol 125:1120–1124

Colten HR (1986) Molecular regulation of complement gene expression. In: Cinader B, Miller RG (eds) Progress in immunology, vol VI. Academic Press, Orlando, pp 314–324

Colten HR, Dowton SB (1986) Regulation of complement gene expression. Biochem Soc Symp 51:37–46

Falus A, Beuscher HU, Auerbach HS, Colten HR (1987) Constitutive and IL-1 regulated murine complement gene expression is strain and tissue specific. J Immunol 138:856–860

Gauldie J, Richards C, Harnish D, Landsdorp P, Baumann H (1987) Interferon β_2/B-cell stimulatory factor type 2 shares identity with monocyte-derived hepatocyte-stimulating factor and regulates the major acute phase protein response in liver cells. Proc Natl Acad Sci USA 84:7251–7255

Gorman JC, Jackson R, Desantola JR, Shreffler DC, Atkinson JP (1980) Development of a hemolytic assay for mouse C2 and determination of its genetic control. J Immunol 125:344–349

Hansen TH, Shreffler DC (1976) Characterization of a constitutive variant of the murine serum protein allotype, Slp. J Immunol 117:1507–1513

Katz Y, Cole FS, Strunk RC (1988) Synergism between γ-interferon and lipopolysaccharide for synthesis of factor B, but not C2, in human fibroblasts. J Exp Med 167:1–14

Kushner I (1982) The phenomenon of the acute phase response. Ann NY Acad Sci 389:39–48
Kushner I, Feldmann G (1978) Control of the acute phase response. Demonstration of C-reactive protein synthesis and secretion by hepatocytes during acute inflammation in the rabbit. J Exp Med 148:466–477
Levi-Strauss M, Tosi M, Steinmetz M, Klein J, Meo T (1985) Multiple duplications of complement C4 gene correlate with H2-controlled testosterone-independent expression of its sex-limited isoform. Proc Natl Acad Sci USA 82:1746–1750
Logan J, Shenk T (1984) Adenovirus tripartite leader sequence enhances translation of mRNAs late after infection. Proc Natl Acad Sci USA 81:3655–3659
Miura N, Prentice H, Schneider PM, Perlmutter DH (1987) Synthesis and regulation of the two human complement C4 genes in stable transfected mouse fibroblasts. J Biol Chem 262:7298–7305
Mollnes TE, Tambs KE, Myreng Y, Engebretsen LF (1988) Acute phase reactants and complement activation in patients with acute myocardial infarction. Complement 5:33–39
Muller U, Stephan D, Phillippsen P, Steinmetz M (1987) Orientation and molecular map position of the complement genes in the mouse MHC. EMBO J 6:369–373
Muller-Eberhard HJ (1975) Complement. Ann Rev Biochem 44:697–724
Nakayama K, Nonaka M, Yokoyama S, Yeul YD, Pattanakitsakul S-A, Takahashi M (1987) Recombination of two homologous MHC class III genes of the mouse (C4 and Slp) that accounts for the loss of testosterone dependence of sex-limited protein expression. J Immunol 138:620–627
Newell SL, Atkinson JP (1983) Biosynthesis of C4 by mouse peritoneal macrophages. II. Comparison of C4 synthesis by resident and elicited cell populations. J Immunol 130:834–838
Newell SL, Shreffler DC, Atkinson JP (1982) Biosynthesis of C4 by mouse peritoneal macrophages. I. Characterization of an in vitro culture system and comparison of C4 synthesis by "low" vs "high" C4 strains. J Immunol 129:653–659
Nonaka M, Gitlin JD, Colten HR (1989a) Regulation of human and murine complement: Comparison of 5′ structural and functional elements regulating human and murine complement factor B gene expression. Mol Cell Biochem (in press)
Nonaka M, Ishikawa N, Passwell J, Natsuume-Sakai S, Colten HR (1989b) Tissue specific initiation of murine complement factor B mRNA transcription. J Immunol 142:1377–1382
Olaisen B, Teisberg R, Jonassen R, Thorsby E, Gedde-Dahl T (1983) Gene order and gene distance in the HLA regions studied by the haplotype method. Am J Hum Genet 47:285–292
Paolucci ES, Shreffler DC (1983) H-2 linked murine factor B phenotypes. Immunogenet 17:67–73
Passwell J, Schreiner G, Nonaka M, Beuscher HU, Colten HR (1988) Local extrahepatic expression of complement genes C3, factor B, C2 and C4 is increased in murine lupus nephritis. J Clin Invest 82:1676–1684
Perlmutter DH, Colten HR (1987) Structure and expression of the complement genes. Pharmacol Ther 34:247–270
Perlmutter DH, Cole FS, Goldberger G, Colten HR (1984) Distinct primary translation products from human liver mRNA give rise to secreted and cell-associated forms of complement protein C2. J Biol Chem 259:10380–10385
Perlmutter DH, Colten HR, Grossberger D, Strominger J, Seidman JG, Chaplin DD (1985) Expression of complement proteins C2 and factor B in transfected L-cells. J Clin Invest 76:1449–1454
Perlmutter DH, Dinarello CA, Punsal PI, Colten HR (1986a) Cachectin/tumor necrosis factor regulates hepatic acute phase gene expression. J Clin Invest 78:1349–1355
Perlmutter DH, Goldberger G, Dinarello CA, Mizel SB, Colten HR (1986b) Regulation of class III major histocompatibility complex (MHC) gene products by interleukin-1. Science 232:850–852
Perlmutter DH, May LT, Sehgal PB (1988) Interferon beta 2/IL-6 modulates synthesis of alpha 1 antitrypsin in human mononuclear phagocytes and in human hepatoma cells. J Clin Invest (in press)
Ramadori G, Sipe JD, Dinarello CA, Mizel SB, Colten HR (1985a) Pretranslational modulation of acute phase hepatic protein synthesis by murine recombinant interleukin (IL-1) and purified human IL-1. J Exp Med 162:930–942
Ramadori G, Sipe J, Colten HR (1985b) Expression and regulation of the murine serum amyloid A (SAA) gene in extrahepatic sites. J Immunol 135:3645–3647
Reid KBM, Bentley DR, Campbell RD, Chung LP, Sim RB, Kristensen T, Tack BF (1986) Complement system proteins which interact with C3b or C4b. A superfamily of structurally related proteins. Immunol Today 7:230–234
Rosa PA, Shreffler DC (1982) Cultured hepatocytes from mouse strains expressing high and low levels of the fourth component of complement differ in rate of synthesis of the protein. Proc Natl Acad Sci USA 80:2332–2336

Sackstein R, Colten HR (1984) Molecular regulation of MHC class III (C4 and factor B) gene expression in mouse peritoneal macrophages. J Immunol 133:1618–1626
Sackstein R, Colten HR, Woods DE (1983) Phylogenetic conservation of class III major histocompatibility complex antigen, factor B: isolation and nucleotide sequence of mouse factor B cDNA clones. J Biol Chem 258:14693–14697
Shreffler DC, Atkinson JP, Chan AC, Karp DR, Killion CC, Ogata RT, Rosa PA (1984) The C4 and SLP genes of the complement region of the murine H-2 major histocompatibility complex. Phil Trans R Soc Lond B 306:395–403
St John Sutton MB, Strunk RC, Cole FS (1986) Regulation of synthesis of the third component of complement and factor B in cord blood monocytes by lipopolysaccharide. J Immunol 136:1366–1372
Strunk RC, Cole FS, Perlmutter DH, Colten HR (1985a) Gamma interferon increases expression of class III complement genes C2 and factor B in human monocytes and in murine fibroblasts transfected with human C2 and factor B genes. J Biol Chem 260:15280–15285
Strunk RC, Whitehead AS, Cole FS (1985b) Pretranslational regulation of the synthesis of the third component of complement in human mononuclear phagocytes by the lipid A portion of lipopolysaccharide. J Clin Invest 76:985–990
Thireos G, Driscoll-Penn M, Greer H (1984) 5′ untranslated sequences are required for the translational control of a yeast regulatory gene. Proc Natl Acad Sci USA 81:5096–5100
Wetsel RA, Colten HR (1988) The molecular biology of complement deficiency syndromes. In: Avner ED (ed) Topics in renal medicine: the genetics of hereditary nephropathies. Kluwer Academic Publishers (in press)
White PC, Grossberger D, Onufer BJ, Chaplin DD, New MI, Dupont B, Strominger JL (1985) Two genes encoding 21-hydroxylase are located near the genes encoding the fourth component of complement in man. Proc Natl Acad Sci USA 85:1089–1093
Whitehead AS, Colten HR, Chang CC, Demars R (1985) Localization of MHC-linked complement genes between HLA-B and HLA-DR by using HLA mutant cell lines. J Immunol 134:641–643
Woods DE, Markham AF, Ricker AT, Goldberger G, Colten HR (1982) Isolation of cDNA clones for the human complement protein factor B, a class III major histocompatibility complex gene product. Proc Natl Acad Sci USA 79:5661–5665
Wu L, Morley BJ, Campbell RD (1987) Cell-specific expression of human complement protein factor B gene: evidence for the role of two distant 5′-flanking elements. Cell 48:331–342
Yu CY, Belt KT, Giles CM, Campbell RD, Porter RR (1986) Structural basis of the polymorphism of human complement components C4A and C4B: gene size, reactivity and antigenicity. EMBO J 5:2873–2881

Chapter 8

Biosynthesis of Acute Phase Proteins by the Liver Cells

G. Feldmann, D. Bernuau, J. Y. Scoazec and M. Maurice

Introduction

Acute phase reactants (APR) are a group of plasma proteins whose plasma concentrations increase during an acute inflammatory reaction (AIR). Their biochemical structure and functions differ markedly, but they are mainly synthesized by the liver (Koj 1974). While our knowledge of the molecular characteristics of APR and the role of these proteins during inflammation is increasing rapidly (Koj 1974; Kushner 1982; Fey and Gauldie 1988), only a few studies have been devoted to the in vivo mechanisms involved in the production of APR by liver cells. More specifically, the problem of the participation of the various cell types of the liver in the synthesis of APR, as well as the influence of cellular interconnections and tissue organization on the expression of the genes coding APR, has been much less studied.

Before discussing these different points, the cellular composition of the liver will be briefly recalled. Two main types of cells, the hepatocytes and the sinusoidal cells, constitute the liver. Hepatocytes represent about 70% of the total hepatic volume. They are arranged in plates, extending from several peripheral portal tracts to one centrilobular vein, constituting a morpho-functional unit known as the hepatic lobule. According to their position in the lobule, hepatocytes are divided into three distinct subpopulations: (i) periportal hepatocytes, which are the cells located around the portal tracts; (ii) centrilobular hepatocytes, those adjacent to the centrilobular vein, and (iii) mediolobular hepatocytes, located in an intermediate zone in between. No anatomical frontier exists between the three zones and all three hepatocyte subpopulations are morphologically identical, at least under light microscopy. However, functional differences exist between hepatocytes of the different lobular zones, not only for many enzyme activities, but also for APR biosynthesis, as we will see later. In contrast with hepatocytes, sinusoidal cells consist of three different cell types: the endothelial cells, the Kupffer cells and the Ito cells. These are all located in the liver capillaries, the sinusoids. Endothelial cells form a discontinuous thin cellular layer along the whole margin of the sinusoid. Kupffer and Ito cells are randomly located in the sinusoid, the first inside the lumen of the vessel, and the second between endothelial cells and hepatocytes in the perisinusoidal space, the space of Disse. The functional zonation described for hepatocytes has not been demonstrated for sinusoidal cells. The Kupffer cell, an immobilized macrophage, and the Ito cell can synthesize some of the APR.

Role of Hepatocytes in the In Vivo Biosynthesis of APR

While it is now established that hepatocytes can synthesize APR and respond to inflammatory mediators in vitro (Baumann et al. 1984; Guillouzo et al. 1984; Bauer et al. 1985; Fuller et al. 1985; Le and Mortensen 1986; Bauer et al. 1986; Perlmutter et al. 1986; Darlington et al. 1986; Baumann et al. 1987; Magielska-Zero et al. 1987; Pos et al. 1988; Andus et al. 1988), the demonstration of such capacities in the whole liver in vivo has been the subject of only a limited number of investigations.

We will describe our findings both in the unstimulated liver and during an AIR. Immunofluorescence or immunoperoxidase cytochemistry using specific antibodies against different APR was employed for the localization of APR in liver cells (for review see Feldmann et al. 1985). More recently, *in situ* hybridization (ISH) was added to immunocytochemical techniques to investigate the transcriptional behaviour of APR in liver cells.

Expression of APR in the Normal Unstimulated Liver

APR were detected by immunocytochemistry in hepatocytes, and also in the Kupffer cells, for some APR. The problems raised by the localization of APR in Kupffer cells will be discussed later. The following APR were demonstrated in hepatocytes: C-reactive protein (CRP) (Kushner and Feldmann 1978), fibrinogen (Courtoy et al. 1981), α2-macroglobulin (α2-M) (Courtoy et al. 1981), α1-antitrypsin (α1-AT) (Lamri et al. 1986), α1-acid glycoprotein (α1-AGP) (Courtoy et al. 1981), ceruloplasmin (Feldmann 1973), haptoglobin (Courtoy et al. 1981) and α1-inhibitor 3 (α1-I3) (Geiger et al. 1987). Serum amyloid protein A (SAA) was also localized in hepatocytes by others (Benson and Kleiner 1980). Most of these proteins were investigated in rat hepatocytes, except for CRP and SAA, which were detected in rabbit and mouse hepatocytes, respectively (Kushner and Feldmann 1978; Benson and Kleiner 1980). Whatever the protein, a similar cellular localization was observed: on light microscopy, APR were visible exclusively in the cytoplasm, and not in the nucleus of the hepatocytes (Fig. 8.1a); on electron microscopy, the proteins were observed in the organelles known to be the site of plasma protein synthesis and secretion, i.e. the rough endoplasmic reticulum and the Golgi apparatus. These findings are in agreement

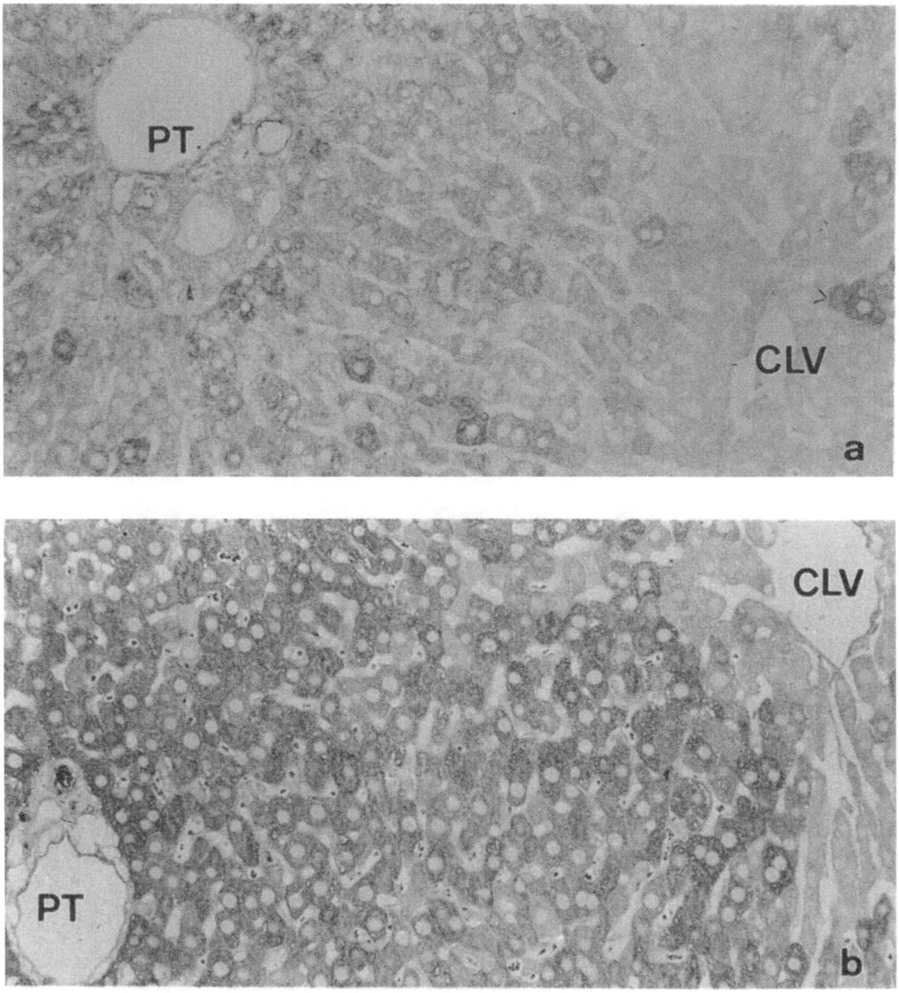

Fig. 8.1a,b. Immunoperoxidase localization of α2-M in liver sections: **a** from a normal rat; **b** from a rat with turpentine-induced acute inflammatory reaction. *PT*: portal tract; *CLV*: centrilobular vein (× 200)

with the results previously obtained for the localization of other plasma proteins of liver origin (Feldmann et al. 1985). The introduction of ISH for the in vivo investigation of APR expression constitutes an important progress. Indeed, with this technique, it has been possible to appreciate the level of transcriptional activity of individual genes within single cells. With ISH, mRNA of some APR, such as α2-M, α1-AT and α1-I3, were localized in the cytoplasm and in the nucleus of hepatocytes (Fig. 8.2a) (Bernuau et al. 1988).

With regard to the distribution of APR-producing hepatocytes in the hepatic lobule, it is now clear that in the normal liver, all or almost all hepatocytes, whatever their location in the lobule, are able to express and synthesize the different APR. As a consequence, it can be assumed that hepatocytes are not

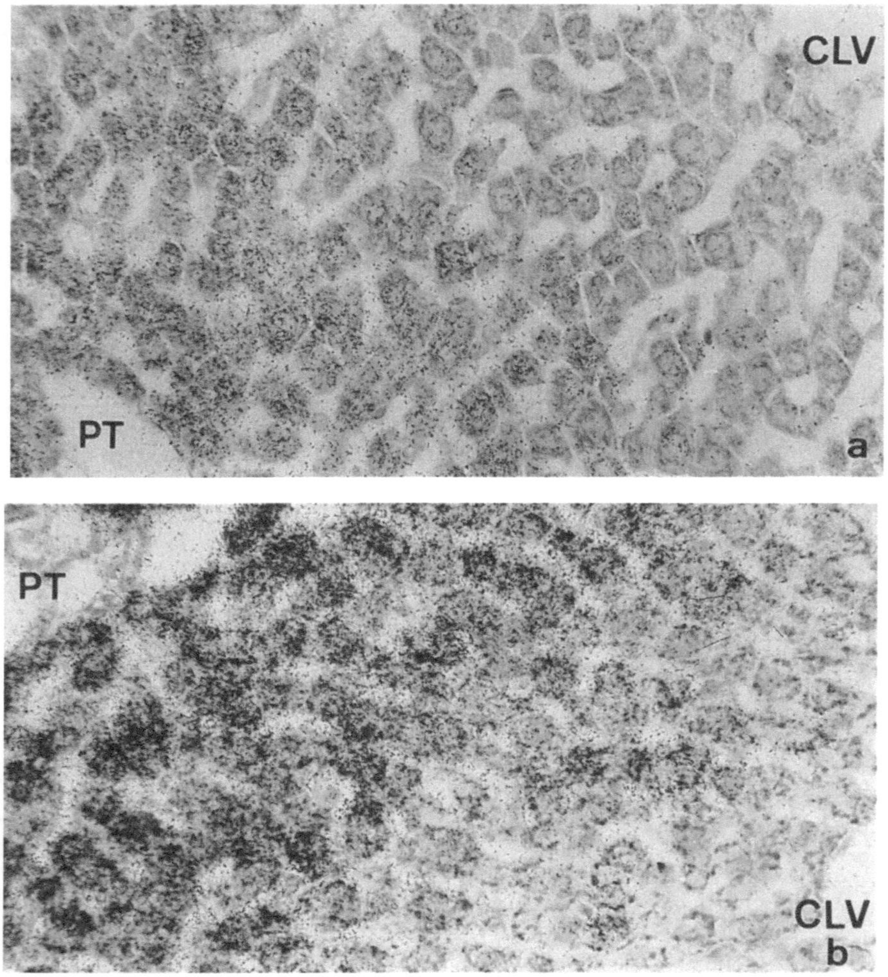

Fig. 8.2a,b. Localization by *in situ* hybridization of α2-M mRNA in liver sections: **a** from a normal rat; **b** from a rat with turpentine-induced acute inflammatory reaction. *PT*: portal tract; *CLV*: centrilobular vein (× 200)

specialized in the synthesis of a given APR, a finding which is also in agreement with the data already observed for other plasma proteins, in particular for albumin (Feldmann et al. 1985). However, it is important to note that the amounts of mRNA and protein in individual hepatocytes differ according to the cell location in the different lobular zones (Bernuau et al. 1989). Various patterns of lobular heterogeneity were recognized for every APR tested: for instance, for the α2-M and α1-I3, the mRNA as well as the corresponding proteins were mainly expressed by periportal and mediolobular hepatocytes, while expression of both genes took place at a very low level in the centrilobular zones (Figs 8.1a, 8.2a) (Bernuau et al. 1989). A different situation was recorded for α1-AT: while equal amounts of mRNA were detected in all three zones of the hepatic lobule, a lobular heterogeneity was observed at the post-translational level, since more protein was present in centrilobular hepatocytes than in hepatocytes of the other zones (Bernuau et al. 1989). Besides a lobular functional heterogeneity, individual variations exist in the level of expression of the APR genes between hepatocytes in a given lobular zone, some individual cells containing or expressing more protein and mRNA than the neighbouring hepatocytes (Figs 8.1a, 8.2a).

Expression of APR During the AIR

The in vivo behaviour of APR-synthesizing hepatocytes varied according to the gene (Bernuau et al. 1989). Induction of APR synthesis was reflected by increasing amounts of mRNA and protein in hepatocytes with, for α2-M and CRP for example, a progressive recruitment of hepatocytes from the periportal to the mediolobular zones (Figs 8.1b, 8.2b). Interestingly, even at the acme of the hepatocyte response, the lobular heterogeneity observed in the unstimulated liver persisted, as we observed for several positive APR, such as CRP (Kushner and Feldmann 1978), fibrinogen (Courtoy et al. 1981), α1-AGP (Courtoy et al. 1981), α2-M and haptoglobin (Courtoy et al. 1981). The lobular heterogeneity also remained visible in the case of α1-I3 (Bernuau et al. 1989) and albumin (Bernuau, unpublished data), two negative APR. In contrast, the lobular heterogeneity observed for α1-AT in the normal liver disappeared during the AIR (Bernuau et al. 1989).

The functional lobular heterogeneity is now well established for hepatocytes (Jungerman 1986; Gumucio and Chianale 1988): while, as is recalled in the Introduction, all hepatocytes share the same morphology, these cells do not function similarly. Examples of this heterogeneity have been reported for numerous hepatic functions, such as, for example, the predominant localization of glycogenolytic enzymes in periportal hepatocytes (Jungermann 1986) or the localization of some cytochromes, particularly the P450 cytochrome, in centrilobular hepatocytes (Gumucio and Chianale 1988). Lobular heterogeneity is not specific for APR, since we have already observed similar findings for α-foetoprotein (Poliard et al. 1986). Therefore, lobular functional heterogeneity applies not only to enzymatic functions of hepatocytes but, as it appears from our results, to plasma protein synthesis as well.

The fact that in the normal liver and during the AIR, APR are not expressed at the same level in the different zones of the hepatic lobule, suggests that as yet unknown factors, linked to the lobular location of hepatocytes, influence the transcription rates and/or the mRNA stabilities of APR. These factors could be either extracellular signals that are unevenly distributed over the liver lobules or intrinsic properties of hepatocytes. While it is now clear that inflammatory mediators released by monocytes and macrophages participate in the regulation of APR genes during the AIR (Koj 1986; Fey and Fuller 1987; Fey and Gauldie 1988), it is not yet clear whether these mediators are also responsible for the uneven expression of APR genes in basal conditions. However, some explanations of the heterogeneity can be provided, based on our results: (i) hepatocytes from the different zones are surrounded by different micro-environments as a result of the unidirectional blood flow, which brings decreasing concentrations of incoming mediators from the portal tract to the centrilobular vein; (ii) the concentration of mediator receptors on the surface of the hepatocytes may differ according to their lobular location. At present, we have no direct evidence to prove that the plasma concentrations of interleukin-1 (IL-1), tumor necrosis factor (TNF) or interleukin-6 (IL-6), the three inflammatory mediators responsible for the increase in APR biosynthesis, decrease along the sinusoid, but this is likely to occur since the plasma concentrations of some hormones, such as insulin, glucagon and corticosteroids have been shown to decrease when the hormones reach the efferent part of the hepatic lobule (Jungermann 1986). In the same way, the distribution of mediator receptors on the plasma membrane of hepatocytes in the different zones of the lobule is unknown. Clearly, additional factors are very likely to play a role in the lobular heterogeneity of APR expression. Indeed, the structure and functions of the intra-hepatocyte components involved in the expression of the APR in response to hormones and external signals probably differ according to the location of hepatocytes within the lobule. Moreover, hepatocytes establish contacts with other hepatocytes, and also with the sinusoidal cells and the liver matrix. Such cellular interrelations are likely to differ according to the lobular topography.

As examples of intra-cellular components which may play a role in APR expression and synthesis by hepatocytes, we will discuss the microtubules and the Golgi apparatus (GA). Relationships between hepatocytes and sinusoidal cells will also be briefly discussed in the next section.

Microtubules localized in the cytoplasm of hepatocytes around the GA are clearly involved in the secretion of APR, as demonstrated by two different experiments. An increase in the volume density of microtubules and of polymerized tubulin, the main constituent of microtubules, was observed in rat liver cells when turpentine was administered to the animals (Maurice and Feldmann 1982). The changes paralleled the increased synthesis of APR by the liver (Maurice and Feldmann 1982). In contrast, an inhibition of APR secretion was observed when colchicine, a microtubule-depolymerizing agent, was given to rats during an experimental AIR (Ritchie et al. 1988). The role of microtubules in the secretion of plasma proteins by the hepatocytes has been well established (for review see Feldmann 1989). This cytoskeleton organelle seems to be essential in the final steps of plasma protein secretion, by preserving the integrity of the GA and by assuring the pathway followed by proteins between the GA and the plasma membrane of the hepatocyte. A delicate equilibrium controlled by many factors (Feldmann 1989) exists between the free cytoplasmic form of tubulin and the polymerized tubulin. How the induction of an AIR can act in hepatocytes to displace this equilibrium towards the polymerized form is not known at the present time, but one can suggest that the answer depends on the basal conditions which control the equilibrium between free and polymerized tubulins.

The GA by itself is also involved in APR secretion by hepatocytes. An increase in the volume density of this organelle has been reported during the AIR (Lombart et al. 1980; Maurice and Feldmann 1982), together with an increase in activity of enzymes located in its membranes, particularly those involved in glycosylation (Lombart et al. 1980). Most APR are glycoproteins. During an AIR, there is not only an enhanced expression of APR themselves, but also an increase in numerous enzymatic proteins involved in APR biosynthesis (Jamieson et al. 1987). This supposes some delicate regulation and coordination at different cellular levels. Up to now, a heterogeneous distribution of microtubules according to the location of hepatocytes in the lobular zones has not been reported. Similarly, it is not known whether glycosyltransferase activity differs between periportal and centrilobular zones.

Role of Sinusoidal Cells in the In Vivo Biosynthesis of APR

At the present time, only one APR, α2-M, has been demonstrated to be synthesized by Ito cells in vitro (Andus et al. 1987). In vivo localization of other APR, such as fibrinogen (Hamashima et al. 1964) or α1-AT (Millward-Sadler 1981) has been reported in Kupffer cells in unstimulated liver. However, as we pointed out (Feldmann et al. 1985), a localization of plasma proteins in Kupffer cells must be interpreted cautiously since this localization could be due to diffusion artefacts, particularly during liver fixation. However, APR biosynthesis by macrophages has been demonstrated by biochemical techniques (Isaacson et al. 1981). On the other hand, we have observed the presence of α1-AT mRNA and its corresponding protein in Kupffer cells during an AIR (Lamri et al. 1986; Bernuau et al. 1988).

Kupffer cells can also produce, like other macrophages, inflammatory mediators, particularly the hepatocyte-stimulating factor (now known as IL-6) as has been demonstrated in vitro by Ritchie and Fuller (1983). However, their role in the induction of the liver acute phase response in vivo, remains unknown. We quantified, by light microscopy, the number of Kupffer cells at different times during an experimental AIR in the liver of two species, the rat and the rabbit (Scoazec and Feldmann 1987). No changes in their number were observed either in periportal or in centrilobular zones. Similarly, no mitosis was observed. Consequently, if Kupffer cells act in vivo on hepatocytes, this role seems to be only functional and is not explained by numerical or morphological changes.

Summary

Hepatocytes are the main site of biosynthesis of acute phase proteins in the liver. These cells are not specialized in APR production. However, in basal conditions, a lobular functional heterogeneity exists for APR synthesis, both at pre- and post-translational levels, with specific patterns for individual genes. During the AIR-induced modulation of APR synthesis, the lobular heterogeneity may, or less often may not, persist, according to the gene. The mechanisms of the functional heterogeneity for APR synthesis are not elucidated, but might depend on variations in intracellular components or cell–cell interactions throughout the lobule. During the AIR, variable expressions of mediator and/or hormone receptors in the hepatic lobule, and lobular hormone gradients, remain to be demonstrated.

Acknowledgments. The authors wish to thank Mrs G. Malet for her help in the preparation of the manuscript.

References

Andus T, Ramadori G, Heinrich PC, Knittel T, Meyer Zum Büschenfelde K-H (1987) Cultured Ito cells of rat liver express the α_2-macroglobulin gene. Eur J Biochem 168:641–646

Andus T, Geiger T, Hirano T, Kishimoto T, Tran-Thi T-A, Decker K, Heinrich P (1988) Regulation of synthesis and secretion of major rat acute-phase proteins by recombinant human interleukin-6 (BSF-2/IL-6) in hepatocyte primary cultures. Eur J Biochem 173:287–293

Bauer J, Kurdowska A, Tran-Thi T-A, Budek W, Koj A, Decker K, Heinrich PC (1985) Biosynthesis and secretion of α_1 acute-phase globulin in primary cultures of rat hepatocytes. Eur J Biochem 146:347–352

Bauer J, Tran-Thi T-A, Northoff H, Hirsch F, Schlayer H-J, Gerok W, Heinrich PC (1986) The acute-phase induction of α_2-macroglobulin in rat hepatocyte primary cultures: action of a hepatocyte-stimulating factor, triiodothyronine and dexamethasone. Eur J Cell Biol 40:86–93

Baumann H, Jahreis GP, Sauder DN, Koj A (1984) Human keratinocytes and monocytes release factors which regulate the synthesis of major acute phase plasma proteins in hepatic cells from man, rat and mouse. J Biol Chem 259:7331–7342

Baumann H, Onorato V, Gauldie J, Jahreis GP (1987) Distinct sets of acute phase plasma proteins are stimulated by separate human hepatocyte-stimulating factors and monokines in rat hepatoma cells. J Biol Chem 262:9756–9768

Benson MD, Kleiner E (1980) Synthesis and secretion of serum amyloid protein A (SAA) by hepatocytes in mice treated with casein. J Immunol 124:495–499

Bernuau D, Legrès L, Lamri Y, Giuily N, Fey G, Feldmann G (1989) Heterogeneous lobular distribution of hepatocytes expressing acute phase genes during the acute inflammation reaction. J Exp Med (in press)

Courtoy PJ, Lombart C, Feldmann G, Moguilevsky N, Rogier E (1981) Synchronous increase of four acute phase proteins synthesized by the same hepatocytes during the inflammatory reaction. A combined biochemical and morphologic kinetics study in the rat. Lab Invest 44:105–115

Darlington GJ, Wilson DR, Lachman LB (1986) Monocyte-conditioned medium, interleukin-1 and tumor necrosis factor stimulate the acute phase response in human hepatoma cells in vitro. J Cell Biol 103:787–793

Feldmann G (1973) Plasma protein synthesis by human liver cells: its morphological demonstration with antibodies labelled with peroxidase. In: Wisse E, Daems WT, Molenaar I, Van Duijn P (eds). Electron microscopy and cytochemistry. North-Holland, Amsterdam, pp 145–149

Feldmann G (1989) The cytoskeleton of the hepatocyte: structure and functions. J Hepatol 8:380–386

Feldmann G, Maurice M, Bernuau D, Rogier E (1985) Morphological aspects of plasma protein synthesis and secretion by the hepatic cells. Int Rev Cytol 96:157–189

Fey G, Fuller GM (1987) Regulation of acute phase gene expression by inflammatory mediators. Mol Biol Med 4:323–338

Fey G, Gauldie J (1988) The acute phase response of the liver in inflammation. In: Popper H, Schaffner F (eds) Progress in liver diseases, vol IX. Grune and Stratton, New York (in press)

Fuller GM, Otto JM, Woloski BM, McGary CT, Adams MA (1985) The effects of hepatocyte stimulating factor on fibrinogen biosynthesis in hepatocyte monolayers. J Cell Biol 101:1481–1486

Geiger T, Lamri Y, Tran-Thi T-A, Gauthier F, Feldmann G, Decker K, Heinrich PC (1987) Biosynthesis and regulation of rat α_1-inhibitor3, a negative acute-phase reactant of the macroglobulin family. Biochem J 245:493–500

Guillouzo A, Delers F, Clement B, Bernard N, Engler R (1984) Long term production of acute-phase proteins by adult rat hepatocytes co-cultured with another liver cell type in serum-free medium. Biochem Biophys Res Commun 120:311–317

Gumucio JJ, Chianale J (1988) Liver cell heterogeneity and liver function. In: Arias IM, Jakoby WB, Popper H, Schachter D, Shafritz DA (eds) The liver: biology and pathobiology, 2nd edn. Raven Press, New York, pp 931–948

Hamashima Y, Harter JG, Coons AH (1964) The localization of albumin and fibrinogen in human liver cells. J Cell Biol 20:271–279

Isaacson P, Jones PB, Millward-Sadler GH, Judd MA, Payne S (1981) Alpha-1-antitrypsin in human macrophages. J Clin Pathol 34:982–990
Jamieson JC, Lammers G, Janzen R, Woloski BMRNJ (1987) The acute phase response to inflammation: the role of monokines in changes in liver glycoproteins and enzymes of glycoprotein metabolism. Comp Biochem Physiol 87B:11–15
Jungermann K (1986) Functional heterogeneity of periportal and perivenous hepatocytes. Enzyme 35:161–180
Koj A (1974) Acute-phase reactants. Their synthesis, turnover and biological significance. In: Allison A (ed) Structure and function of plasma proteins, vol 1. Plenum Press, London, pp 72–131
Koj A (1986) Biological functions of acute phase proteins and the cytokines involved in their induced synthesis. In: Reutter W, Heinrich PC, Popper H, Keppler D, Arias IM, Landmann L (eds) Modulation of liver cell expression. Falk Symposium No 43. MTP Press, Lancaster, pp 331–342
Kushner I (1982) The phenomenon of the acute phase response. In: C-reactive protein and the plasma protein response to tissue injury. Ann NY Acad Sci 389, pp 39–48
Kushner I, Feldmann G (1978) Control of the acute phase response. Demonstration of C-reactive protein synthesis and secretion by hepatocytes during acute inflammation in the rabbit. J. Exp Med 148:466–477
Lamri Y, Eriksson S, Moreau A, Gauthier F, Feldmann G (1986) Cellular and subcellular distribution of alpha-1-antitrypsin in the normal rat liver and during acute inflammatory reaction. A light and electron microscopic study with peroxidase labelled antibodies. Cell Mol Biol 32:691–699
Le PT, Mortensen RF (1986) Induction and regulation by monokines of hepatic synthesis of the mouse serum amyloid P-component (SAP). J Immunol 136:2526–2533
Lombart C, Sturgess J, Schachter H (1980) The effect of turpentine-induced inflammation on rat liver glycosyltransferases and Golgi complex ultrastructure. Biochim Biophys Acta 629:1–12
Magielska-Zero D, Rokita H, Cieszka K, Kurdowska A, Koj A, Sipe JD, Gauldie J (1987) Comparison of the acute phase response of cultured Morris hepatoma 7777 cells and of rat hepatocytes. Br J Exp Pathol 68:485–492
Maurice M, Feldmann G (1982) Quantitative biochemical and ultrastructural studies of microtubules in rat liver during the acute inflammatory reaction. Exp Mol Pathol 36:193–203
Millward-Sadler GH (1981) Alpha-1-antitrypsin deficiency and liver disease. Acta Med Port 2 (suppl): 91–102
Perlmutter DH, Dinarello CA, Punsal PI, Colten HR (1986) Cachetin/tumor necrosis factor regulates hepatic acute-phase gene expression. J Clin Invest 78: 1349–1354
Poliard AM, Bernuau D, Tournier I, Legrès LG, Schoevaert D, Feldmann G, Sala-Trepat JM (1986) Cellular analysis by in situ hybridization and immunoperoxidase of alpha-fetoprotein and albumin gene expression in rat liver during the perinatal period. J Cell Biol 103:777–786
Pos O, Van Dijk W, Ladiges N, Linthorst C, Sala M, Van Tiel D, Boers W (1988) Glycosylation of four acute-phase glycoproteins secreted by rat liver cells in vivo and in vitro. Effects of inflammation and dexamethasone. Eur J Cell Biol 46:121–128
Ritchie DG, Fuller GM (1983) Hepatocyte-stimulating factor: a monocyte-derived acute-phase regulatory protein. In: Mosesson MW, Doolittle RF (eds) Molecular biology of fibrinogen and fibrin. Ann NY Acad Sci 408, pp 490–502
Ritchie DG, Carter KC, Papaconstantinou J (1988) Microtubule involvement in the HSF mediated regulation of hepatic acute phase reactant mRNA levels. In: Proceedings of the IV international congress of cell biology, Montreal, p 160
Scoazec JY, Feldmann G (1987) Quantitation of hepatic sinusoidal macrophages during the acute systemic inflammatory reaction in two animal species. Proc Soc Exp Biol Med 184:389–395

Chapter 9

The Plasma Serine Protease Inhibitors (Serpins): Structural Modifications in Inflammation

P. E. Stein and R. W. Carrell

The Serpins

The serine protease inhibitor (serpin) superfamily comprises more than 20 glycoproteins that show primary sequence homology at protein or cDNA level with the archetype of the family, α1-antitrypsin (abbreviated hereafter to antitrypsin). Members of the family are presumed to be derived by divergent evolution from a common ancestral protein over a period of more than 500 million years (Hunt and Dayhoff 1980; Carrell and Boswell 1986). Although the serpins are widely distributed in nature, and have been identified in viruses and plants, the best studied members of the family are the human plasma serpins (Table 9.1). Most plasma serpins function as inhibitors of serine proteases and are principally involved in inflammation, controlling the activities of enzymes that trigger the inflammatory cascades including coagulation, fibrinolysis, complement activation and kinin release. However some serpins have apparently lost their antiprotease activity and have evolved other functions, such as hormone binding in the case of thyroxine binding globulin (TBG) and cortisol binding globulin (CBG).

Structure of the Serpins

The serpins show primary sequence homology of approximately 30%, with an even distribution of conserved residues indicating a common secondary and tertiary structure. The greatest variation in amino acid sequence occurs at the

Table 9.1. Human serpins

	Concentration (μM)	Mass (kDa)	
Inhibitors			*Target protease*
α1-Antitrypsin	25	51	Neutrophil elastase
α1-Antichymotrypsin	7	69	Cathepsin G
Antithrombin III	2	58	Thrombin, Factor Xa
C_1-inhibitor	2	104	C1s, kallikrein
α2-Antiplasmin	1	70	Plasmin
Heparin cofactor II	1	66	Thrombin
PC inhibitor	10^{-2}	57	Activated protein C
PA inhibitor I	10^{-4}	52	Plasminogen activator
Protease nexin I	—	43	Thrombin
Non-inhibitors			*Function*
Angiotensinogen	10^{-2}	50	Donor of angiotensin II
TBG	0.3	54	Transport of thyroid hormones
CBG	0.5	58	Transport of cortisol, progesterone

amino (N) terminus prior to the start of ordered structure, and near the carboxy (C) terminus in the region immediately surrounding the reactive centre.

Cleaved Tertiary Structure

The only serpin for which a crystallographic three dimensional structure has been determined is antitrypsin, in an inactive form in which the peptide bond at the reactive site has been cleaved by chymotrypsin (Fig. 9.1). This structure (Loebermann et al. 1984) shows a highly ordered, globular molecule 30% of which is arranged as α-helices (A-I) and 40% as β-pleated sheets (A-C). Unexpectedly, the two amino acids forming the reactive site peptide bond in the native inhibitor (Met 358 and Ser 359) are found at opposite poles of the molecule separated by a distance of 67 Å and indicating that a major structural transition has taken place on proteolytic cleavage of the native protein.

Native Tertiary Structure

To reconstruct the reactive site, as it is presumed to exist in native antitrypsin, the peptide strand terminating with the reactive centre methionine (strand A4) must be withdrawn from the major (A) β-pleated sheet and bent back on itself to join this methionine with the reactive centre serine of the C-terminal fragment of the molecule. This forms an externally situated, stressed peptide loop of some 15 residues that presents the reactive centre in an exposed position where it is accessible to target proteases. Support for this model of native antitrypsin has been obtained from study of the limited proteolysis of the active, native inhibitor. All the proteolytic cleavages investigated with antitrypsin occur within a stretch of nine residues on the N-terminal side of the reactive centre (Fig. 9.2). These residues are inaccessible to proteolytic attack in the crystallographic structure of the cleaved inhibitor since they lie on the central A4 strand of the A β-sheet. However, in the proposed model of native antitrypsin with the loop intact, the

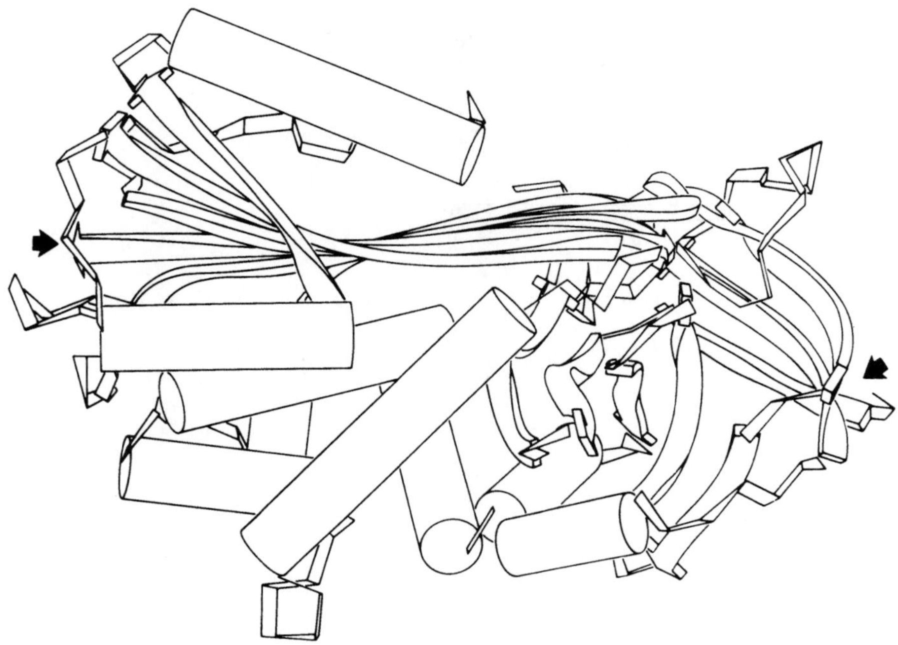

Fig. 9.1. Structure of antitrypsin cleaved at its reactive centre. α-Helical regions are represented as *cylinders*, β-pleated sheets as *broad ribbons*, and regions of indeterminate secondary structure as *flat ribbon* segments. The *arrows* indicate the residues which in the native inhibitor form the reactive site. Figure prepared by Dr Arthur Lesk from coordinates supplied by Prof. Robert Huber.

same residues are externally situated on the loop and therefore susceptible to cleavage. No cleavage sites for antitrypsin have been mapped on the C-terminal side of the reactive centre, partly due to the presence here of a double prolyl sequence, but also consistent with the model which indicates that the sequence becomes inaccessible within a few residues of the reactive centre. The exact configuration of the reactive centre will remain unknown until crystallographic data is available for a native, uncleaved serpin. Work with crystals of native antithrombin is in progress (Samama et al. 1987).

Synthesis of Serpins in Inflammation

The liver is the major site of synthesis of the plasma serpins, some of which undergo changes in their rate of synthesis as part of the acute phase response associated with inflammation. Antitrypsin and α1-antichymotrypsin are acute phase proteins (Ganrot 1974). Plasma levels of immunologically reactive antitrypsin increase more than 3-fold during the acute phase response, although at an inflammatory site functional activity of the inhibitor may be conversely reduced (pp. 113 and 115). Induction of angiotensinogen mRNA in rat liver has been reported in response to an acute inflammatory stimulus (Kageyama et al. 1985), suggesting that angiotensinogen is also an acute phase protein with a possible role in inflammation. In vitro studies in the rat indicate that antithrombin

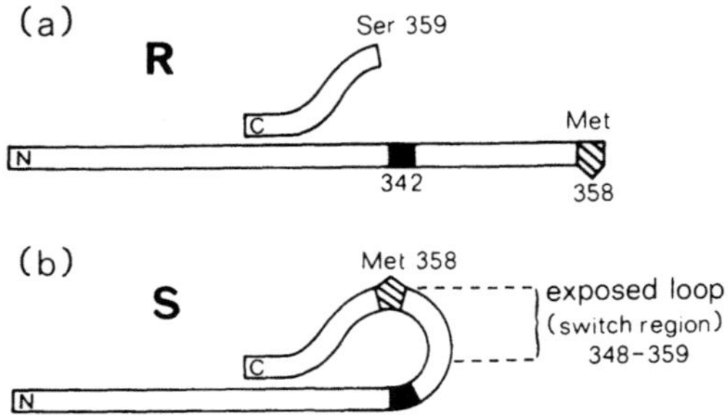

Numbers (related to P_1-P_1' reactive centre)	15	14	13	12	11	10	9	8	7	6	5	4	3	2		1	1'	2'	3'	4'	5'		
Antitrypsin	G	T	E	A	A	G	A ↑1	M	F ↑3, 18	L	E ↑4	A	I	P	- ↑16, 17	M ↑5, 6	S	I	P	P	E	-	-
Antithrombin	G	S	E	A	A	A ↑2	S	T	A ↑2	V	V	I ↑7	A	G	-	R	S ↑8	L	N	P	N	-	-
C_1-Inhibitor	G	V	E	A	A	A	A	S	A ↑9	I ↑10	S ↑11	V ↑10, 9	A	-	-	R	T	L	L	V	-	-	-
α_2 - Antiplasmin	G	V	E	A	A	A	A	T	S	I	A ↑21	M	S	-	-	R	M	S	L	S	S	-	-
TBG	G	T	E	A	A	A	V	P	E	V	E ↑12	L	S	D	-	Q	P	E	N	T ↑12, 13	F	L	H
CBG	G	V	D	T	A	G	S	T	G	V ↑14	T	L	N	L	-	T	S	K	P	I	I	-	-
Angiotensinogen	E	R	E	P	T	E	S	T ↑15	Q	Q	L ↑15	N	K	P	-	-	-	-	-	-	-	-	E
Ovalbumin	G	R	E	V	V	G	S	A	E ↑19	A	G	V	D ↑19	A	-	A ↑19, 20	S	V	S	-	E	E	-

1, 2	C. adamanteus venom proteinase II
3	S. aureus metalloproteinase
4	S. aureus serine and cysteine proteinases
5	Papain
6	S. Marcescens Metalloproteinase
8, 11, 16	Ps. aeruginosa elastase
7, 10, 13, 14, 15, 21	Human neutrophil elastase
9, 12	Bitis arietan venom
17	Macrophage elastase
18	Human neutrophil metalloproteinase
19	Subtilisin BPN
20	Pancreatic elastase

Fig. 9.2. The structure of the serpins is based on the cleaved, relaxed (*R*) form of antitrypsin shown diagrammatically (**a**). The native, stressed (*S*) form has to be reconstructed by a deduced exposed loop that hinges near residue 342 (**b**). Sequences of the exposed loop have been aligned on the basis of conserved hinge residues and show (↑) the sites of cleavage by different enzymes. References: Cleavage site 1 (Kress et al. 1979); 2 (Kress and Catanese 1981); 3, 4 (Potempa et al. 1986); 5 (Johnson and Travis 1977); 6 (Virca et al. 1982); 7 (Carrell and Owen 1985); 8 (P. Christey, unpublished work); 9, 10 (Pemberton et al. 1989); 11 (Salvesen et al. 1985); 12 (P. Pemberton, unpublished work); 13, 14 (Pemberton et al. 1988); 15 (P. Stein, unpublished work); 16 (Morihara et al. 1984); 17 (Banda et al. 1987); 18 (Desrochers and Weiss 1988); 19, 20 (Linderstrom-Lang 1952; Wright 1984); 21 (Potempa et al. 1988).

III is a negative acute phase protein (Owens and Miller 1980), but in the human, plasma levels remain remarkably constant in inflammation.

Antitrypsin

Antitrypsin is the inhibitor present in highest concentration in normal plasma and is the best characterized member of the serpin family. The mature protein (Mr 51000) comprises a single polypeptide chain of 394 amino acids with three carbohydrate side chains (Carrell et al. 1982). Although inhibition of trypsin was the first function ascribed to antitrypsin (Schultze et al. 1962), the name is inappropriate, because the primary physiological target of this inhibitor has now been identified as neutrophil elastase (Travis and Salvesen 1983).

Antitrypsin Deficiency

Severe deficiency of antitrypsin has an incidence of 1 in 1000 in the northern European population and is associated with S (264 Glu→Val) and Z (342 Glu→Lys) variants of the normal (M) inhibitor.

In the S variant, loss of a glutamic acid which in the normal protein takes part in an important, stabilizing salt-bridge results in reduced stability of the inhibitor (Lieberman 1973). A false splice site in the gene coding for the mutant (S) protein may lead to loss of mRNA contributing to deficiency of the protein (Schindler 1984). In the Z variant, the defect appears to lie in the transport of the newly synthesized polypeptide through the endoplasmic reticulum prior to final processing of the carbohydrate side chains. Abnormal (Z) antitrypsin accumulates as microscopic inclusions in the periportal hepatocytes, accompanied by enhanced lysosomal activity associated with proteolytic degradation (Bathurst et al. 1985).

Emphysema in Antitrypsin Deficiency

The role of antitrypsin is illustrated by considering the most important consequence of the deficiency associated with ZZ and SZ phenotypes, that of premature emphysema (Laurell and Eriksson 1963). The basic pathogenic mechanism in emphysema is believed to be protease–antiprotease imbalance (Lieberman 1976), resulting either from prolonged exposure to neutrophil proteases released in chronic inflammation or from lack of appropriate inhibitors as in antitrypsin deficiency. In each case, destruction of lung connective tissue results from excessive exposure to neutrophil enzymes, particularly elastase which hydrolyses a wide range of proteins including elastin and collagen.

Liver Disease in Antitrypsin Deficiency

The Z antitrypsin variant carries a predisposition to liver disease, particularly in males. In approximately 10% of ZZ individuals liver disease occurs in childhood, in the form of neonatal hepatitis which may progress to juvenile cirrhosis (Sveger and Thelin 1981). A further 20% develop cirrhosis in adult life with increased risk of primary liver cell cancer (Eriksson et al. 1986). The intracellular accumulation of antitrypsin, rather than plasma deficiency of the inhibitor, is now believed to be the mechanism underlying the liver disease associated with the ZZ phenotype (Carrell et al. 1982). The accumulation of abnormal protein appears to induce the heat shock response, involving a major alteration in the pattern of gene expression in the hepatocyte. Specific heat shock proteins (hsp 70 and 90), that may have functions in limiting or repairing cell damage, have been identified in hepatocytes of ZZ individuals (Perlmutter et al 1988). The recent finding in transgenic mice (Carlson et al 1988), that insertion of the abnormal (Z) human gene results in hepatocyte inclusions and associated liver pathology, supports the hypothesis that liver disease is indeed a consequence of the intracellular accumulation of antitrypsin. On this basis, the use of plasma replacement therapy or stimulants of antitrypsin synthesis (e.g. danazol) in treatment of this liver disease, are clearly inappropriate.

The Reactive Centre

Mechanism of Protease Inhibition

Serpins are believed to act as ideal substrates for their target proteases (Travis and Salvesen 1983), resulting in formation of a stable, 1 : 1 complex between enzyme and inhibitor that is subsequently catabolized by the cells of the reticuloendothelial system. The exposed loop of the native serpin is under tension, holding the molecule in a stressed conformation and distorting the reactive centre bond, probably towards tetrahedral geometry, such that it precisely fits the active site pocket of the target protease. Hence maintenance of inhibitory activity is critically dependent on the integrity of the exposed loop.

Specificity of Inhibition

The structure of the reactive centre of a serpin enables prediction of its probable target protease. Specificity is substantially determined by the residue at the N-terminal side of the reactive centre (P_1), which is believed to act as a bait, providing a potential cleavage site for the target enzyme (Laskowski and Kato 1980). The P_2 residue also contributes to specificity.

Antitrypsin has a methionine in the P_1 position (Metserpin) and inhibits proteases that cleave specifically at small hydrophobic residues, principally neutrophil elastase and cathepsin G. Complexes are formed with other serine proteases, although in many cases reaction rates are too slow to be of physiological significance (Travis et al. 1987). Antithrombin III, C1-inhibitor and $\alpha 2$-antiplasmin all have an arginine as the P_1 residue (Argserpins) with potential to inhibit a range of trypsin-like proteases including thrombin, kallikrein, factor XIa and plasmin. In the Argserpins, features distant from the reactive centre may modulate specificity, enabling more specific targeting to different proteases. For example, binding of the glycosaminoglycan heparin to externally situated, basic residues of antithrombin III may result in a steric change that exposes the reactive centre, making it accessible to the target enzyme, thrombin.

Antitrypsin Pittsburgh

Evidence for the dependence of specificity on the P_1 residue is provided by the changes in inhibitory activity seen in both natural and genetically engineered reactive centre mutants. The importance of the P_1 residue (Met 358) at the reactive centre of antitrypsin was particularly well illustrated by the identification of a unique natural mutation at this site. The patient, a boy from Pittsburgh, presented with a severe bleeding disorder. Electrophoresis of his plasma showed an abnormal antitrypsin that, on analysis (Owen et al. 1983), revealed a substitution of the normal reactive centre methionine for an arginine residue. As predicted, the mutant protein, antitrypsin Pittsburgh, showed an altered pattern of inhibitory activity in comparison to normal antitrypsin, with a 10-fold reduction in elastase inhibition and increases in inhibition of thrombin, kallikrein, factor XIa and plasmin, so in effect the mutant protein resembled antithrombin III. Unlike normal antithrombin III, antitrypsin Pittsburgh showed full inhibitory activity in the absence of heparin. Since the normal molar concentration of antitrypsin in plasma is an order of magnitude higher than that of antithrombin III, the mutant protein was present in great excess over normal antithrombin III. Furthermore the reaction to tissue haemorrhage induced an acute phase response with even greater synthesis of the abnormal protein and aggravation of the condition.

Reactive Centre Oxidation of Antitrypsin

During inflammation, stimulation of neutrophils results in degranulation with release of enzymes, including elastase (Weissmann et al. 1980), and oxidant metabolites (Babior 1978). The methionine at the reactive centre of antitrypsin is particularly labile and readily oxidized to methionine-sulphoxide, a derivative that is too large to fit the reactive centre of elastase, resulting in virtually complete loss of elastase inhibitory activity (Johnson and Travis 1979). At a centre of inflammation it has been suggested that antitrypsin is oxidatively inactivated within close proximity of stimulated neutrophils by reactive oxidants generated by neutrophil NADPH oxidase or myeloperoxidase systems (Carp and Janoff 1980). This inactivation of antitrypsin allows unopposed elastase activity within a limited area, facilitating breakdown of connective tissue and discrete localization of the inflammatory focus.

In contrast to oxygen radicals which have only a small radius of diffusion, elastase can diffuse away from the inflammatory focus with the potential to cause tissue destruction at distant sites. However, beyond the zone of oxidation at the inflammatory centre, antitrypsin is active and effectively inhibits elastase, thus limiting tissue damage.

Evidence for the oxidative inactivation of antitrypsin was provided by studies of a genetically engineered mutant of antitrypsin in which an oxidation resistant valine residue was substituted for the methionine at the reactive centre (Rosenberg et al. 1984). This mutant inhibitor retained its anti-elastase activity and was not inactivated by oxidation. Normal and mutant inhibitors were tested (George et al. 1984) in an experimental model of inflammatory tissue damage in which stimulated neutrophils degraded glomerular basement membrane collagen. Proteolysis was only minimally inhibited by physiological concentrations of normal antitrypsin, but the valine mutant showed a 10-fold increase in inhibition at the same concentration. However when normal neutrophils were replaced by those from a patient with chronic granulomatous disease, which are unable to produce oxygen free radicals, the normal antitrypsin remained active and effectively inhibited proteolysis.

It has been reported that oxidized, inactivated antitrypsin can be partially reactivated by addition of mild reducing agents (Carp and Janoff 1980). An enzyme (methionine sulphoxide peptide reductase) has been identified in human lung homogenates and in lysates of human neutrophils and type 2 alveolar cells that can regenerate active inhibitor from some forms of oxidized antitrypsin (Carp et al. 1983).

Role of Oxidation in the Pathogenesis of Emphysema

It has been proposed that oxidative inactivation of antitrypsin by components of tobacco smoke may contribute to the development of chronic lung damage in emphysema (Carp et al. 1982). The extent of this contribution is controversial, but it seems likely that neutrophils which accumulate in the lungs of the smoker, together with components of tobacco smoke, provide the oxidant capacity for local inactivation of antitrypsin in the vicinity of the neutrophil, allowing focal elastase mediated damage to lung tissue.

The SR Conformational Change

The predicted structure of native antitrypsin may be reconstructed from the known cleaved structure by withdrawal of the A4 strand from the centre of the major β-sheet to form an exposed, stressed loop, a transition that would require disruption of hydrogen bonds and considerable energy cost (Loebermann et al. 1984). This is consistent with the observation (Carrell and Owen 1985) that cleavage of the exposed loop of native antitrypsin is associated with a marked increase in thermal stability of the protein accompanying the profound molecular transition from the native, active, stressed (S) structure to a cleaved, inactive, relaxed (R) form of lower energy. This increase in heat stability on cleavage of the exposed loop sequence was also demonstrated for antithrombin III and subsequently for C_1-inhibitor (Pemberton et al. 1989), in each case associated with loss of inhibitory activity. These observations suggest that the SR transition is a general feature, at least of the inhibitor members of the serpin family and is of physiological significance.

Physiological Significance of the SR Change

Local Inactivation of Inhibitors

It has been proposed that the susceptibility of the exposed peptide loop to proteolytic cleavage provides a physiological switch that allows catalytic inactivation of inhibitors, associated with the SR conformational change at sites of inflammation (Carrell and Owen 1985).

Although antitrypsin is not cleaved by neutrophil elastase, a specific site has been demonstrated on the exposed loop (between Phe 352 and Leu 353) that is susceptible to proteolytic cleavage by a metalloprotease released from stimulated neutrophils (Desrochers and Weiss 1988; Vissers et al. 1988). This catalytic cleavage is associated with inactivation of antitrypsin (as measured by loss of elastase inhibition) and is presumably accompanied by the SR transition described above. There is evidence to suggest that the, as yet uncharacterized, metalloprotease responsible for this effect is released from neutrophils in a latent form and is activated by a process linked to generation of neutrophil oxidants (Desrochers and Weiss 1988). The catalytic inactivation of antitrypsin by a neutrophil metalloenzyme, together with oxidation of the reactive centre methionine by neutrophil oxidants (p. 113) provide neutrophil-dependent physiological mechanisms for the structural modification and local inactivation of antitrypsin at an inflammatory site, resulting in a focus of unopposed neutrophil elastase activity.

Antithrombin III, C1-inhibitor and α2-antiplasmin are each susceptible to cleavage of the reactive centre loop by neutrophil elastase (Fig. 9.2), associated with inactivation of the inhibitor and the presumed SR change. Inactivation of inhibitors in this way would provide the potential for localized modification of the proteolytic cascades at an inflammatory site.

Neutrophil Chemotaxis

Forms of structurally modified antitrypsin (and perhaps other serpins), although inactive as protease inhibitors, may have a role in stimulating neutrophil migration and thus amplifying the inflammatory response. It has been reported (Banda et al. 1987) that antitrypsin proteolytically inactivated by mouse macrophage elastase which cleaves in the exposed loop region (between Pro 357 and Met 358), acts as a potent chemotactic factor for human neutrophils.

Modulation of Hormone Transport

TBG and CBG are serpins without known inhibitory activity, but are nevertheless vulnerable to proteolytic cleavage, by neutrophil elastase, in the peptide segment corresponding to strand A4 (Pemberton et al. 1988). In each case cleavage is associated with the marked increase in heat stability characteristic of the SR transition (Fig. 9.3) and providing evidence that these hormone binding proteins have retained the typical serpin tertiary structure. The ability of the serpin molecule to undergo such a profound structural change during inflammation may explain its selection during evolution for the role of hormone transport.

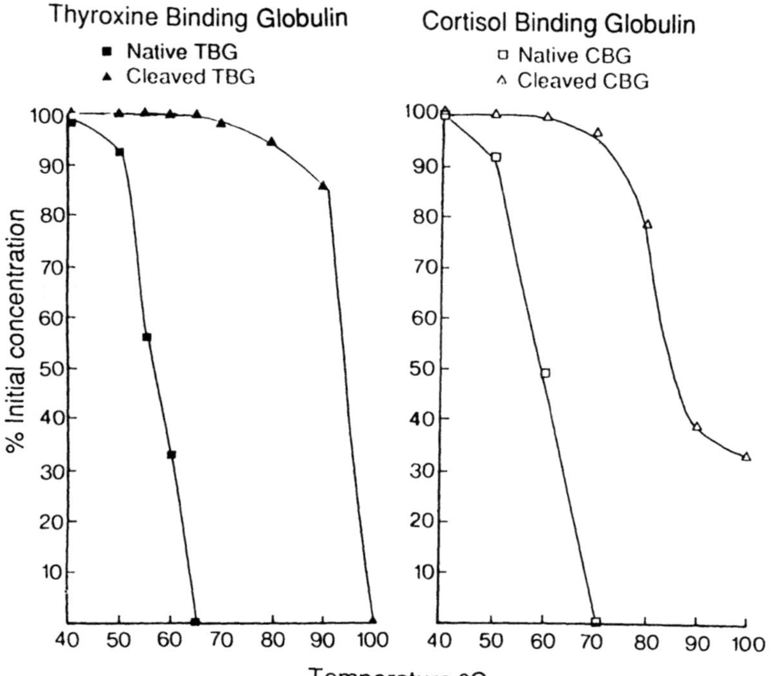

Fig. 9.3. Thermal stability of TBG and CBG showing the marked increase in stability that follows cleavage of the exposed loop sequence by neutrophil elastase. Reprinted by permission from *Nature* vol. 336 pp. 257–258. © 1988 Macmillan Magazines Ltd.

While native and cleaved forms of TBG show no difference in their affinities for thyroxine, CBG shows a 10-fold decrease in affinity for cortisol subsequent to cleavage by neutrophil elastase (Fig. 9.4). Such cleavage of CBG will result in direct release of cortisol at a site of inflammation with the potential to locally modify and contain the inflammatory response.

Pathological Significance of the SR Change

Inactivation of Serpins by Pathological Proteases

A disadvantage of the exposure of the reactive centre on a stressed loop is its vulnerability to proteolytic cleavage, with subsequent inactivation of the serpin as a pathological process. Serpins may be catalytically inactivated by various snake venoms and bacterial metalloproteinases, and in each case inactivation is associated with cleavage of the reactive centre loop (Kress 1986). Antitrypsin is susceptible to loop cleavage and inactivation by proteases of *Pseudomonas aeruginosa*, *Staphylococcus aureus* and *Serratia marcescens*. Unopposed activity

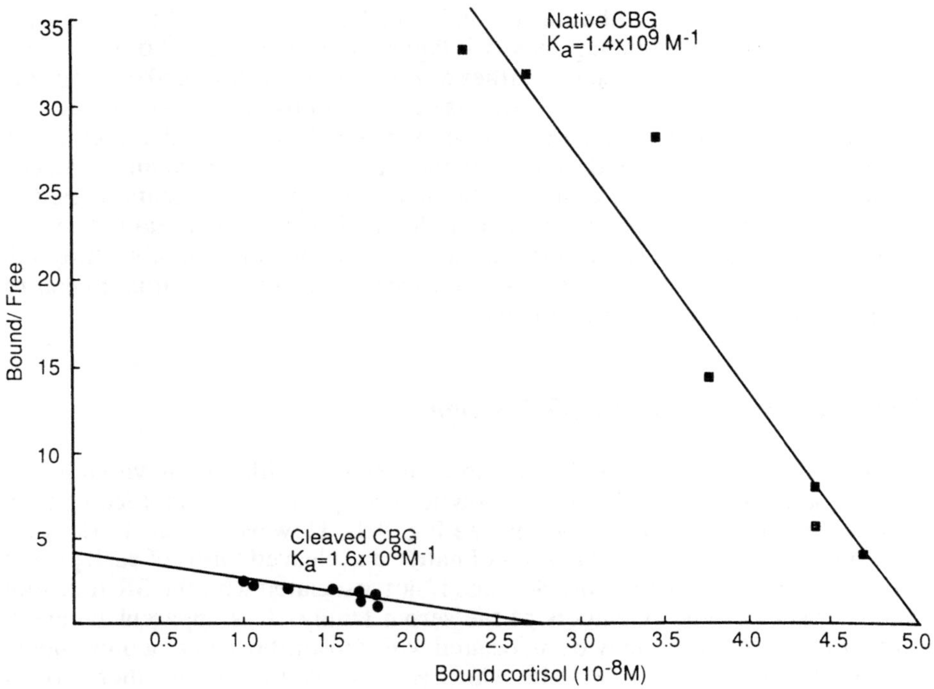

Fig. 9.4. Scatchard plot for CBG showing the marked decrease in cortisol binding affinity subsequent to cleavage by neutrophil elastase. Reprinted by permission from *Nature* vol. 336 pp. 257–258. © 1988 Macmillan Magazines Ltd.

of neutrophil elastase associated with infection of the lungs by such bacteria may contribute to the pathogenesis of the chronic pulmonary disease in cystic fibrosis.

Shock Syndromes

The inactivation of antitrypsin by stimulated neutrophils, either through oxidation of the reactive centre methionine or by proteolytic cleavage of the exposed loop, results in unopposed elastase activity. While this may be advantageous in a local inflammatory reaction, it becomes harmful if there is massive activation of neutrophils as seen in the shock syndromes associated with septicaemia or endotoxaemia (Duswald et al. 1985), or following neutrophil sequestration in cardiopulmonary bypass procedures (Westaby 1983) or in myeloid leukaemias. In these syndromes, uncontained and unopposed elastase activity results in widespread cleavage and inactivation of antithrombin III, C1-inhibitor and *α*2-antiplasmin with triggering of coagulation, fibrinolysis, complement activation and kinin release. This fulminant activation of all the plasma proteolytic cascades contributes to life-threatening disseminated intravascular coagulation and shock lung.

It has been reported that sera from patients with severe shock of bacterial (Zouaghi et al. 1983) or fungal (Savu et al. 1981) origin are virtually depleted of cortisol binding activity characteristic of CBG, although cortisol binding was not significantly altered in other acute inflammatory disorders. Such profound impairment of cortisol binding activity is typical of the cleaved, relaxed (R) form of CBG and this may represent a further example of the widespread cleavage of a circulating serpin by uncontrolled elastase activity in the shock syndrome.

Several genetically engineered serpin variants have potential therapeutic application in the shock syndromes. Mutants of antitrypsin with valine or leucine substituted for the reactive centre methionine would provide oxidation resistant elastase inhibition. The leucine variant additionally retains inhibition of cathepsin G. Arginine substituted antitrypsin (antitrypsin Pittsburgh) inhibits thrombin and kallikrein and is resistant to elastase cleavage, with the potential to control both coagulation and contact systems.

Evolutionary Loss of the SR Change

Human angiotensinogen and hen ovalbumin, serpins without known inhibitory activity, are each susceptible to proteolytic cleavage at sites homologous to the exposed peptide loop of antitrypsin (see Fig. 9.2). However recent work in our laboratory shows that heat stabilities of native and cleaved forms of each protein are similar, suggesting that this cleavage is not associated with the SR transition characteristic of other members of the serpin family. This apparent failure to undergo the SR change may be associated with substitutions in angiotensinogen and ovalbumin at the P_{10}–P_{12} site which is highly conserved in other serpins. Structural evidence, such as the distribution of other conserved residues and the position of glycosylation and phosphorylation sites, supports the prediction that these two proteins do in fact share the tertiary structure of antitrypsin. Angiotensinogen and ovalbumin appear to be examples of members of the serpin family in which the ability to undergo the SR transition has provided no selective advantage and has therefore been lost by evolutionary change.

Other Structural Modifications in Inflammation

N-terminal Cleavage: Angiotensinogen

The N-terminus of the serpin molecule provides a further region of exposed, unordered sequence (in addition to the reactive centre loop) that is vulnerable to proteolytic cleavage. N-terminal cleavage appears to occur more rapidly than cleavage of the reactive centre loop and, in contrast to loop cleavage, is not associated with loss of inhibitory activity (Kress 1986). The physiological significance of N-terminal cleavage is unknown except in the case of angiotensinogen.

The N-terminus of angiotensinogen is susceptible to cleavage between residues 10(Leu) and 11(Val), releasing the decapeptide angiotensin I. This peptide undergoes further modification at its C-terminus to form the biologically active octapeptide angiotensin II, with roles in blood pressure control and sodium metabolism.

Two pathways generating angiotensin II have been described. The first involves cleavage of circulating angiotensinogen by renin to release angiotensin I (Skeggs et al. 1967), which is subsequently converted to angiotensin II by angiotensin converting enzyme in the lungs (Skeggs et al. 1954). The second pathway is neutrophil-dependent and mediated by cathepsin G, which may use either angiotensinogen or angiotensin I as a substrate for generation of angiotensin II (Wintroub et al. 1984). This latter pathway is a potential mechanism for generation of angiotensin II at an inflammatory site, where it may have a role in modulating local vascular responses.

Modification of Carbohydrate

The microheterogeneity shown by antitrypsin is partly explained by the presence of isoforms with different proportions of tri- and bi-antennary oligosaccharide side chains (Vaughan and Carrell 1981). A marked change in the pattern of these isoforms has been described in inflammation (Vaughan et al. 1982) with increased production of tri-antennary forms. The physiological significance of this observation is unclear, although it has been suggested that desialylated tri-antennary forms of antitrypsin are more susceptible to hepatic catabolism than the corresponding bi-antennary forms.

Conclusions

Various mechanisms have been described whereby structural modification of serpins at a site of inflammation results in altered functional activity and ultimately in modulation of the inflammatory response. These events are summarized as follows:

1. Stimulated neutrophils release proteolytic enzymes (including elastase, cathepsin G and metalloproteinases) and reactive oxygen metabolites.
2. Antitrypsin is locally inactivated by:
   (i) Direct oxidation of the reactive centre methionine.
   (ii) Selective catalytic cleavage by an oxidation-dependent neutrophil metalloproteinase.

This results in unopposed activity of target proteases, elastase and cathepsin G, within the zone of oxidation.

3. Local effects of neutrophil elastase include:
   (i) Cleavage of exposed loop sequences of C1-inhibitor, antithrombin III and $\alpha 2$-antiplasmin, triggering the SR transition with loss of inhibitory activity. Target serine proteases are released from inhibition with activation of the proteolytic cascades at the inflammatory centre.
   (ii) Selective cleavage of CBG with release of cortisol.
   (iii) Breakdown of connective tissue.
4. Neutrophil cathepsin G cleaves angiotensin II from the N-terminus of angiotensinogen. Angiotensin II mediates local vascular effects.

5. Structurally modified and inactivated serpins may stimulate neutrophil migration, thus augmenting the inflammatory response.

6. Neutrophil enzymes diffuse away from the inflammatory centre with the potential to cause tissue damage at distant sites. Antitrypsin also diffuses freely through the extracellular fluid and, beyond the zone of oxidation at the inflammatory centre, remains an active and effective inhibitor of elastase and cathepsin G.

Acknowledgements. This work was supported by the Medical Research Council and the Wellcome Trust. Human angiotensinogen was a gift from Dr Duane Tewksbury.

References

Babior BM (1978) Oxygen-dependent microbial killing by phagocytes. N Engl J Med 298:659–668

Banda MJ, Clark EJ, Sinha S, Travis J (1987) Interaction of mouse macrophage elastase with native and oxidised human α_1-proteinase inhibitor. J Clin Invest 79:1314–1317

Banda MJ, Rice AG, Griffin GL, Senior RM (1988) α_1-Proteinase inhibitor is a neutrophil chemoattractant after proteolytic inactivation by macrophage elastase. J Biol Chem 263:4481–4484

Bathurst IC, Errington DM, Foreman RC, Judah JD, Carrell RW (1985) Human Z α_1-antitrypsin accumulates intracellularly and stimulates lysosomal activity when synthesised in the xenopus oocyte. FEBS Lett 183:304–308

Carlson JA, Rogers BB, Sifers RN et al. (1988) P_iZ α_1-antitrypsin causes liver disease in transgenic mice. J Clin Invest 82:26–36

Carp H, Janoff A (1980) Potential mediator of inflammation. Phagocyte-derived oxidants suppress the elastase-inhibitory capacity of α_1-proteinase inhibitor in vitro. J Clin Invest 66:987–995

Carp H, Miller F, Hoidal JR, Janoff A (1982) Potential mechanism of emphysema: α_1-proteinase inhibitor recovered from lungs of cigarette smokers contains oxidised methionine and has decreased elastase inhibitory capacity. Proc Natl Acad Sci USA 79:2041–2045

Carp H, Janoff A, Abrams W et al. (1983) Human methionine sulfoxide-peptide reductase, an enzyme capable of reactivating oxidised α_1-proteinase inhibitor in vitro. Am Rev Respir Dis 127:301–305

Carrell RW, Boswell DR (1986) Serpins: a family of serine proteinase inhibitors. In: Barrett A, Salvesen G (eds) Protease inhibitors. Elsevier, Amsterdam, pp 403–420

Carrell RW, Owen MC (1985) Plakalbumin, α_1-antitrypsin, antithrombin and the mechanism of inflammatory thrombosis. Nature 317:730–732

Carrell RW, Jeppsson JO, Laurell CB et al. (1982) Structure and variation of human α_1-antitrypsin. Nature 298:329–334

Desrochers PE, Weiss SJ (1988) Proteolytic inactivation of α_1-proteinase inhibitor by a neutrophil metalloproteinase. J Clin Invest 81:1646–1650

Duswald KH, Jochum M, Schramm W, Fritz H (1985) Released granulocytic elastase: an indicator of pathobiochemical alterations in septicaemia after abdominal surgery. Surgery 98:892–899

Eriksson S, Carlson J, Velez R (1986) Risk of cirrhosis and primary liver cancer in α_1-antitrypsin deficiency. N Engl J Med 314:736–739

Ganrot K (1974) The plasma protein patterns in acute infectious diseases. Scand J Clin Lab Invest 34:75–81

George PM, Vissers MCM, Travis J, Winterbourn CC, Carrell RW (1984) A genetically engineered mutant of α_1-antitrypsin protects connective tissue from neutrophil damage and may be useful in lung disease. Lancet ii:1426–1428

Hunt LT, Dayhoff MO (1980) A surprising new protein superfamily containing ovalbumin, antithrombin-III and α_1-proteinase inhibitor. Biochem Biophys Res Commun 95:864–871

Johnson D, Travis J (1977) Inactivation of human α_1-proteinase inhibitor by thiol proteinases. Biochem J 163:639–641

Johnson D, Travis J (1979) The oxidative inactivation of human α_1-proteinase inhibitor. Further evidence for methionine at the reactive center. J. Biol Chem 254: 4022–4026

Kageyama R, Ohkubo H, Nakanishi S (1985) Induction of rat liver angiotensinogen mRNA following acute inflammation. Biochem Biophys Res Commun 129:826–832

Kress LF (1986) Inactivation of human plasma serine proteinase inhibitors (serpins) by limited proteolysis of the reactive site loop with snake venom and bacterial metalloproteinases. J Cell Biochem 32:51–58

Kress LF, Catanese JJ (1981) Identification of the cleavage sites resulting from enzymatic inactivation of human antithrombin III by Crotalus adamanteus proteinase II in the presence and absence of heparin. Biochemistry 20:7432–7438

Kress LF, Kurecki T, Chan SK, Laskowski M (1979) Characterisation of the inactive fragment resulting from limited proteolysis of human α_1-proteinase inhibitor by Crotalus adamanteus proteinase II. J Biol Chem 254:5317–5320

Laskowski M, Kato I (1980) Protein inhibitors of proteinases. Annu Rev Biochem 49:593–626

Laurell CB, Eriksson S (1963) The electrophoretic α_1-globulin pattern of serum in α_1-antitrypsin deficiency. Scand J Clin Lab Invest 15:132–140

Liebermann J (1973) Heat lability of α_1-antitrypsin variants. Chest 64:579–584

Liebermann J (1976) Elastase, collagenase, emphysema and α_1-antitrypsin deficiency. Chest 70:62–67

Linderstrom-Lang KU (1952) The enzymatic breakdown of ovalbumin. Med Sci 6:73–92

Loebermann H, Tokuoka R, Deisenhofer J, Huber R (1984) Human α_1-proteinase inhibitor: crystal structure analysis of two crystal modifications, molecular model and preliminary analysis of the implications for function. J Mol Biol 177:531–556

Morihara K, Tsusuki H, Harada M, Iwato T (1984) Purification of human plasma α_1-proteinase inhibitor and its inactivation by Pseudomonas aeruginosa elastase. J Biochem 95:795–804

Owen MC, Brennan SO, Lewis JH, Carrell RW (1983) Mutation of antitrypsin to antithrombin. α_1-Antitrypsin Pittsburgh (358 Met→Arg), a fatal bleeding disorder. N Engl J Med 309:694–698

Owens MR, Miller LL (1980) Net biosynthesis of antithrombin III by the isolated rat liver perfused for 12–24 hours. Biochim Biophys Acta 627:30–39

Pemberton PA, Stein PE, Pepys MB, Potter JM, Carrell RW (1988) Hormone binding globulins undergo serpin conformational change in inflammation. Nature 336:257–258

Pemberton PA, Harrison RA, Lachmann PJ, Carrell RW (1989) The structural basis for neutrophil inactivation of C_1-inhibitor. Biochem J 258:193–198

Perlmutter DH, Schlesinger MJ, Pierce JA, Schwartz AL (1988) Intracellular accumulation of α_1-antitrypsin (α_1AT) in P_iZZ α_1AT deficiency with constitutive expression of heat shock genes. Clin Res 36:611A

Potempa J, Watorek W, Travis J (1986) The inactivation of human plasma α_1-proteinase inhibitor by proteinases from *Staphylococcus aureus*. J Biol Chem 261:14330–14334

Potempa J, Shieh BH, Travis J (1988) α_2-Antiplasmin: a serpin with 2 separate but overlapping reactive sites. Science 241:699–700

Rosenberg S, Barr PJ, Najarian RC, Hallewell RA (1984) Synthesis in yeast of a functional, oxidation resistant active centre variant of human α_1-antitrypsin. Nature 312:77–80

Salvesen GS, Catanese JJ, Kress LF, Travis J (1985) Primary structure of the reactive site of human C_1-inhibitor. J Biol Chem 260:2432–2436

Samama JP, Delarue M, Moras D, Petitou M, Lormeau JC, Choay J (1987) Crystallographic investigation of antithrombin III (Abstract 966). Thromb Haemost 58:264

Savu L, Zouaghi H, Carli A, Nunez EA (1981) Serum depletion of corticosteroid binding activities, an early marker of human septic shock. Biochem Biophys Res Commun 102:411–419

Schindler D (1984) In: Ahmad S, Black S, Schulz J, Scott W, Whelan J (eds) 16th Miami winter symposium. Advances in gene technology, human genetic disorders, p 32

Scultze HE, Heide K, Haupt H (1962) α_1-Antitrypsin aus Humanserum. Klin Wochenschr 8:428–434

Skeggs LT, Marsh WH, Khan JR, Shumaray NP (1954) The existence of two forms of hypertension. J Exp Med 99: 275–282

Skeggs LT, Lentz KE, Gould AB et al. (1967) Biochemistry and kinetics of the renin-angiotensin system. Fed Proc 26:42–47

Sveger T, Thelin T (1981) Four-year-old children with α_1-antitrypsin deficiency. Acta Paediatr Scand 70:171–177

Travis J, Salvesen GS (1983) Human plasma proteinase inhibitors. Annu Rev Biochem 52:655–709

Travis J, Salvesen GS, Matheson NR (1987) The use of association rate constants in determining the specificity of proteinase inhibitors. In: Cunningham DD, Long GL (eds) Proteases in biological control and biotechnology. Alan R. Liss, New York, pp 307–310

Vaughan L, Carrell R (1981) α_1-antitrypsin isoforms: structural basis of microheterogeneity. Biochem Int 2:461–467

Vaughan L, Lorier MA, Carrell RW (1982) α_1-antitrypsin microheterogeneity. Isolation and physiological significance of isoforms. Biochim Biophys Acta 701:339–345

Virca GD, Lyerly D, Kreger A, Travis J (1982) Inactivation of human plasma α_1-proteinase inhibitor by a metalloproteinase from *Serratia marcescens*. Biochim Biophys Acta 704:267–271

Vissers MCM, George PM, Bathurst IC, Brennan SO, Winterbourn CC (1988) Cleavage and inactivation of α_1-antitrypsin by metalloproteinases released from neutrophils. J Clin Invest 82:706–711

Weissmann G, Smolen JE, Korchak HM (1980) Release of inflammatory mediators from stimulated neutrophils. N Engl J Med 303:27–34

Westaby S (1983) Complement and the damaging effects of cardiopulmonary bypass. Thorax 38:321–325

Wintroub BU, Klickstein LB, Dzau VJ, Watt KWK (1984) Granulocyte–angiotensin system. Identification of angiotensinogen as the plasma protein substrate of leukocyte cathepsin G. Biochemistry 23:227–232

Wright HT (1984) Ovalbumin is an elastase substrate. J Biol Chem 259:14335–14336

Zouaghi H, Savu L, Delorme J, Kleinknecht D, Nunez EA (1983) Loss of serum transcortin in human shock associated with severe infection by *Candida albicans*. Acta Endocrinol (Copenh) 102:277–283

Chapter 10

The Three Dimensional Structure of SAP

H. E. White, B. P. O'Hara, G. Oliva, T. L. Blundell, M. B. Pepys and S. P. Wood

---

Serum amyloid P (SAP) component and C-reactive protein (CRP) are members of a closely related family of pentameric or decameric proteins called pentraxins which are involved in amyloidosis and the acute phase response (Pepys and Baltz 1983). Although CRP was crystallised in 1947 (McCarty 1947), no detailed three dimensional structural information has been obtained by X-ray analysis. However, several recent reports of CRP (DeLucas et al. 1987) and SAP (Wood et al. 1988) crystals have demonstrated that this is now a realistic objective. In this paper we review progress made on the X-ray analysis in our laboratory, and describe for the first time preliminary, medium resolution electron density maps of a pentameric form of SAP.

As isolated from human serum in calcium free media, SAP has a molecular weight of 254 000 and comprises 10 subunits each of 204 amino acids. Each subunit contains a single intrachain disulphide bridge, and carries a single asparagine linked complex oligosaccharide which contributes about 8% of the total mass. CRP on the other hand has a molecular weight of 115 000 and comprises five subunits, each of 206 amino acids with no attached carbohydrate. The amino acid sequences are included in Fig. 10.11.

The pentagonal arrangement of subunits, in disks 85–110 Å diameter and 30–40 Å deep have been reported by various researchers (Osmand et al. 1977; Painter et al. 1976; Pepys et al. 1977; Baltz et al. 1982a) for both proteins. Human SAP appears as two disks interacting through their planar faces (Fig. 10.1).

Our attempts to crystallize these proteins began in 1985 and quickly led to a crystal form of human SAP suitabie for high resolution structure analysis (Oliva et al. 1986). Large prismatic crystals could be grown in 4–5 weeks from protein solutions containing calcium ions, at pH 5.5 (Fig. 10.2). The preliminary crystallographic details for this form and a number of others grown subsequently (O'Hara et al. 1988) are summarized in Table 10.1.

X-ray diffraction data for the original crystal form were measured to 5.5 Å resolution and the non-crystallographic symmetry was examined using the fast

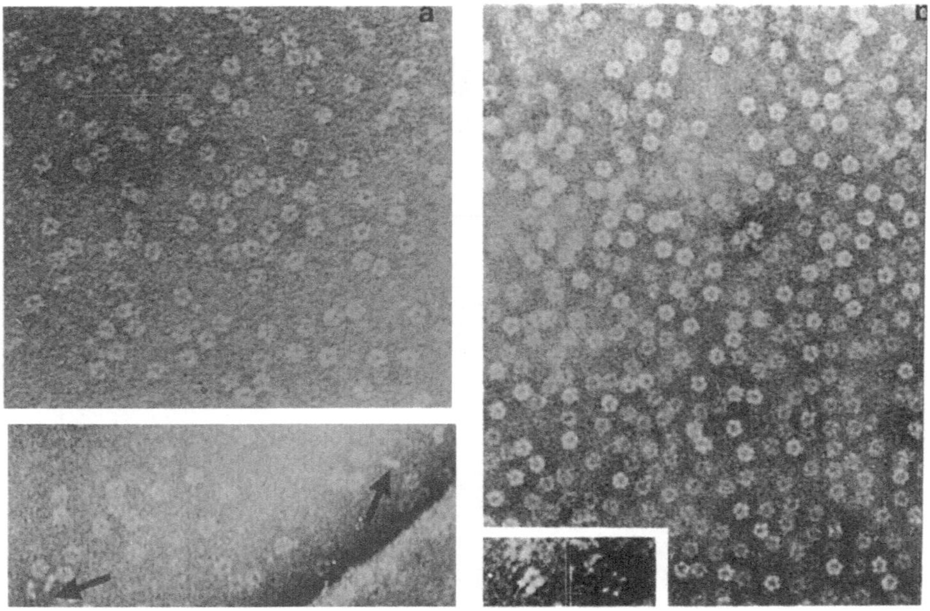

Fig. 10.1a,b. Electron micrographs of negatively stained preparations of **a** human C-reactive protein and **b** human serum amyloid P component. Most molecules are seen face-on but *arrows* in **a** and *insets* in **b** (at higher magnification) show views of the disk-like molecules side-on. Magnification, × 320 000. Micrography performed by Dr E. A. Munn, Department of Biochemistry, ARC Institute of Animal Physiology, Babraham.

a

Fig. 10.2a. Prismatic crystals of SAP (dimension 0.5–1.0 mm) grown at pH 5.5 in the presence of calcium ions. (Figure continued on p. 125.)

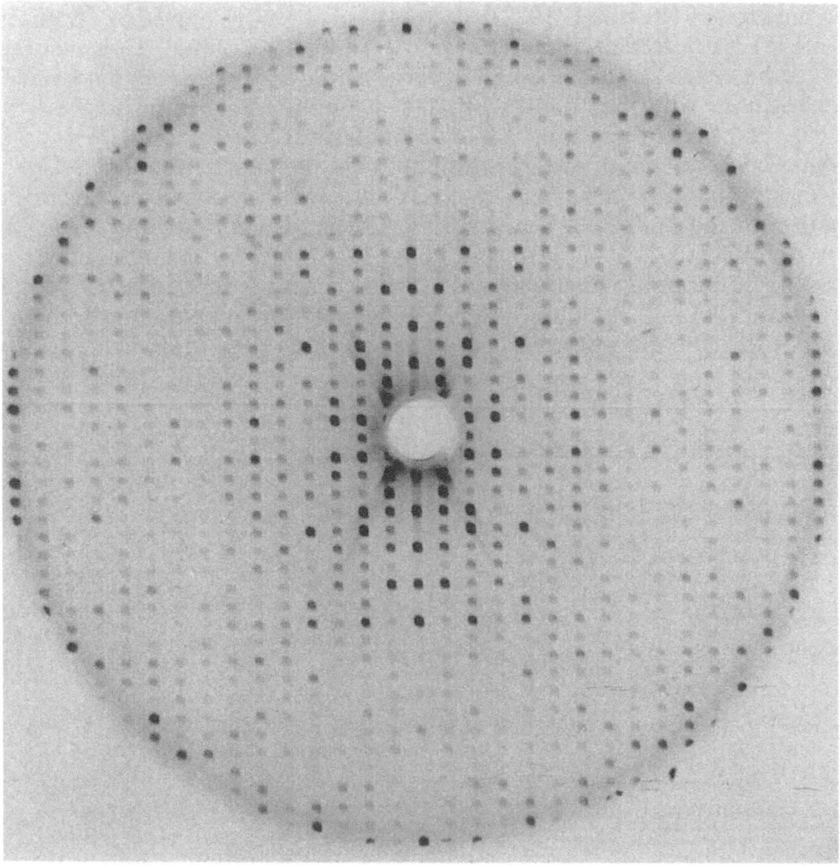

b

Fig. 10.2b. Zero layer (hk0) X-ray precession photograph (μ = 12°).

Table 10.1. Crystal forms of SAP

Shape	Conditions	Space group	Cell			
			a	b	c	β°
1. Prismatic	pH 5.5	$P2_1$	69.0	99.3	96.8	96.1
2. Prismatic	pH 5.5/Ca/MOβDG	$P2_1$	69.0	99.3	96.8	96.1
3. Rods/plates	pH 8/Ca	?				
4. Rods	pH 8/Ca/MOβDG	$P2_1$	121.6	108.9	117.2	94.5
5. Rods	pH 5.5/Sr		Very soft			
6. Prismatic	pH 5.5/Ba	$P2_1$	60.7	106.5	103.0	94.6
7. Prismatic	pH 5.5/La	$P2_1$	74.0	164.5	185.0	95.0

rotation function of Crowther (1972). The rotation function showed 52 symmetry as expected for a decamer in the asymmetric unit of the crystal. However, the volume of the crystal unit cell was not large enough to accommodate such a decamer together with the quantity of solvent usually found in protein crystals (Wood et al. 1988). The apparent 2-fold symmetry was subsequently found to be an attribute of the calculation at low resolution but the implication that SAP was present in a pentameric form was not consistent with the expected decameric organisation of the molecule. Low angle scattering of neutrons and X-rays by solutions of SAP and CRP (Perkins and Pepys 1986) showed that, at neutral pH, SAP exists as a monodisperse population of decamers independent of protein concentration, while CRP shows evidence of aggregation of pentamers to higher oligomers with increasing protein concentration (Fig. 10.3). However, further

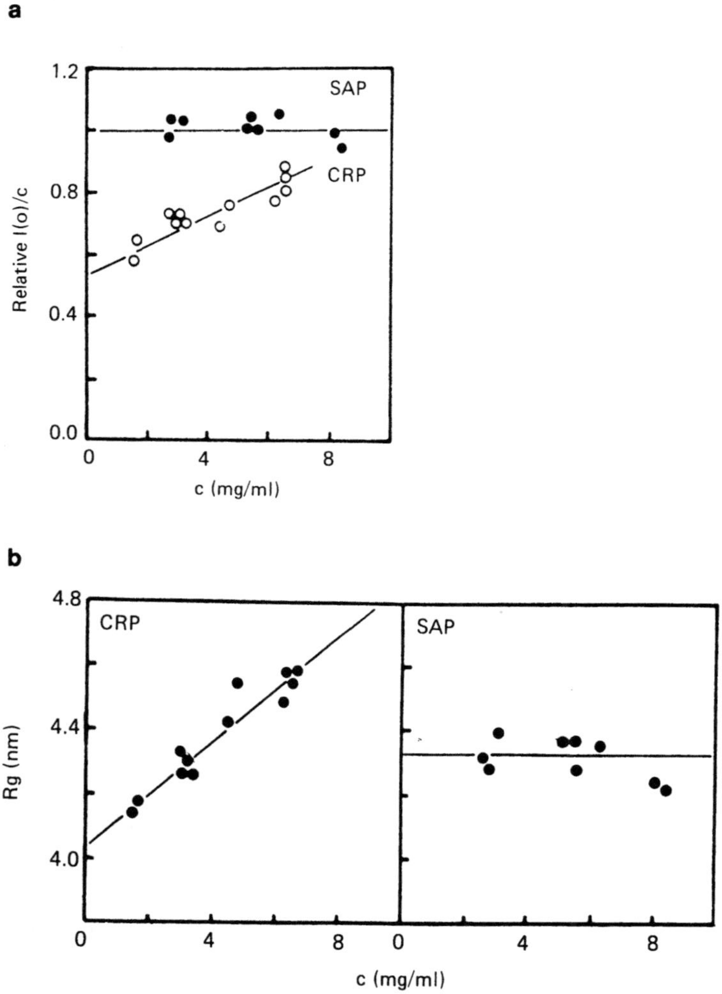

Fig. 10.3a,b. Plots of the **a** relative molecular weights (I(o)/c – the forward scattering X-ray intensity) and **b** radius of gyration (Rg) for SAP and CRP as a function of protein concentration, determined by low angle X-ray scattering of protein solutions, show the aggregation tendency for CRP and the larger more stable molecular form for SAP (Perkins and Pepys 1986).

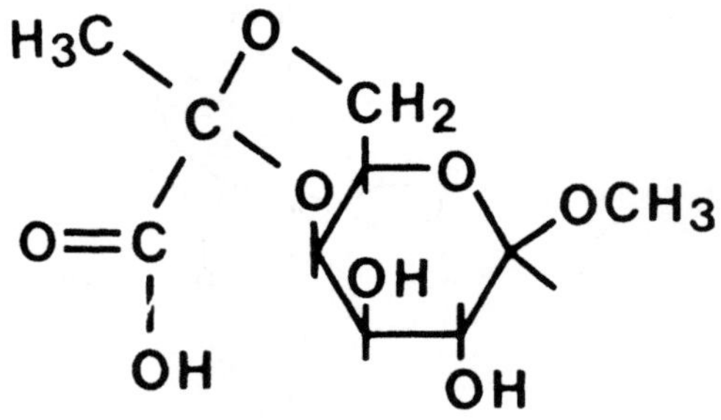

Fig. 10.4. The synthetic monosaccharide, methyl 4,6-O-(1-carboxyethylidene)-β-D-galactopyranoside (MOβDG) which has been used to control the rate of SAP crystal growth and modify the crystal form.

measurements carried out in the crystallization solvent buffer showed that the SAP had dissociated into pentamers. Gel filtration experiments confirmed the pentameric nature of SAP under these conditions and, furthermore, demonstrated that SAP could be readily converted between pentameric and decameric forms simply by suitable buffer exchange (Wood et al. 1988).

In Table 10.1 are listed several additional crystal forms of SAP which were grown subsequently and which raise further questions about the aggregation state of the molecule. At pH 5.5 we found that the galactose derivative (MOβDG) (Fig. 10.4) shown by Pepys and co-workers (Hind et al. 1984a), to be the constituent of agarose to which SAP binds in the presence of calcium ions, dramatically slows the rate of crystal growth but does not affect the final product. From precipitates of SAP which rapidly develop on adding calcium ions at pH 8 (Baltz et al. 1982b), distinctive plates and prisms grew together. When MOβDG was included at pH 8 the normal precipitation was not observed, but overnight at 4° C rod shaped crystals grew. The volume of the asymmetric unit of these crystals was large enough to accommodate a decamer of SAP (O'Hara et al. 1988). Slow cooling of this mixture, however, gave rise to the crystal form seen at pH 5.5. These experiments again indicate that the equilibrium between pentameric and decameric SAP populations can be readily perturbed to provide different crystalline forms. Further recent neutron scattering experiments at pH 8 in the presence of calcium ions and MOβDG indicate a pentameric population of SAP molecules (S. Perkins, unpublished observations). We have successfully substituted other metals such as strontium, barium and lanthanum for calcium in the crystallization of SAP but there is clearly disturbance to the crystal lattice with these other metals.

In Fig. 10.5 some of these observations are summarized together with some implications in cartoon form. The central point seems to be that in the presence of calcium ions the pentamer–decamer equilibrium is displaced towards pentamers which at near neutral pH can only be retained in solution by co-ordination of MOβDG. In its absence, calcium-mediated intermolecular interactions lead to precipitation (Baltz et al. 1982b; Hind et al. 1984b). At lower pH, calcium binding via sidechain carboxylates would be somewhat less favourable and calcium does not precipitate SAP at pH 5.5. However, MOβDG can still compete for the calcium-mediated intermolecular interaction and retard crystal growth. This proposed role for calcium involvement in the interactions between pentamers seems to be confirmed by the packing changes observed with other metals

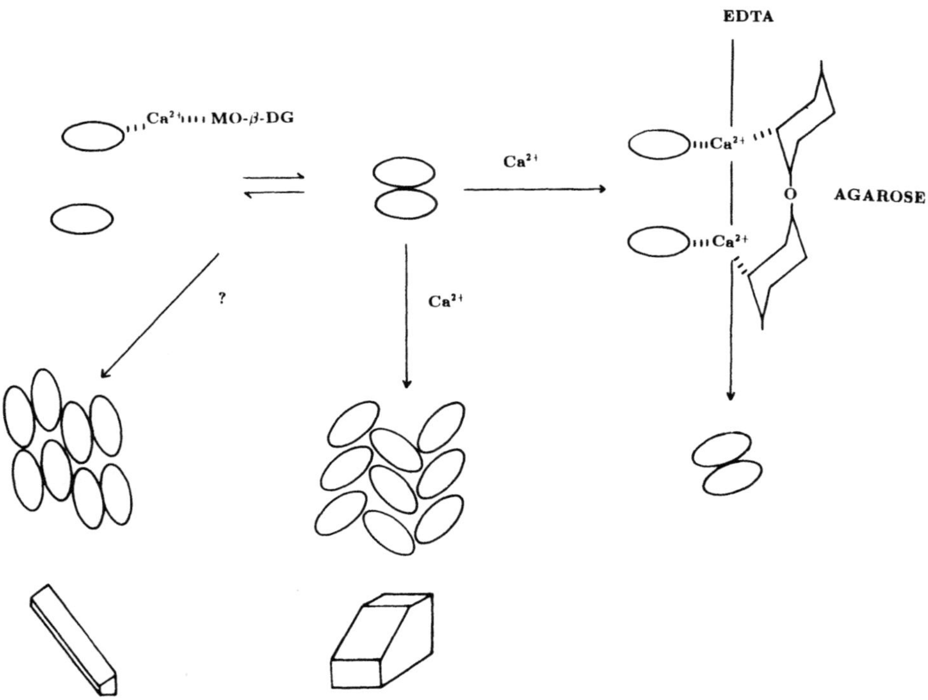

Fig. 10.5. Each oval signifies a pentameric disk of SAP. Two of these are shown interacting face-to-face in the centre of the figure and represent the molecular species (decameric) encountered in the absence of calcium ions. To the left and right pentameric forms are favoured in the presence of calcium and these may be the species involved in binding agarose or the related synthetic galactose derivative (MOβDG). EDTA sequesters calcium to release a decameric product (*right*) from affinity columns of agarose while suitable manipulation of the equilibrium (*left*) by change of protein concentration pH, temperature, precipitant or MOβDG addition during crystallization leads to the production of crystal lattices where pentamers (*centre*) or decamers (*left*) are packed.

such as barium. Calcium pentamers form prismatic crystals rather slowly while rapid growth at pH 8 traps what appears to be a decameric form. The ease with which the SAP aggregation state can change, in particular in response to calcium ions, raises the question of what form is involved in ligand binding. The cartoon in Fig. 10.5 implies that pentamers associate with agarose in the presence of calcium ions, and following elution of SAP with EDTA the protein reassociates to form decamers. In serum, SAP sediments as a decamer in spite of calcium ion concentrations capable of causing precipitation in the isolated protein (Baltz et al. 1982b). More quantitative data on the aggregation behaviour of SAP is required in order to determine the species that interacts with natural ligands.

A variety of different X-ray data sets have now been collected (summarized in Table 10.2). The native protein data measured by oscillation photography at the Daresbury synchrotron source extend to 2 Å resolution (Fig. 10.6). Many attempts have been made to produce suitable heavy atom derivatives of the SAP crystals. Most of the more commonly used reagents have not been successful. Of the derivative data sets collected so far, only those for thorium nitrate and tungsten-cluster soaked crystals has been of use in phase calculation. Two binding

Table 10.2. X-ray data sets collected for SAP

Crystal	Collection method	Resolution (Å)	Refls. measured	Refls. unique	R symmetry (%)
Native	Diffractometer	5.6	10341	4112	4.5
K_2PtCl_4	Diffractometer	6.0	7387	3372	5.0
Na_2WO_4	Diffractometer	6.0	6574	3294	5.1
Native	Fast	3.8	43175	11802	5.3
$Th(NO_3)_4$	Fast	4.0	34176	9521	6.7
$Na_3PO_4.12WO_3$	Fast	4.0	38419	12107	7.0
$Th(NO_3)_2$	Fast	2.8	96444	29551	5.0
Native	Synchrotron	2.0	345000	86000	7.7
K_2PtI_6	Fast	3.5	41778	15119	8.4
NAT/PE[a]	Diffractometer	5.0	6272	5659	5.0
NAT/Ba[b]	Fast	3.5	77468	17078	9.8

[a] PE: crystals soaked in phosphoethanolamine.
[b] Ba: co-crystallisation product with Ba^{2+} ions.

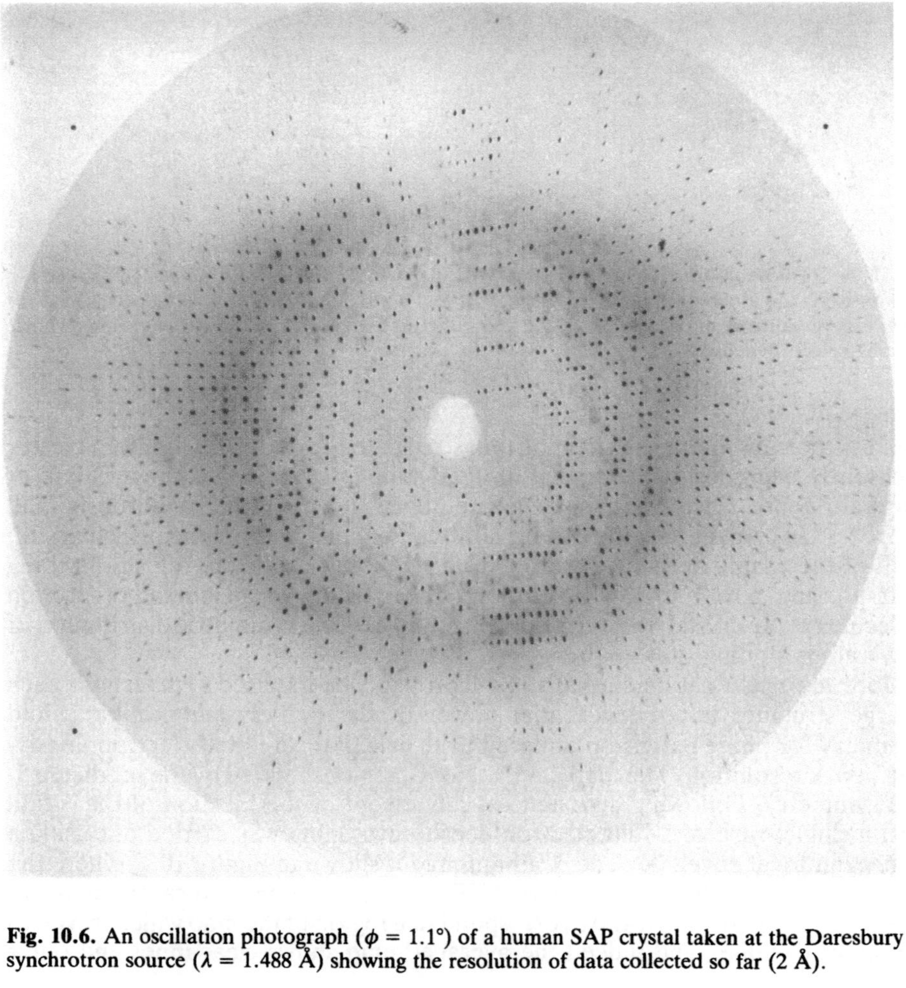

Fig. 10.6. An oscillation photograph ($\phi = 1.1°$) of a human SAP crystal taken at the Daresbury synchrotron source ($\lambda = 1.488$ Å) showing the resolution of data collected so far (2 Å).

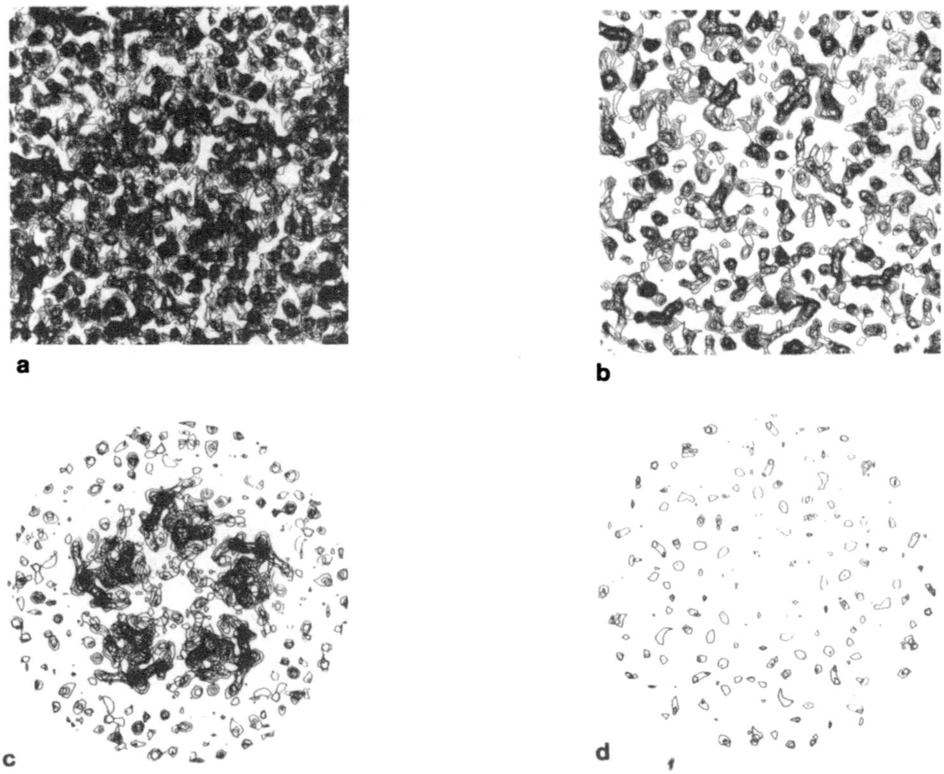

Fig. 10.7a Corresponds to a region of the 6 Å MIR map which traverses the molecular disk, and the averaged map in the position shows clear inner and outer boundaries **c**. **b** and **d** correspond to sections perpendicular to the 5-fold axis but above the molecular disk, and after averaging **d** show no strong electron density.

sites for thorium and three sites for tungsten were located. The tungsten cluster apparently degrades in the crystal mother liquor, to provide a slow release of tungstate ions, since sodium tungstate alone reacts rather vigorously and disorders the crystals to some extent, although binding at the same principal site. Both of the usable derivatives suffer from severe lack of isomorphism, limiting their phasing power to low resolution. The packing arrangement of protein molecules in the crystal has so far not permitted a 5-fold symmetric distribution of heavy atom binding sites.

More good heavy atom derivatives will probably be required to determine such a large structure but the potential power of the non-crystallographic 5-fold symmetry for phase extension tempted us to calculate an initial electron density map at 6 Å resolution. The map has distinctive regions related by the predicted 5-fold symmetry. Following an extended refinement of the location of the 5-fold axis, a single cycle of 5-fold electron density averaging was carried out using a large cylindrical envelope. The resultant map is shown in Figure 10.7, where the same sections of the original 6 Å MIR map and of the averaged map are represented. Figure 10.7a and c correspond to a region of the map which traverses the molecular disk, and indeed the averaged map shows clear inner and outer

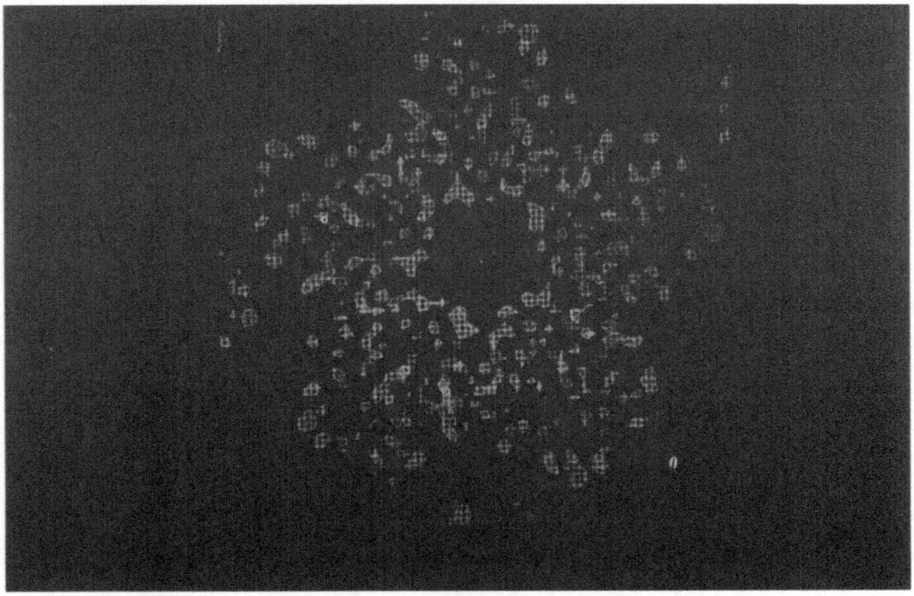

a

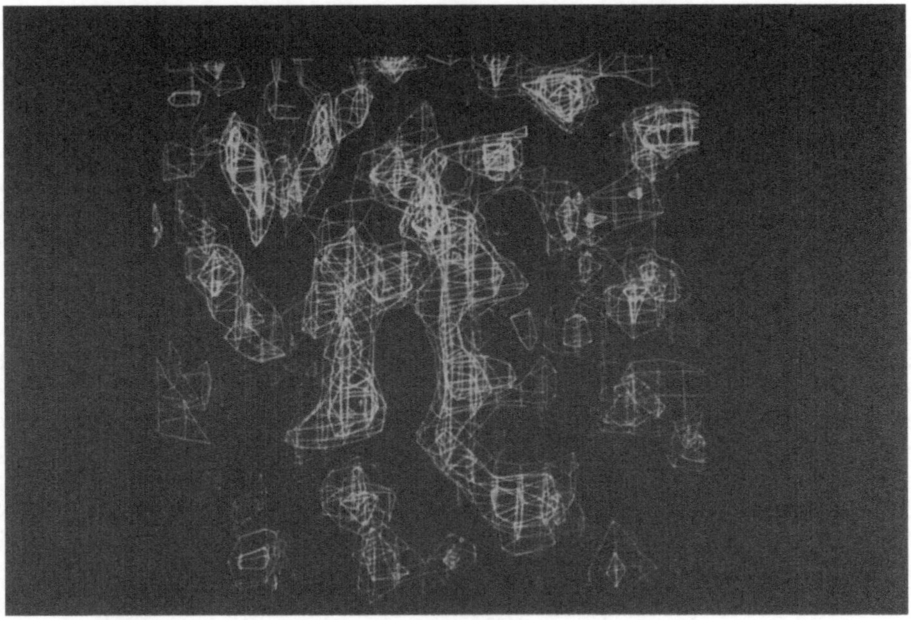

b

Fig. 10.8a. A shallow slice of electron density for the SAP molecule perpendicular to the 5-fold axis calculated at 3.5 Å resolution revealing a molecule about 100 Å across. Closer examination of a small region of one subunit (**b**) shows extended regions of density to which a polypeptide fragment could be built.

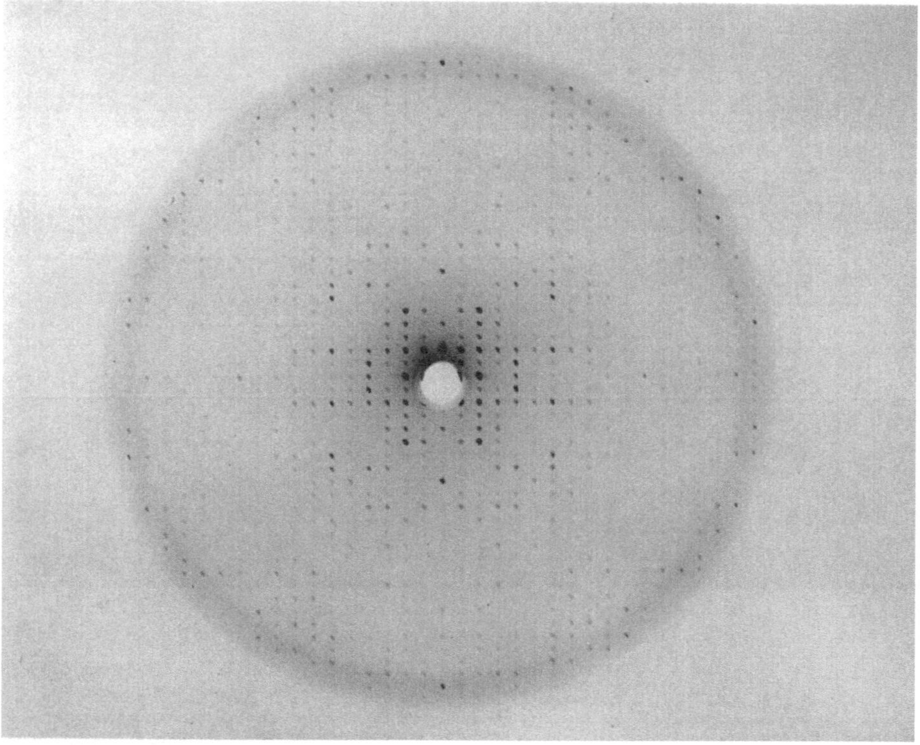

Fig. 10.9. X-ray precession photograph ($\mu = 10°$) for an orthorhombic crystal of human CRP (Birkbeck).

molecular boundaries. Figure 10.7b and d correspond to sections perpendicular to the molecular 5-fold axis but above the molecular disk, and the averaged sections, as expected, do not show any strong electron density. As a convincing test, the phases from the averaged map were capable of locating the heavy atom positions and these are situated at the exposed periphery of the subunits. We proceeded to extend the resolution of the phase information using the 5-fold symmetry to constrain the phases by the method of Bricogne (1976). At intervals we checked that the heavy atoms were still detectable. Solvent flattening was also employed (Wang 1985). At 3.5 Å resolution the electron density map shows rather clear subunits with dimensions in accord with those seen by electron microscopy (Fig. 10.8). Some extended regions of interconnecting electron density are clearly visible. However, at this resolution it is not possible to follow reliably the complete polypeptide chain, and a detailed interpretation must await further improvement and extension of the phases.

Structural studies on CRP are somewhat less advanced but are proceeding steadily in three laboratories on different crystal forms, in research by Greenhough and co-workers (Keele/Alabama, DeLucas et al. 1987), Turnell, Bloomer and Pepys (LMB, Cambridge/RPMS, unpublished results) and Wood et al. (Birkbeck/RPMS, unpublished results; orthorhombic a = 102.5, b = 106.0, c = 118.4 Å; Fig. 10.9). Clearly, all these crystallographic analyses of CRP are closely

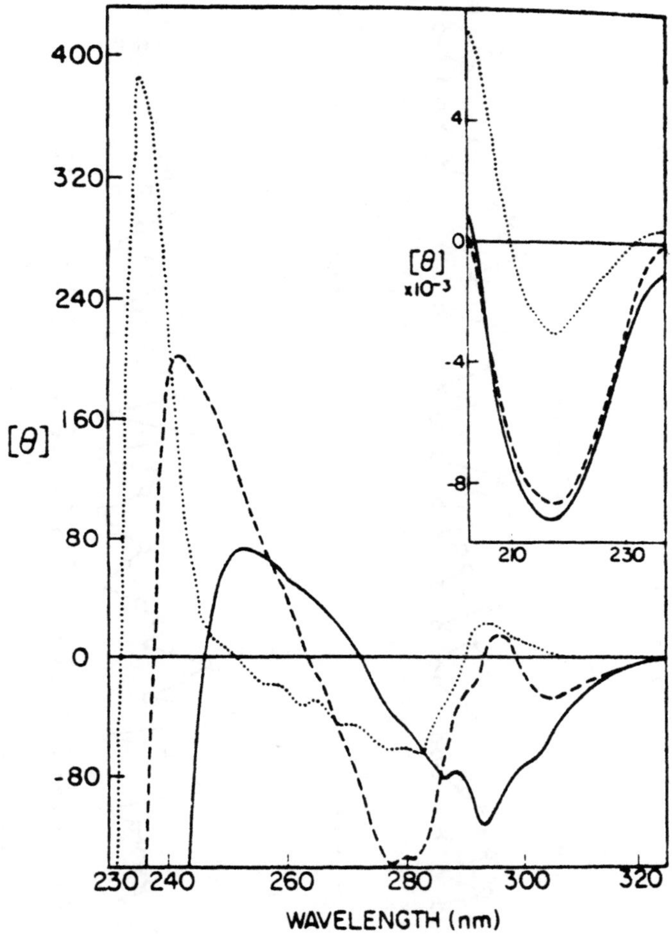

Fig. 10.10. CD spectra for SAP (········) and CRP (____) in the near and far UV measured at pH 7.5 and 1 mg ml^{-1} protein concentration, showing close similarity to the peptide region but more distinct features in the aromatic region (Young and Williams 1978).

related and a successful structure determination for any one of them, or for SAP, should be catalytic for solving the various other crystal forms.

What sort of structural features do we expect to find in these molecules? Both circular dichroism and sequence based prediction methods have provided indications of the amount and types of secondary structure to be expected. Young and Williams (1978) showed that SAP and CRP were very similar in the far ultraviolet region of the CD spectrum and estimated that there was 34% helix and 45% sheet (Fig. 10.10). In the near UV region the spectra are more distinct. We have measured the near UV spectrum of SAP in conditions which favour pentamers or decamers at the same protein concentration. For some proteins the trapping of aromatic sidechains at subunit interfaces during association can give rise to very convenient changes in rotational strength which can provide rough measure of the degree of association (Wood et al. 1975). Apart from a small change in the tryptophan region no major differences were observed. Aromatic

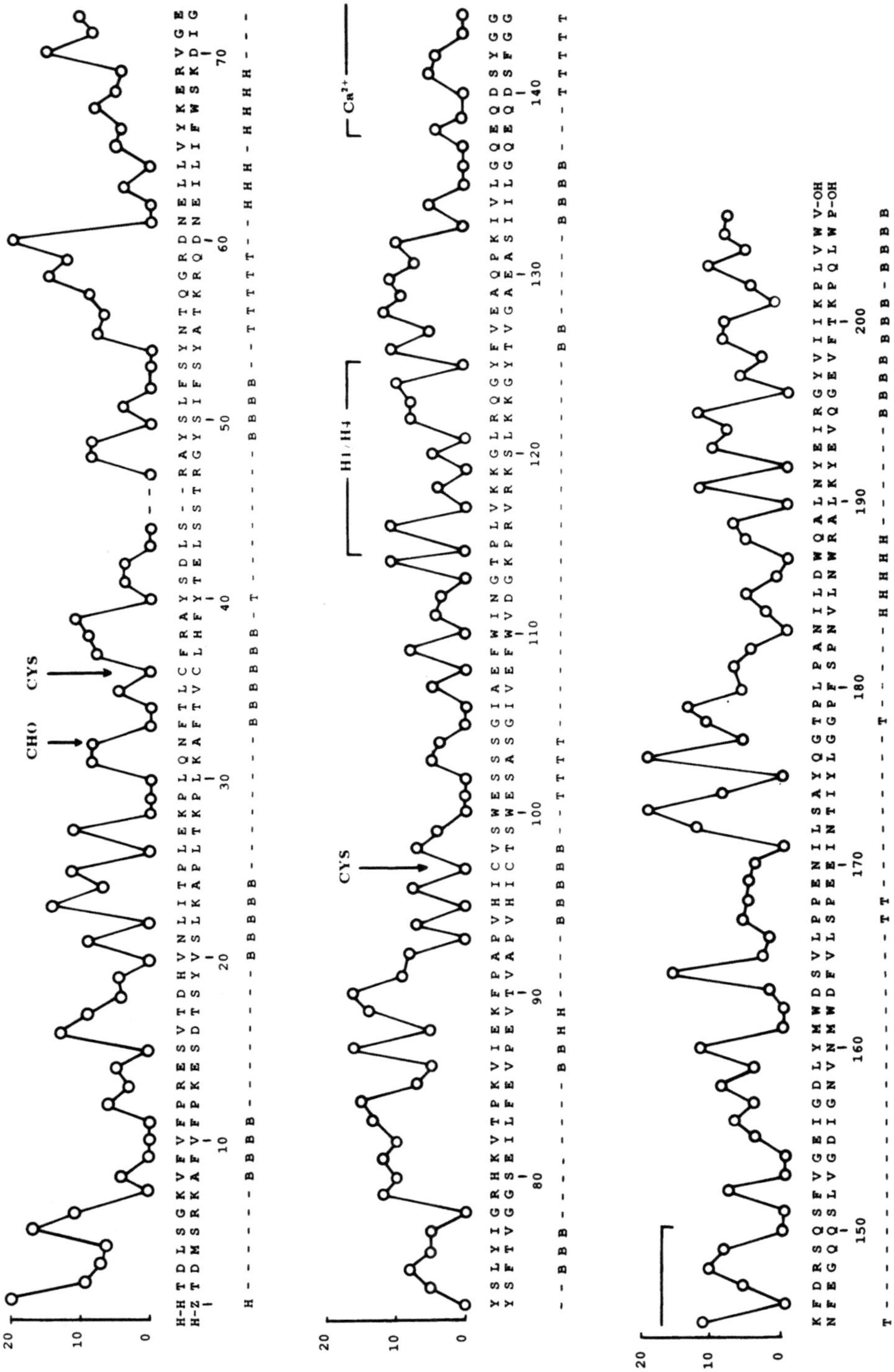

Fig. 10.11. Summary of some amino acid sequence based predictions of secondary structure in the pentraxins. Eight available sequences from Anderson et al. (1989) were aligned and the residue "likeness" at each position (see text) is plotted on a scale from 1 to 20 against sequence number and the common secondary structure predicted from seven different prediction algorithms (*H*: helix, *B*: sheet, *T*: turn). The complete sequences for human SAP (top) and CRP (bottom) are included. A region homologous to histones H1 and H4, and possibly responsible for DNA binding, and a putative calcium binding region (Turnell et al. 1988) are shown.

residues may therefore not be buried at the pentamer–pentamer interface (B. O'Hara, unpublished results).

Secondary structure prediction methods generally confirm the estimate from circular dichroism of the amount of helix and sheet. Figure 10.11 shows a combined prediction summary. The ordinate of the graph is a residue "likeness" index scored at each sequence position from a comparison of seven aligned pentraxin sequences, using a relatedness Venn diagram based on features of size, polarity, aromacity, etc., for each amino acid, as proposed by Taylor (1986a,b). Beneath the abscissa (residue number) the consensus secondary structure prediction for human SAP, derived from seven different prediction algorithms (available within the Leeds University sequence data base), is plotted. This figure provides an overview of the conservation of residue type in the pentraxin family, likely position (buried or exposed) and possible secondary structure. The sequences shown for human SAP (upper) and CRP are much more alike than some other members of the set analysed. Some regions of predicted sheet are clearly quite well conserved, particularly if they are hydrophobic and possibly buried. Some regions are consistently predicted to be helical (e.g. 60–70). Also included in the prediction summary are regions proposed by Turnell et al. (1988) to be homologous with histones H1 and H4 and to be a putative helical DNA binding region adjacent to a calcium binding region (Nguyen et al. 1986; Turnell 1989).

The physiological and pathophysiological role of SAP is of increasing interest in view of the recent discovery that it is the major specific calcium-dependent DNA and chromatin binding protein of the serum (Pepys and Butler 1987), and the development of labelled SAP as a specific targeting agent for clinical diagnosis of amyloidosis (Hawkins et al. 1988). It is evident that these properties and uses of SAP are determined by its specific ligand binding properties and it is hoped that elucidation of the protein structure to high resolution will both explain these interactions and open the way for manipulation of protein function.

References

Anderson JK, Taylor JA, Baltz ML et al. (1989) Primary structure of murine serum amyloid P component (submitted)

Baltz ML, de Beer FC, Feinstein A et al. (1982a) Phylogenetic aspects of C-reactive protein and related proteins. Ann NY Acad Sci 389:49–75

Baltz ML, de Beer FC, Feinstein A, Pepys MB (1982b) Calcium-dependent aggregation of human serum amyloid P component. Biochim Biophys Acta 701:229–236

Bricogne G (1976) Methods and programs for direct-space exploitation of geometric redundancies. Acta Cryst A32:832–847

Crowther RA (1972) In: Rossman MG (ed) The molecular replacement method. Gordon and Breach, pp 173–175

DeLucas LJ, Greenhough TJ, Rule SA (1987) Preliminary X-ray study of crystals of human C-reactive protein. J Mol Biol 196:741–742

Hawkins PN, Myers MJ, Lavender JP, Pepys MB (1988) Diagnostic radionuclide imaging of amyloid: biological targeting by circulating human serum amyloid P component. Lancet i:1413–1418

Hind CRK, Collins PM, Renn D et al. (1984a) Binding specificity of serum amyloid P-component for the pyruvate acetal of galactose. J Exp Med 159:1058–1069

Hind CRK, Collins PM, Pepys MB (1984b) Calcium-dependent aggregation of human serum amyloid P component. Inhibition of the cyclic 4,6-pyruvate acetal of galactose. Biochim Biophys Acta 802:148–150

McCarty MJ (1947) The occurrence during acute infections of a protein not normally present in the blood IV. Crystallization of the C-reactive protein. J Exp Med 85:491–498

McCarty MJ (1982) Historical perspective on C-reactive protein. Ann NY Acad Sci 389:1–9

Ngkuyen NY, Suzuki A, Boykins RA, Liu T-Y (1986) The amino acid sequence of *Limulus* C-reactive protein. Evidence of polymorphism. J Biol Chem 261:10456–10465

O'Hara BP, Wood SP, Oliva G et al. (1988) Crystallizations of human serum amyloid P component (SAP). J Crystal Growth 90:209–212

Oliva G, O'Hara BP, Wood SP et al. (1986) Preliminary crystallographic studies of human serum amyloid P component (SAP). In: Peeters H (ed) Protides of the biological fluids, vol 34. Pergamon Press, Oxford, pp 371–374

Osmand AP, Freidenson B, Gewurz H (1977) The characteristics of C-reactive protein and the complement subcomponent C1t as homologous proteins displaying cyclic pentameric symmetry (pentraxins). Proc Natl Acad Sci USA 74:739–743

Painter RH, Pintenc L, Hoffman T, Kells DIC, Katz A (1976) Ultrastructure and chemistry of C1t subcomponent of C1: Similarities to amyloid P-component. J Immunol 116:1745

Painter RH, de Escallion I, Massey A et al. (1982) The structure and binding characteristics of serum amyloid protein (9.5S α1-glycoprotein). Ann NY Acad Sci 389:199–215

Pepys MB, Baltz ML (1983) Acute phase proteins with special reference to C-reactive protein and related proteins and serum amyloid A protein. Adv Immunol 34:141–212

Pepys MB, Butler PJG (1987) Serum amyloid P component is the major calcium-dependent specific DNA binding protein of the serum. Biochem Biophys Res Commun 148:308–313

Pepys MB, Dash AC, Munn EA, Feinstein A, Skinner M, Cohen AS, Gewurz H, Osmand AP, Painter RH (1977) Isolation of amyloid P-component (protein AP) from normal serum as a calcium-dependent binding protein. Lancet i:1029–1031

Perkins SJ, Pepys MB (1986) X-ray and neutron scattering studies on CRP and SAP. In: Peeters H (ed) Protides of the biological fluids colloquium XXXIV. Pergamon Press, Oxford, pp 323–326

Taylor WR (1986a) The classification of amino acid conservation. J Theor Biol 119:205–218

Taylor WR (1986b) Identification of protein sequence homology by consensus template alignment. J Mol Biol 188:233–258

Turnell WG (1989) Searches for calcium-binding motifs in protein sequences. In: Reid E, Cook GMW, Luzio JP (eds) Biochemical approaches to cellular calcium. Royal Soc Chem, London (Methodological surveys in biochemistry and analysis, vol 19) (in press)

Turnell WG, Satchwell SC, Travers AA (1988) Hypothesis: a decapeptide motif for binding to the minor groove of DNA—a proposal. FEBS Lett 232:263–268

Wang BC (1985) Resolution of phase ambiguity in macromolecular crystallography. Meth Enzymol 115:90–112

Wood SP, Blundell TL, Wollmer A et al. (1975) The relation of conformation and association of insulin to receptor binding; X-ray and circular dichroism studies on bovine and hystriomorph insulins. Eur J Biochem 55:531–542

Wood SP, Oliva G, O'Hara BP et al. (1988) A pentameric form of human serum amyloid P component. Crystallization, X-ray diffraction and neutron scattering studies. J Mol Biol 202:169–173

Young NM, Williams RE (1978) Comparison of the secondary structures and binding of C-reactive protein and the phosphorylcholine-binding murine myeloma proteins. J Immunol 121:1893–1898

Chapter 11

Structure, Metabolism and Function of Acute Phase High Density Lipoprotein

F. C. de Beer, A. F. Strachan and E. G. Shephard

Introduction

The acute phase response is a multi-system adaptation induced by autologous cell death or by products of exogenous invasion (Kushner 1982). These influence macrophages to secrete various monokines that, among other functions, affect hepatic gene expression for many proteins (Ramadori et al. 1985). This response consists of a variety of biochemical, cellular, hormonal and metabolic changes with characteristic increases in many plasma proteins. Serum amyloid A protein (apo-SAA) is particularly notable in that it is normally present in trace amounts in human serum, but the concentration increases up to 100-fold within 48 hours of initiation of the acute phase. (Eriksen and Benditt 1984; McAdam et al. 1978; de Beer et al. 1982). Segrest et al. (1976) first noted that a 45-residue fragment of the apo-SAA cleavage product, amyloid A protein (AA), formed stable complexes with phospholipid. This finding preceded the discovery, by Benditt and Eriksen (1977), that apo-SAA is transported in human plasma mainly with high density lipoproteins (HDL) and is thus classified as an apolipoprotein. Although HDL (expressed as HDL cholesterol) is a significant negative acute phase reactant, it can play a role in inflammation. HDL has been reported to function as a vehicle to transfer damaged cellular constituents to the liver and to bind bacterial lipopolysaccharides (Ulevitch et al. 1981) and neutrophil elastase (Jacob et al. 1981). The function of apo-SAA rich HDL in response to injury is, however, unknown. The phylogenetic conservation of the association of apo-SAA with HDL suggests that it plays an important role.

In rare instances during chronic inflammation cleavage of apo-SAA (Mr 11 500; 104 amino acids) at the 76–77 serine–leucine bond yields amyloid A protein (residues 1–76 of apo-SAA) (Parmelee et al. 1982). During an ill-understood process these AA peptides associate in an anti-parallel fashion to form the fibrils of reactive systemic amyloidosis.

Further studies on the cellular interaction and metabolism of apo-SAA are essential to elucidate its function. Such studies would be complex when one considers the dynamic state of individual HDL components and the polydispersity of HDL particles. A first prerequisite for such studies is the definition of the exact physical/chemical nature of the acute phase HDL particle. Biological experiments using purified apo-SAA in a lipid free environment are more difficult to interpret because of solubility problems and aggregation of apo-SAA.

Apo-SAA in Plasma

Monokine-induced hepatic secretion of apo-SAA appears to be followed by rapid association of this protein with HDL in plasma (Hoffman and Benditt 1982). This association with HDL was confirmed for all animals in which apo-SAA was identified, such as mouse (Benditt et al. 1979), rabbit (Skogen et al. 1979), monkey (Parks and Rudel 1983) and mink (Marhaug et al. 1984). The situation in the rat is of particular interest as no apo-SAA-like protein is demonstrable in rat HDL. This may explain why rats are resistant to the induction of AA amyloidosis (Baltz et al. 1987). However, there is evidence for the existence of apo-SAA-related genes in the rat. Sipe et al. (1986) have reported the presence of RNA species in acute phase rat liver that have 74% (34/46) amino acid identity with the C-terminal portion of human apo-SAA. As the NH_2-terminal residues are important in lipid binding (Turnell et al. 1986) it was proposed by Baltz et al. (1987) that, in the rat, apo-SAA-like molecules would be coded for in such a manner as to lack lipid binding NH_2-terminal residues. Thus in the rat it is not an apolipoprotein, and behaves differently from apo-SAA from other species.

Apo-SAA, like other apolipoproteins of HDL, exists partially in the free form. This equilibrium is heavily directed towards HDL, as only a small proportion of apo-SAA appears to circulate in a lipoprotein-free form in human plasma. Binding to HDL is also relatively specific as only small amounts (<10%) of apo-SAA are associated with either very low density lipoprotein (VLDL), intermediate density lipoprotein (IDL) or low density lipoprotein (LDL) (Benditt and Eriksen 1977; Coetzee et al. 1986). About 20% of the total protein of lymph chylomicrons in the vervet monkey was found to be apo-SAA (Parks and Rudel 1983).

Structure of Apo-SAA

Apo-SAA is a polymorphic protein with at least six isoforms that can exist in three different patterns in individual humans (Strachan et al. 1988) (Fig. 11.1, Table 11.2). The amino acid sequence of one of the major isoforms has been determined and consists of 104 amino acids (Parmelee et al. 1982), and this corresponds to the pI 6.4 isoform (Strachan et al. unpublished work). The basis

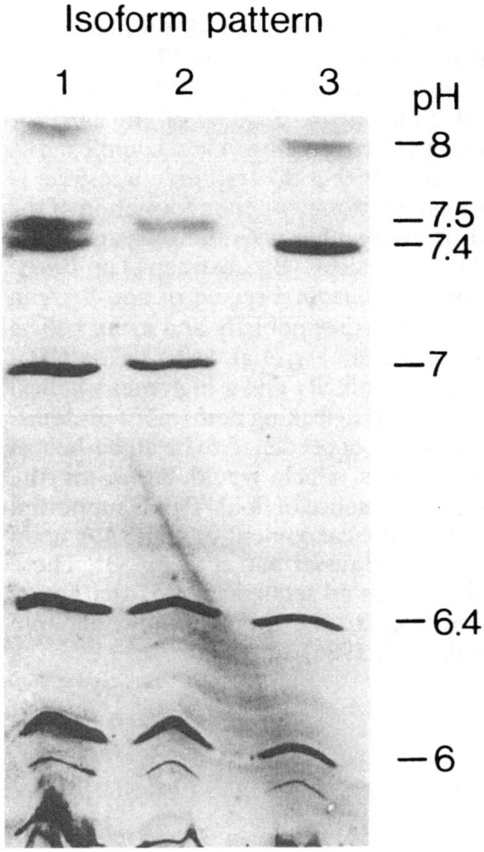

Fig. 11.1. The three patterns of human apo-SAA isoforms: electrofocusing of delipidated acute phase HDL samples was performed on 0.3 mm polyacrylamide gels containing 7 M urea (Strachan et al. 1988).

for the different patterns could be at the level of the structural genes, could reflect different rates of synthesis or metabolism of the same gene products in different subjects, or could be the result of variable post-translational modifications other than glycosylation.

Table 11.1. The apo-SAA isoform complement of patients with different conditions

Patient details[a]		Apo-SAA isoform quantification[b]					
Condition	Sex	pH 6.0	pH 6.4	pH 7.0	pH 7.4	pH 7.5	pH 8.0
Pneumonia	M	26	34	23	ND	17	ND
Pneumonia	M	9	25	10	14	12	30
Pneumonia	F	31	36	17	ND	16	ND
Pneumonia	F	36	34	20	ND	10	ND
Rheumatoid arthritis	M	33	50	9	ND	8	ND
Rheumatoid arthritis	F	24	27	ND	35	ND	14
Trauma, surgery	M	31	34	9	14	3	9
Trauma, surgery	F	22	38	7	12	5	16

[a] Some data from Strachan et al. (1988).
[b] Apo-SAA isoforms were quantified as previously described (Strachan et al. 1988).
ND: Not detectable. Patients with isoform patterns 2 and 3 do not express certain isoforms (see Fig. 11.1).

A mathematical prediction of the secondary structure of apo-SAA (Turnell et al. 1986) predicts two alpha-helical regions between residues 12 and 27 and between residues 72 and 87. A lipid binding domain of apo-SAA is probably located in or around the NH_2-terminal alpha helix. Studies of the intrinsic fluorescence spectrum of amyloid A protein both in free form and complexed to dimyristol-phosphatidylcholine (DMPC) suggests that the tryptophan residue at position 18 moves from a polar to a non-polar environment upon formation of the apo-SAA–DMPC complex. This would be compatible with the burying of the tryptophan reside in fatty acyl chains of DMPC vesicles (Bausserman et al. 1983). Further support for the implication of the NH_2-terminal region of apo-SAA in binding to HDL comes from analyses of mean hydrophobicity and mean helical hydrophobic moment using the technique of Eisenberg et al. (1982). The NH_2-terminal amino acids have a high mean hydrophobicity and a high mean helical hydrophobic moment which is characteristic of lipid-seeking domains of proteins. Although the initial NH_2-terminal residues were not predicted to be alpha-helical (Turnell et al. 1986) they have characteristics which would allow for the possibility of a coil-helix transformation in the presence of lipid. This is supported by the observation of an increase in the alpha-helical content of apo-SAA upon association with DMPC vesicles, as reported by Bausserman et al. (1983). There is another possible lipid binding site in the conserved sequence 48–51 that forms part of a predicted surface loop with homology to the Ca^{2+} and phospholipid binding sites of phospholipase A_2 (Turnell et al. 1986).

Structure of Acute Phase HDL

Lipoproteins are spherical particles consisting of a hydrophobic core surrounded by a hydrophilic shell (Eisenberg 1984). The core is rich in cholesteryl esters and triglycerides, while phospholipids, unesterfied cholesterol and apolipoproteins constitute the shell. The hydrophobic domains of these shell components are directed towards the core while the hydrophilic regions are directed towards the aqueous environment. This makes the complex water soluble (Eisenberg 1984). HDL is the lipoprotein that sediments at density 1.063–1.21 g ml^{-1} in the ultracentrifuge and has apo-AI and apo-AII as major apolipoproteins. Within this density range, two major subclasses classified as HDL_2 and HDL_3 exist. HDL_2 with density 1.063–1.125 g ml^{-1} is larger (~10 nm) and differs in lipid and protein composition from HDL_3 (~8 nm) with density 1.25–1.21 g ml^{-1}. Within each subclass, considerable heterogeneity exists which is demonstrated by density gradient and rate zonal ultracentrifugation (Patsch et al. 1980) as well as gradient polyacrylamide gel electrophoresis (Blanche et al. 1981). Immunoaffinity chromatography reflects the heterogeneity that exists with respect to relative content of major apoproteins in both acute phase and normal HDL (Coetzee et al. 1986; Cheung and Albers 1984). More than 75% of HDL associated apo-SAA is associated with HDL_3 while only 20% is associated with HDL_2 (Coetzee et al. 1986). Apart from HDL-associated apo-SAA, the only other significant apo-SAA-containing fraction recovered following density gradient ultracentrifugation is in the density >1.21 layer. Whether apo-SAA exists as free SAA at this

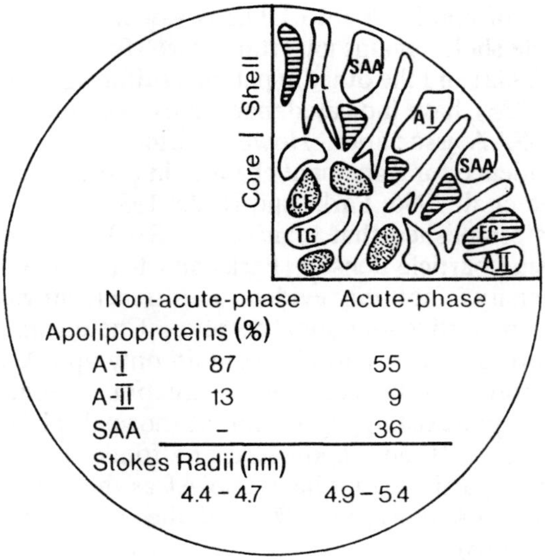

Fig. 11.2. The HDL particle: phospholipids (*PL*), free cholesterol (*FC*) and apolipoproteins occupy the shell whilst cholesteryl ester (*CE*) and triglyceride (*TG*) occupy the core. Typical apolipoprotein composition and Stokes radii of normal and acute phase HDL particles are shown in the *lower portion* of the figure.

density or whether it is associated with a very high density lipoprotein fraction has not been established.

Studies of the structure of apo-SAA-enriched HDL, created by in vitro reconstitution, indicate that apo-SAA exhibits a high affinity for HDL and can displace the bulk of apo-AI from the particles (Coetzee et al. 1986; Clifton et al. 1985). It seems that apo-SAA is secreted by hepatocytes in a monomeric and lipid free form and then associates with existing HDL in the circulation (Hoffman and Benditt 1982). The normal HDL that apo-SAA associates with in vivo represents a heterogeneous spectrum of particles and subclasses identified with a variety of techniques.

Using immunoassays for plasma apo-SAA standardized with apo-SAA-containing HDL_3 with known apo-SAA content, plasma apo-SAA is shown to reach levels of 1200 mg l^{-1} during the acute phase response (Godenir et al. 1985). On acute phase HDL apo-SAA can even become the major apolipoprotein. Generally, values of relative apo-SAA content of acute phase HDL_3 range from 20% to 50% or more of total protein (Coetzee et al. 1986; Eriksen and Benditt 1984; Bausserman et al. 1982; Strachan et al. 1986).

The acquisition of apo-SAA by HDL results in a loss of normal apolipoproteins from HDL. When one considers the respective molecular weight of the two apoproteins (apo-AI, 28 000 Da and apo-SAA, 11 500 Da), the loss of one apo-AI molecule with the concomitant gain of three apo-SAA molecules would lead to a small increase in relative protein content and density of the particle. The loss of apo-AI from the HDL particle could be a result of displacement, as has been shown in vitro (Coetzee et al. 1986). Another factor that needs to be considered is that apo-AI synthesis during the acute phase is likely to be decreased.

Whether or not the incorporation of apo-SAA into HDL is associated with displacement of lipid from the particle shell remains uncertain. Two studies, one in the mouse (Hoffman and Benditt 1982) and the other in patients suffering from myocardial infarction (Clifton et al. 1985), have indicated that some displacement of phospholipid does occur with apo-SAA association. However, other studies in man and monkey have shown no evidence for significant changes in protein and lipid composition of HDL_3 (Coetzee et al. 1986; Parks and Rudel 1983).

In mouse, man and monkey it is established that relative apo-SAA content increases progressively with increasing particle density (Parks and Rudel 1985; Coetzee et al. 1986). This polydispersity is not only evident on ultracentrifugal gradients, but is also confirmed by competitive immunoabsorption. Cheung and Albers (1984) have shown that whereas certain particles contain only apo-AI, none have apo-AII as the only apoprotein. In the acute phase situation it seems that apo-SAA associates with this pre-existing spectrum of normal HDL particles. No particles with only apo-AII and apo-SAA are found but a polydisperse spectrum exists including particles that have apo-AI as the major apoprotein and particles where apo-SAA constitutes 80% of the apoprotein (Strachan et al. 1986; Coetzee et al. 1986).

When normal and acute phase HDL are compared with respect to size using gradient gel electrophoresis under non-denaturing conditions, or gel filtration, it becomes apparent that apo-SAA-bearing particles are larger than normal HDL particles of similar density. This increase has led Clifton et al. (1985) to suggest that these particles may represent HDL_2-like particles that gained apo-SAA with resultant increased density but retained their larger size. This increase in particle size is probably the result of an increase in the width of the shell of the HDL particle which comprises about 80% of particle volume (Coetzee et al. 1986).

Agarose electrophoresis of acute phase HDL_3 and of reconstituted acute phase particles showed a lower mobility than that for normal HDL_3. The degree of retardation, however, showed individual variation that was not solely a function of the apo-SAA content (Strachan et al. 1986).

The question of polydispersity with respect to apo-SAA isoforms associated with particles of varying density has been studied. It was found that all isoforms remained equally distributed amongst the various particles (Strachan et al. unpublished work).

In summary, the acquisition of apo-SAA by normal HDL results in the displacement of normal apolipoproteins, increases the density of the HDL, increases the size of the particles and, finally, results in a spectrum of particles which are polydisperse with respect to apolipoprotein content.

Metabolic Function of Normal HDL

When considering the metabolism of apo-SAA, cognisance needs to be taken of certain facts with respect to normal HDL metabolism. In vitro, every component of HDL turns over a different rate (Eisenberg 1984). In the circulation, HDL is continually remodelled by plasma enzymes, lipid transfer proteins and lecithin cholesterol acyl transferase. As originally pointed out by Glomset (1968), HDL is involved in the movement of cholesterol from the periphery to the liver – reverse cholesterol transport. The precise mechanism by which HDL interacts with different cells and the precise roles played by the different HDL apolipoproteins remain in dispute. Various studies have pointed to roles for apo-AI, and possibly apo-AII, as well as HDL surface lipids as the interactive entities with cells (Fidge and Nestel 1985; Hoeg et al. 1985; Havekes et al. 1984; Tabas and Tall 1984).

Another important fact is that HDL represents a polydisperse group of particles in continual dynamic flux (Eisenberg 1984). Studies on HDL thus represent a significantly more complex problem than, for instance, studies on LDL, which has a stable apolipoprotein composition.

In addition to involvement in cholesterol metabolism, HDL has been shown to specifically bind endotoxin in vivo and in vitro (Ulevitch et al. 1979; Freudenberg et al. 1980). Ulevitch has also shown that endotoxin is biologically altered by this association. This could be important in modulating the induction of the acute phase response. The influence of apo-SAA association on endotoxin binding by HDL still needs elucidation.

Synthesis of Apo-SAA

Hepatocytes are believed to be the major site of synthesis for apo-SAA (Benson and Kleiner 1980). Following an acute phase inducing stimulus, random hepatocytes in the hepatic lobules can be immunochemically shown to synthesize apo-SAA. More prolonged stimulation results in the hepatic cords lining the space of Disse secreting most apo-SAA (Shirahama et al. 1984). The question whether significant extra hepatic secretion of apo-SAA occurs and whether this could be of pathogenetic importance in the development of amyloidosis remains unclear. Studies that utilized sensitive radio-immunoassays to show apo-SAA in homogenized liver and kidney but not in spleen, lung or heart suffer from the limitation of not being able to distinguish between absorption and synthesis. More convincing evidence comes from the immunohistochemically identified apo-SAA in foetal endothelial cells and embryonal fibroblasts in culture (Linder et al. 1976; Linder et al. 1977). Following in vivo administration of endotoxin or recombinant interleukin-1 (IL-1) to mice, mRNA specific for murine apo-SAA was detected in liver, heart, spleen, lung, intestine, kidney and peritoneal macrophages (Ramadori et al. 1985). Whether this mRNA is in fact translated and secreted as apo-SAA remains to be proved. Analysis of human and murine cDNA and genomic clones indicate that apo-SAA is encoded by multiple genes (Sipe et al. 1985; Sack 1983; Yamamoto et al. 1986; Woo et al. 1987). The mechanisms for control and expression of the various apo-SAA genes are not known but transcriptional control seems to be an important regulatory point (Baumann et al. 1984). Of particular importance would be the elucidation of how the various isoforms in humans arise – the role of individual structural genes, different rates of synthesis or catabolism or the influence of post-translational modifications. This is exemplified in the mouse, where, it has been suggested, resistance to amyloid deposition in SJL mice arises from the relative low level of production of SAA_2

mRNA – the amyloidogenic isoform (Yamamoto et al. 1986). Whether amyloidogenic isoforms of apo-SAA exist in humans or whether differential synthesis rates for the various isoforms can occur is unknown.

Plasma Clearance of Apo-SAA

Hoffman and Benditt (1983) studied comparative clearance kinetics of acute phase apolipoproteins in BALB/c mice using ^{125}I-apo-SAA-rich HDL and unlabelled plasma apo-SAA. Very rapid clearance of apo-SAA (half-life 75–80 min) when compared to ±11 hours for mouse apo-AI was shown. They noted the complicating fact that the rapid plasma clearance of apo-SAA was lost when apo-SAA was purified or labelled with ^{125}I and then used to reconstitute normal HDL. ^{125}I-apo-SAA was not selectively excreted and the distribution among the major organs presented no evidence that the metabolic fate of ^{125}I-apo-SAA was any different to that of other HDL apolipoproteins.

One of the two secreted apo-SAA isoforms in the mouse, apo-SAA_2 is selectively deposited as amyloid fibrils after degradation. Meek et al. (1986) studied the kinetics of the respective mouse isoforms during casein induction of amyloidosis in CBA mice. Serum apo-SAA levels peaked 24 hours after casein was administered and then gradually declined. The decline was entirely accounted for by a dramatic fall in apo-SAA_2 with apo-SAA_1 remaining constant. At the same time the ratio of apo-SAA_1:apo-SAA_2 mRNA remained constant and no mRNA production could be detected in the spleen, which is a major site of fibril deposition. The authors' findings are explained on the basis of the rapid and selective removal of apo-SAA_2 from the circulation.

Parks and Rudel (1983) performed similar metabolic studies in the vervet monkey (*Cercopithecus aethiops*), a species with a close phylogenetic relationship to man. Chair restraint induced an acute phase response in these animals and the metabolism of apo-SAA rich ^{131}I-labelled chylomicrons and ^{125}I-labelled acute phase HDL was compared. The clearance was biphasic with the initial phase similar for the two sources of apo-SAA (half-life of 0.39–0.48 days). The second phase was significantly faster for the apo-SAA associated with HDL. These data pointed to differences between apo-SAA of HDL and of chylomicron origin, intestinal synthesis being a possibility that cannot be excluded. These studies, however, again confirmed that apo-SAA is more rapidly catabolized than apo-AI or apo-AII.

Bausserman et al. (1984) studied the clearance kinetics of two human apo-SAA isoforms (so-called apo-SAA_1 and apo-SAA_4) with identical NH_2-terminal sequence but with different physical/chemical properties. The purified isoforms were iodinated and then combined with HDL and administered to cynomolgus monkeys. Both had similar die-away curves, remained associated with HDL and remained unaltered, indicating that there was no precursor to product relationship. Human apo-AI cleared from plasma at a slower rate than apo-C-III-1 with apo-SAA_4 clearing most rapidly. In contrast to the mouse, where isolation and iodination of apo-SAA followed by re-association with HDL seem to slow the clearance of apo-SAA, this seems not to be the case in humans and would considerable facilitate further studies on the metabolism of apo-SAA in man.

In a subsequent study, Bausserman et al. (1987) studied the clearance and degradation of human apo-SAA in rats in an isolated perfused rat liver model. Important conclusions reached from this study are that apo-SAA is catabolized independently from other HDL apoproteins in what is likely to be, in part, a specific process.

All of the preceeding studies can be explained either by apo-SAA associating with an HDL subpopulation that turns over more rapidly than the bulk of HDL or that apo-SAA dissociated from HDL before clearance from plasma.

Cellular Association and Degradation of Apo-SAA

Purified apo-SAA was shown to bind to the surface of mononuclear cells at 4°C, but at 37°C association could only be shown when cells were preincubated with diisopropyl-fluorophosphate (DFP) or specific elastase inhibitors. No internalized apo-SAA was demonstrated (Zucker-Franklin et al. 1981). Lavie et al. (1980) further characterized the surface enzymes that have the capacity to degrade apo-SAA. These enzymes are elastases but differ from secreted macrophage elastase in molecular weight. Lavie et al. suggested that two groups of enzymes exist: one group has the capacity to degrade apo-SAA to small peptides and the other degrades apo-SAA to AA-like intermediates.

Neutrophil leucocytosis occurs concomitantly with increases in plasma apo-SAA during the acute phase response. This led Shephard et al. (1987) to study the comparative neutrophil association and degradation of acute phase and normal HDL_3. Results showed that more ^{125}I-labelled acute phase HDL_3 associated with neutrophils than ^{125}I-labelled normal HDL_3 and that the presence of α1-antitrypsin in the incubation medium further increased the association (Fig. 11.3). This is probably due to proteolytic alteration of the apolipoproteins in the

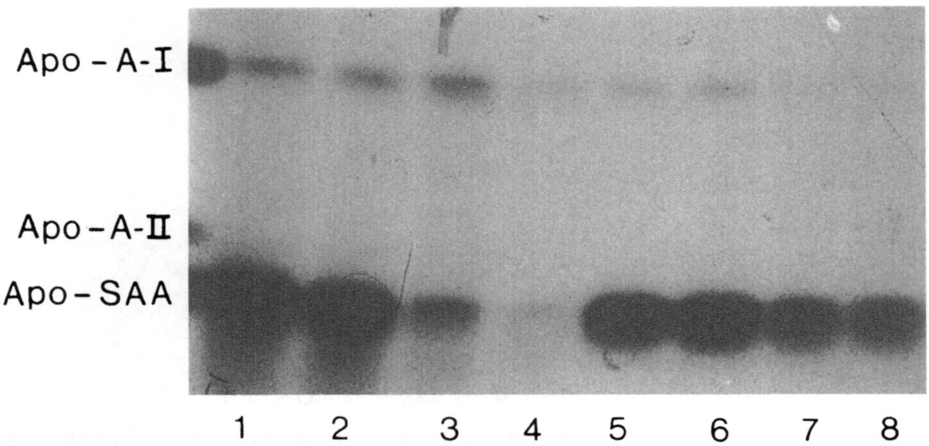

Fig. 11.3. Nature of apolipoproteins associated with 250 μg neutrophil or lymphocyte protein after incubation of 25 μg ^{125}I-labelled acute phase HDL_3 ml^{-1} with 10 × 10^6 cells (60 min, 37°C). Neutrophil associated apolipoproteins in the presence (*Tracks 1 and 2*) and in the absence (*Tracks 3 and 4*) of α1-antitrypsin (1.5 mg ml^{-1}). Lymphocyte associated apolipoprotein in the presence (*Tracks 5 and 6*) and in the absence (*Tracks 7 and 8*) of α1-antitrypsin (1.5 mg ml^{-1})

absence of α1-antitrypsin, which affects binding to cells. Qualitative analysis using SDS-electrophoresis showed that apo-A-I is the major apoprotein associating with neutrophils when normal HDL_3 was bound. This is expected as it is believed that apo-AI is the major apoprotein involved in cellular binding although the nature of this process has not been resolved (Eisenberg 1984). More remarkable, however, was that when ^{125}I-labelled acute phase HDL_3 was associated with neutrophils, apo-SAA was the predominant associated apolipoprotein saturating at 150 μg of total acute phase apolipoprotein per ml. Analogous results were obtained with lymphocytes, bovine endothelial cells and Hep G2 cells (Figs. 11.3 and 11.4). A fundamental question that remains unresolved is whether the apo-SAA that preferentially associates with the neutrophils dissociated from acute phase HDL particles, or whether the bound apo-SAA results from an apo-SAA enriched subset of particles that preferentially bind to cells. The latter possibility needs to be entertained because in the polydisperse nature of acute phase HDL_3, particles have been shown that have apo-SAA as the major apoprotein (Strachan et al. 1986). When trypsinization of cells was employed, it became obvious that the associated apo-SAA is not internalized but is bound on the cell surface. The apo-SAA bound on the cell membrane is rapidly degraded. As it was reported that normal HDL_3 mediates the egress of elastase from neutrophils with resultant degradation of apo-AII (Byrne et al. 1984), conditioned medium degradation was compared to cellular (neutrophil) degradation. A relatively α1-antitrypsin resistant cellular apo-SAA degradation of considerably larger magnitude than that achieved by conditioned medium was to exist. These findings correspond with the findings of Lavie et al. (1980) using monocytes, that apo-SAA degradation is a cell surface event where membrane associated serine proteases play an important role. However, using neutrophils from many individuals and a variety of batches of acute phase HDL_3,

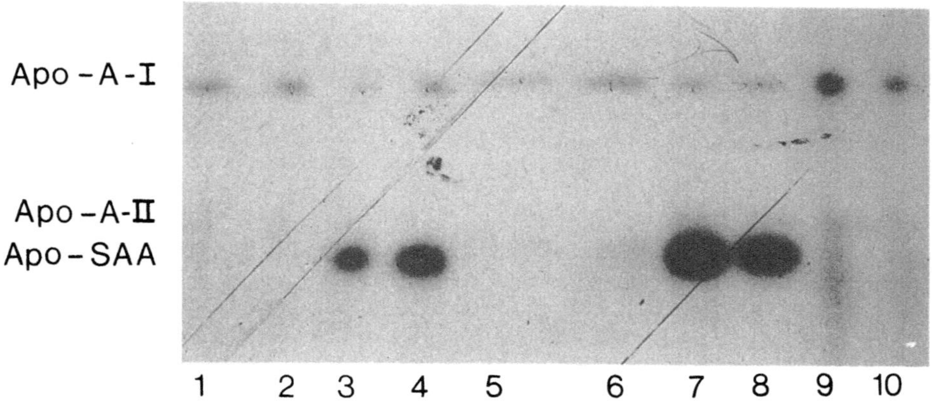

Fig. 11.4. Endothelial cells, Hep G2 cells, or lymphocytes were incubated (60 min, 37°C) with either normal HDL_3 or acute phase HDL_3 (25 μg ml^{-1}) and the nature of apolipoproteins associated with 250 μg cellular protein analysed by 5%–20% SDS-polyacrylamide gel electrophoresis. Apolipoprotein associated with endothelial cells. *Tracks 1 and 2*: normal HDL_3; *Tracks 3 and 4*: acute phase HDL_3, Hep G2 cells; *Tracks 5 and 6*: normal HDL_3; *Tracks 7 and 8*: acute phase HDL_3, lymphocytes; *Tracks 9 and 10*: normal HDL_3.

the degradation to AA-like intermediates was never detected – cell degradation rapidly yielded small peptides.

Using an isolated perfused rat liver system, Bausserman et al. (1987) showed that the degradation of human apo-SAA appeared to be mediated by cell associated enzymes, as recirculated perfusate accounted for less than 15% of the observed degradation.

The precise mechanisms involved in apo-SAA degradation need to be further elucidated to promote understanding of these events in pathophysiological processes.

Functions of Apo-SAA

Of all acute phase proteins, least is known about the function of apo-SAA. The consequences and possible functions of the radically altered HDL_3 during the acute phase response are unknown. The phylogenetic conservation of the association of apo-SAA with HDL suggests an important role.

The most important function of normal HDL is its involvement in reverse cholesterol transport from the periphery to the liver. Whether the association of apo-SAA with HDL and the resultant physical/chemical changes induced alter the ability of HDL to facilitate the egress of cholesterol from peripheral cells is unknown. Apo-SAA–HDL_3 with its higher binding would be an interesting tool to study this event.

The rapid clearance of apo-SAA from the circulation in comparison to other HDL apoproteins has led to suggestions that the function of apo-SAA may be to confer a rapid clearance ability on HDL particles when complexed to toxic or unwanted material. This suggestion would only be valid if apo-SAA is associated with a rapidly clearing group of particles, but it is quite possible that apo-SAA dissociates from particles before being cleared. Although lipopolysaccharide has been shown to complex to apo-SAA bearing HDL in the rabbit, the ^{125}I-labelled lipopolysaccharide was not cleared more rapidly (Tobias et al. 1982).

Purified human and murine apo-SAA have been shown to suppress T-cell-dependent in vitro antibody responses to sheep erythrocytes (Benson and Aldo-Benson 1979; Aldo-Benson and Benson 1982). Benson et al. (1986) subsequently studied apo-SAA levels in carcinoma of the lung and showed a significant correlation between apo-SAA serum concentrations and peripheral blood lymphocytes (PBL) mitogenic response to Con. A. In addition, both high apo-SAA concentrations and depressed PBL responses to Con. A correlated with shortened survival.

These findings all suggest that apo-SAA might in some way alter cellular activation. During the acute phase response, protein kinase C (PKC) is often activated in several tissues and cells involved in inflammation, e.g. neutrophils and lymphocytes. PKC is a key enzyme in cell-surface signal transduction in a variety of mammalian tissues (Nishizuka 1984). Receptor mediated stimulation of membrane lipid turnover with generation of 1,2-diacylglycerol activates PKC (Nishizuka 1984). It was shown that the apolipoprotein moiety of normal HDL_3 can activate PKC in bovine endothelial cells (Darbon et al. 1986). As these are all cell membrane events, and as apo-SAA is the major acute phase apolipoprotein associating with neutrophil and lymphocyte membranes (Shephard et al. 1987; De Beer et al., unpublished observations), Nel et al. (1988) studied the interaction between PKC and apo-SAA. Apo-SAA was the only apolipoprotein substrate phosphorylated by PKC (K_m = 12 μM). Phosphorylation of the apo-SAA containing HDL_3 was selective for the more basic isoforms of apo-SAA with the relatively more acidic isoforms only phosphorylating when the HDL was delipidated.

The precise function of apo-SAA remains an enigma and an extensive programme of comparative testing of acute phase and normal HDL_3 in various promising biological systems would be needed to elucidate its function.

Concluding Remarks

Recent years have brought considerable advances in our knowledge of apo-SAA. Many essential features of apo-SAA enriched HDL_3 have become available, studies on the isoforms of apo-SAA have progressed and much information concerning the genes for apo-SAA has unfolded. Data on the precise molecular mechanism that results in apo-SAA synthesis and the biological function of the secreted proteins remain limited. Further studies on cellular interaction and metabolism of apo-SAA are essential to elucidate its function and would assist materially in understanding the mechanism(s) involved in the pathogenesis of amyloidosis.

Acknowledgements. The financial support of the South African Medical Research Council and the South African Arthritis and Rheumatism Association is acknowledged.

References

Aldo-Benson MA, Benson MD (1982) SAA suppression of immune response in vitro: evidence for an effect on T cell-macrophage interaction. J Immunol 128:2390–2392

Baltz ML, Rowe IF, Caspi D, Turnell W, Pepys MB (1987) Acute-phase high-density lipoprotein in the rat does not contain serum amyloid A protein. Biochem J 242:301–303

Baumann H, Held WA, Berger FG (1984) The acute phase response of mouse liver. Genetic analysis of the major acute phase reactants. J Biol Chem 259:566–573

Bausserman LL, Saritelli AL, Herbert PN, McAdam KPWJ, Shulman RS (1982) NH_2-terminal analysis of four of the polymorphic forms of human amyloid A proteins. Biochim Biophys Acta 704:556–559

Bausserman LL, Herbert PN, Forte T et al. (1983) Interaction of serum amyloid A proteins with phospholipid. J Biol Chem 258:10681–10688

Bausserman LL, Herbert PN, Rodgers R, Nicolosi RJ (1984) Rapid clearance of serum amyloid A from high density lipoproteins. Biochem Biophys Acta 792:186–191

Bausserman LL, Saritelli AL, Van Zuiden P, Gollaher CJ, Herbert PN (1987) Degradation of serum amyloid A by isolated perfused rat liver. J Biol Chem 262:1583–1589

Benditt EP, Eriksen N (1977) Amyloid protein SAA is associated with high density lipoprotein from human serum. Proc Natl Acad Sci USA 74:4025–4028

Benditt EP, Eriksen N, Hanson RH (1979) Amyloid protein SAA is an apoprotein of mouse plasma high density lipoprotein. Proc Natl Acad Sci USA 76:4092–4096
Benson MD, Aldo-Benson MA (1979) Effect of purified protein SAA on immune response in vitro: mechanisms of suppression. J Immunol 122:2077–2082
Benson MD, Kleiner E (1980) Synthesis and secretion of serum amyloid protein A (SAA) by hepatocytes in mice treated with casein. J Immunol 124:495–499
Benson MD, Eyanson S, Fineberg NS (1986) Serum amyloid A in carcinoma of the lung. Cancer 57:1783–1787
Blanche PJ, Gong EL, Forte TM, Nichols AV (1981) Characterization of human high-density lipoproteins by gradient gel electrophoresis. Biochim Biophys Acta 665:408–419
Byrne RE, Polacek D, Gordon JI, Scanu AM (1984) The enzyme that cleaves apolipoprotein A-II upon in vitro incubation of human plasma high-density lipoprotein 3 with blood polymorphonuclear cells in an elastase. J Biol Chem 259:14537–14543
Cheung MC, Albers JJ (1984) Characterization of lipoprotein particles isolated by immunoaffinity chromatography—particles containing A-I and A-II and particles containing A-I but no A-II. J Biol Chem 259:12201–12209
Clifton PM, MacKinnon AM, Barter PJ (1985) Effects of serum amyloid A protein (SAA) on composition, size and density of high density lipoproteins in subjects with myocardial infarction. J Lipid Res 26:1389–1398
Coetzee GA, Strachan AF, Van der Westhuyzen DR, Hoppe HC, Jeenah MS, De Beer FC (1986) Serum amyloid A-containing human high density lipoprotein 3. Density, size and apolipoprotein composition. J Biol Chem 261:9644–9551
Darbon JM, Tournier JF, Tauber JP, Bayard F (1986) Possible role of protein phosphorylation in the mitogenic effect of high density lipoproteins on cultured vascular endothelial cells. J Biol Chem 261:8002–8008
de Beer FC, Mallya RK, Fagan EA, Lanham JG, Hughes GRV, Pepys MB (1982) Serum amyloid A protein concentration in inflammatory disease and its relationship to the incidence of reactive systemic amyloidosis. Lancet ii:231–234
Eisenberg D, Weiss RM, Terwilliger TC (1982) The helical hydrophobic moment: a measure of the amphiphilicity of a helix. Nature 299:371–374
Eisenberg S (1984) High density lipoprotein metabolism (Review). J Lipid Res 25:1017–1058
Eriksen N, Benditt EP (1984) Trauma, high density lipoproteins and serum amyloid A protein. Clin Chim Acta 140:139–149
Fidge NH, Nestel PJ (1985) Identification of apolipoproteins involved in the interaction of human high density lipoprotein 3 with receptors on cultured cells. J Biol Chem 260:3570–3575
Freudenberg MA, Bog-Hansen TC, Back U, Galanos C (1980) Interaction of lipopolysaccharides with plasma high density lipoprotein in rats. Infect Immunol 28:373–380
Glomset JA (1968) The plasma lecithins: cholesterol acyltransferase reaction. J Lipid Res 9:155–167
Godenir NL, Jeenah MS, Coetzee GA, Van der Westhuyzen DR, Strachan AF, de Beer FC (1985) Standardisation of the quantitation of serum amyloid protein (SAA) in human serum. J Immunol Methods 83:217–225
Havekes L, Schouten D, Van Hinsbergh V, De Wit E (1984) Characterisation of the binding of apolipoprotein E-free high density lipoprotein to cultured human endothelial cells. Biochem Biophys Res Commun 122:785–790
Hoeg JM, Demosky SJ Jr, Edge SB, Gregg RE, Osborne JC Jr, Brewer HB Jr (1985) Characterization of a human hepatic receptor for high density lipoproteins. Arteriosclerosis 5:228–237
Hoffman JS, Benditt EP (1982) Secretion of serum amyloid protein and assembly of serum amyloid protein-rich high density lipoprotein in primary mouse hepatocyte culture. J Biol Chem 257:10518–10522
Hoffman JS, Benditt EP, (1983) Plasma clearance kinetics of the amyloid-related high density lipoprotein apoprotein, serum amyloid protein (apo-SAA) in the mouse. J Clin Invest 71:926–934
Jacob MP, Bellon G, Robert L et al. (1981) Elastase-type activity associated with high density lipoproteins in human serum. Biochem Biophys Res Commun 103:311–318
Kushner I (1982) The phenomenon of the acute phase response. Ann NY Acad Sci 389:39–48
Lavie G, Zucker-Franklin D, Franklin EC (1980) Elastase type proteases on the surface of human blood monocytes—possible role in amyloid formation. J Immunol 125:175–180
Linder E, Anders RF, Natvig JB (1976) Connective tissue origin of the amyloid-related protein SAA. J Exp Med 144:1336–1346
Linder E, Lehto VP, Virtanen I, Stenman S, Natvig JB (1977) Localization of amyloid-related serum protein SAA-like material to intermediate (10 nm) filaments of cultured human embryonal fibroblasts. J Exp Med 146:1158–1163

Marhaug G, Borresen AL, Husby G, Nordstoga K (1984) A comparative study of serum amyloid A protein (SAA) from mink and man. Comp Biochem Physiol 78B:401–406

McAdam KPWJ, Elin RJ, Sipe JD, Wolff SM (1978) Changes in human serum amyloid-A and C-reactive protein after etiocholanolone-induced inflammation. J Clin Invest 61:390–394

Meek RL, Hoffman JS, Benditt EP (1986) Amyloidogenesis. One serum amyloid A isotype is selectively removed from the circulation. J Exp Med 163:499–510

Nel AE, de Beer MC, Shephard EG, Strachan AF, Vandenplas ML, de Beer FC (1988) Phosphorylation of human serum amyloid A protein by protein kinase C. Biochem J 255:26–34

Nishizuka Y (1984) The role of protein kinase C in cell-surface signal transduction and tumor promotion. Nature 308:693–698

Parks JS, Rudel LL (1983) Metabolism of the serum amyloid A proteins (SAA) in high-density lipoproteins and chylomicrons of non-human primates (Vervet monkey). Amer J Pathol 112:243–249

Parmelee DC, Titani K, Ericsson LH, Eriksen N, Benditt EP, Walsh KA (1982) Amino acid sequence of amyloid-related apoprotein (apo-SAA_1) from human high density lipoprotein. Biochem 21:3298–3303

Patsch W, Schonfeld G, Gotto AM Jr, Patsch JR (1980) Characterization of human high-density lipoproteins by zonal ultracentrifugation. J Biol Chem 255:3178–3185

Ramadori G, Sipe JD, Dinarello CA, Mizel SB, Colten HR (1985) Pretranslation modulation of acute phase hepatic protein synthesis by murine recombinant interleukin I (IL-1) and purified human IL-1. J Exp Med 162:930–942

Sack GH (1983) Molecular-cloning of human genes for serum amyloid-A. Gene 21:19–24

Segrest JP, Pownall HJ, Jackson RL, Glenner GG, Pollack PS (1976) Amyloid A: amphipathic helixes and lipid binding. Biochemistry 15:3187–3191

Shephard EG, de Beer FC, de Beer MC, Jeenah MS, Coetzee GA, Van der Westhuyzen DR (1987) Neutrophil association and degradation of normal and acute-phase high-density lipoprotein 3. Biochem J 248:919–926

Shirahama T, Skinner M, Cohen AS (1984) Heterogenous participation of the hepatocyte population in amyloid protein AA synthesis. Cell Biol I 8:849–856

Sipe JD, Colten HR, Goldberger G et al. (1985) Human serum amyloid A (SAA): biosynthesis and postsynthetic processing of pre-SAA and structural variants defined by complementary DNA. Biochemistry 24:2931–2936

Sipe JD, Rokita H, Shirahama T, Cohen AS, Koj A (1986) Analysis of SAA gene expression in rat and mouse during acute inflammation and accelerated amyloidogenesis. In Peeters H (ed) Protides of the biological fluids, vol 34. Pergamon Press, Oxford, pp 331–334

Skogen B, Borresen AL, Natvig JB, Berg K, Michaelson TE (1979) High-density lipoprotein as carrier for amyloid-related protein SAA in rabbit serum. Scand J Immunol 10:39–45

Strachan AF, de Beer FC, Coetzee GA, Hoppe HC, Jeenah MS, Van der Westhuyzen DR (1986) Characteristics of apo-SAA containing HDL_3 in humans. In Peeters H (ed) Protides of the biological fluids, vol 34. Pergamon Press, Oxford, pp 359–362

Strachan AF, de Beer FC, Van der Westhuyzen DR, Coetzee GA (1988) Identification of three isoform patterns of human serum amyloid A protein. Biochem J 250:203–207

Tabas I, Tall AR (1984) Mechanisms of the association of HDL_3 with endothelial cells, smooth muscle cells and fibroblasts – evidence against the role of specific ligand and receptor proteins. J Biol Chem 259:13897–13905

Tobias PS, McAdam KPWJ, Ulevitch RJ (1982) Interactions of bacterial lipopolysaccharide with acute phase rabbit serum and isolation of two forms of rabbit serum amyloid A. J Immunol 128:1420–1427

Turnell W, Sarra R, Glover ID et al. (1986) Secondary structure prediction of human SAA_1 – presumptive identification of calcium and lipid binding sites. Mol Biol Med 3:387–407

Ulevitch RJ, Johnston AR, Weinstein DB (1979) New function for high density lipoproteins – their participation in intravascular reactions of bacterial lipopolysaccharide. J Clin Invest 64:1516–1524

Ulevitch RJ, Johnston AR, Weinstein DB (1981) New function for high density lipoproteins – isolation and characterization of a bacterial lipopolysaccharide high density lipoprotein complex formed in rabbit plasma. J Clin Invest 67:827–837

Woo P, Sipe JD, Dinarello CA, Colten HR (1987) Structure of a human serum amyloid A gene and modulation of its expression in transfected L-cells. J Biol Chem 262:15790–15795

Yamamoto K, Shiroo M, Migita S (1986) Diverse gene expression for isotypes of murine serum amyloid A protein during acute phase reaction. Science 232:227–229

Zucker-Franklin D, Lavie G, Franklin EC (1981) Demonstration of membrane-bound protolytic activity on the surface of mononuclear leukocytes. J Histochem Cytochem 29 (3A Suppl):451–456

Chapter 12

Clinical Measurement of Acute Phase Proteins to Detect and Monitor Infectious Diseases

P. E. Laurent

Introduction

The acute phase (AP) response includes fever, leukocytosis with neutrophilia of the blood, changes in certain heavy metals in the blood and increased uptake of amino acids with secretion of acute phase proteins by the liver (Koj 1983). The bacterial growth associated with access of endotoxin to tissue and its dissemination produces a systemic inflammatory response, and the acute phase proteins are one of the most studied biological markers.

In the absence of illness, plasma proteins constitute 3.5% of the total body protein, and plasma turnover accounts for 10% of the daily protein turnover. During the acute phase response of systemic inflammation, the incorporation of amino acids into plasma proteins overall may increase by 2- to 10-fold, mostly reflecting acute phase protein response (Powenda and Moyer 1981).

The acute phase reactant (APR) proteins are a heterogeneous group of plasma proteins which alter their blood concentration in response to a wide variety of stimuli, including tissue damage such as necrosis, acute and chronic inflammation and neoplasia (Fisher et al. 1976; Koj 1974; Owen 1967). In man, the most studied acute phase (AP) proteins are α1-acid glycoprotein (AGP), α1-antitrypsin (α1-AT) α1-antichymotrypsin (ACH), serum amyloid A (SAA) protein, haptoglobin (HPT), ceruloplasmin (CER), fibrinogen (FIB) and C-reactive protein (CRP). In acute inflammation, CRP, SAA and ACH become elevated in the first 12 hours after the onset of tissue damage. AGP, α1-AT, HPT and FIB increase their blood level some 24 hours later. In general, the blood levels of these proteins change in parallel. However, both the rate of change and peak concentration vary according to the rate of synthesis and catabolism. Most of the acute phase proteins have half-lives of 2–4 days. The exceptions are CRP and

SAA which have half-lives of 5–7 hours. From these different data, Kushner proposed to classify AP proteins in three groups:

Group I: concentration increases about 50% such as CER.

Group II: concentration increases about 2- to 4-fold: AGP, α1-AT, ACH, HPT, FIB.

Group III: concentration increases several hundred fold: CRP, SAA.

Because AP protein changes occur promptly after tissue damage, they are especially useful in detecting and monitoring inflammation processes. However, the AP protein response is non-specific because a wide variety of types of tissue damage and different infections promote exactly the same pattern.

Table 12.1. The general characteristics of AP proteins

Protein	Electrophoretic mobility	Molecular weight (Da)	Serum content in adults ($g\,l^{-1}$)
CER	α2	151 000	0.3–0.6
C3	β1	180 000	0.8–1.7
AGP	α1	40 000	0.5–1.0
AT	α1	54 000	1.5–3.0
ACH	α1	68 000	0.2–0.5
HPT	α2	86 000	0.8–2.0
FIB	β	341 000	2.0–4.0
CRP	γ	120 000	10×10^{-3}
SAA	α1	12 000	10×10^{-3}

Microbial Growth and AP Protein Response

The aggressiveness of bacteria in invading host tissue results in rapid and extensive tissue destruction. *Staphylococcus aureus*-induced inflammation is presented as a typical example. *S. aureus* continues to be a major cause of infection morbidity and mortality both in the community and in hospitals. In addition, *S. aureus* is becoming more and more aggressive. Furthermore, when compared with most other microbiol pathogens, *S. aureus* produces inflammation through a wide variety of mechanisms (see Fig. 12.1) such as:

Toxic substances with proteolytic activities which may directly injure the host tissues.

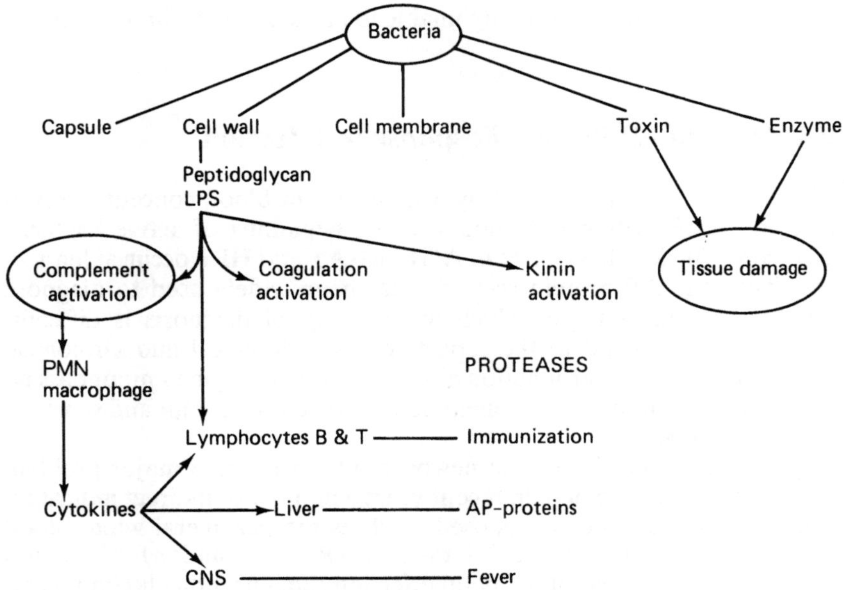

Fig. 12.1. Microbial growth and AP protein response.

Very intense complement activation by endotoxin-like substances (peptidoglycan) and other related substances which induce polymorphonuclear and mononuclear activation.

Activation of coagulation and kinin systems.

The AP proteins are predominantly synthesized in the liver, which is also the main site of their destruction. The exact nature of the factors and mediators which signal to the hepatocytes to increase the rate synthesis of AP proteins remains unclear. A great deal of published data strongly suggest that interleukin-1 (IL-1) plays a key role in the induction of biosynthesis of AP proteins (Dinarello 1986). In addition, cortisol, neurohumoral mediators and prostaglandins may participate significantly in the induction and regulation of AP proteins synthesis.

It is well recognized today that, after tissue injury, leukocytes accumulate at the injury site and infiltrated polymorphonuclears and monocytes secrete numerous cytokines which act as modulators of the inflammation response. These cytokines, and among them interleukin-1, seem to be the main bridge between bacterial growth and the AP response, since peptidoglycans and endotoxins are very powerful leukocyte activators (Kampschmidt et al. 1978). Within an hour of injection of IL-1-related molecules or some other cytokines such as ETAF, there is a flux of amino acids into the liver (Wannemacher et al. 1972), which suggests that this may be a necessary prelude to increased synthesis of acute phase proteins (Powenda et al. 1973; Wannemacher et al. 1975; Koj 1983).

In contrast to bacterial infection the main AP proteins such as CRP, AGP and HPT show only a slight increase during viral infection unless tissue damage resulting from viral pathogenesis or superimposed bacterial infection is also present as an efficient trigger to induce AP protein response. This clinical observation may suggest that viruses are unable to initiate signals for induction of AP protein synthesis.

General Pattern of AP Protein Response in Infection

AP protein response is characterized by a rapid rise in blood concentration of CRP, SAA and ACH within 8–12 hours of the beginning of active bacterial growth in tissues. The kinetics of rise in AGP, α1-AT and HPT occur at least 20 hours later. Thus the AP protein response has been widely used to diagnose infection in circumstances under which microbiological diagnosis is difficult. Because of the delay related to the reports of bacteriological and virological investigation, AP protein determination either with rapid enzyme immunoassay, or fluoroimmunoassay or immunonephelometry provides a prompt and sensitive guide for the diagnosis.

The early identification of sepsis in newborn infants is still a major problem with neonatal infection. Although in recent years a number of indirect indicators of bacterial infection have been proposed, such as the peripheral white blood count, the immature: total neutrophil ratio, CRP, HPT and α1-AT serum determination, early diagnosis of neonatal infection remains difficult (Philip and Hewitt 1980; Manroe et al. 1977; Speer et al. 1983). Thus, this question has been re-evaluated recently by Sann et al. (1984).

The blood concentration of AP protein at birth differs from levels in adults, as indicated in Table 12.2.

Table 12.2. Normal values of the most useful AP proteins for diagnosing infection in newborn infants

Protein	Birth ($g\,l^{-1}$)	First week of life ($g\,l^{-1}$)	Adult ($g\,l^{-1}$)
CRP	<0.001	<0.001	0.001
AGP	0.18	0.30–0.50	0.50–1
HPT	0.10	0.50	0.80–2
AT	2.20		1.50–3

In 85% of well-documented cases of bacterial septicaemia, meningitis and peritonitis, CRP and AGP blood levels are elevated with values up to $0.3\,g\,l^{-1}$ for CRP and $1.5–2\ g\ l^{-1}$ for AGP. Interestingly, the magnitude of the increase in AGP is higher in neonatal infections (× 10) than in adults (Sann et al. 1984).

Surgery increases AP protein blood concentration, but a decrease toward normal values is expected after three to five days if wound infection does not occur. Consequently, changes in the serum AP protein levels, mostly CRP and AGP, or remaining high levels of these two proteins indicate changes in the inflammatory process or occurrence of tissue damage. The literature provides numerous data which confirm the usefulness of CRP in the serial post-operative assessment of infection in surgical patients, but they also point out its inability to help clinicians localize the site and exact nature of infection, since CRP is a non-specific biological marker.

Since CRP levels in the blood intimately follow the course of the acute phase of systemic inflammation, monitoring CRP seems to provide a valuable clinical barometer of illness. In this respect, CRP measurement can be considered in many ways a refined quantitative alternative or supplement to erythrocyte sedimentation rate (ESR). Indeed, it is now appreciated that an elevated ESR is largely a phenomenon secondary to enhanced activity of certain AP proteins.

Monitoring Disease Activity Using AP Protein Response

The acute phase proteins may be used in a number of well-defined clinical contexts. However, since they are essentially a non-specific biological marker of tissue injury, changes must be interpreted in the light of clinical data. The most well-documented clinical conditions in which AP-protein measurement may be of great value are listed in Table 12.3.

Table 12.3. Clinical situations where AP protein measurements are useful

Monitoring disease activity and response to anti-inflammatory and anti-infection therapy
Diagnosing and monitoring infections in neonates, after trauma, after surgery, during cancer therapy (immunosuppressed host)
To diagnose pyrexia in patients suffering from systemic lupus erythematosus, Crohn's disease, AIDS

AP protein serial measurements are most widely used to monitor therapy in infectious diseases. CRP and AGP blood concentrations are two reliable indicators to assess the effectiveness of anti-microbial therapy, since therapeutic efficacy is associated with a decreased blood level toward normal values.

The non-steroidal anti-inflammatory agents such as aspirin, indomethacin and niflumic acid do not modify the AP protein response, despite causing a clinical response. In contrast, steroidal anti-inflammatory drugs strongly suppress AP protein response.

The value of sequential measurements of CRP and C4 component of the complement system in the management of seriously injured patients, such as burn patients, has been reported (Laurent and Marichy 1982). CRP level is very high after thermal injury because of the skin necrosis associated with other skin and muscle damage. In the case of superimposed bacterial infection such as septicaemia, there a consumption of C4 with a resulting increase in the CRP : C4 ratio. When this ratio reaches a value of 1 the prognosis is bad. The use of ratio, such as CRP : prealbumin, has been developed in the prognosis of advanced bowel cancer and in patients with ulcerative colitis.

It must be noted that in spite of very large tissue damage, such as that after thermal injury, paradoxical AGP normal values may be seen. When increased microvascular permeability occurs, as attested by AGP fall, the prognosis is usually poor with a fatal outcome in most cases.

AP Protein Response in Viral Infection

AP protein blood level elevations associated with viral infections have been reported to be much lower than those associated with bacterial infections. Most viral diseases such as paralytic poliomyelitis, serous meningitis, viral pneumonia and hepatitis enhance CRP level marginally, if at all (Sabel and Wadsworth 1979; Wager and Jannson 1957; Peltola 1982; McCarthy et al. 1978). Thus, in the differentiation of viral from bacterial meningitis, CRP is more sensitive than cerebrospinal fluid (CSF) leukocyte count, CSF polymorphonuclear count, CSF glucose, CSF total protein and Gram staining of the CSF, since the elevated CRP level in the serum or CSF is high in the case of a bacterial cause. However, Clarke and Cost (1983) recommended that only the serum CRP values should be measured and that these values provide enough information to make a differential diagnosis between viral and bacterial meningitis in children.

The presence or absence of high blood level of CRP has also been proposed in the differentiation between bacterial tracheitis and viral laryngotracheitis (Shabino 1983). Again, CRP level rises only in the bacterial type, unless inflammation associated with tissue damage occurs in the viral type.

HPT, AGP and CRP present low-to-normal levels in all types of acute viral hepatitis. In some cases, α1-AT blood concentration is elevated chiefly in acute hepatitis type B. On the other hand, in chronic viral hepatitis only α2-macroglobulin is significantly increased while the level of α1-AT is often normal (Meliconi 1986).

Sera from 30 patients with AIDS and sera from 31 patients with AIDS-related complex (ARC) were investigated by Marhaug et al. (1986). In the AIDS group the mean value of SAA was 30 mg l^{-1}, significantly higher than in patients with ARC syndrome, who had a mean value of 10 mg ml^{-1}. The highest SAA values in the AIDS group were found when *Pneumocystis carinii* infection occurred, and the lowest ones in patients with Kaposi's sarcoma without superimposed infection. A fall in SAA level was also a good indicator of effective antibiotic treatment. The same results have been yielded by serial measurement of CRP. Interestingly, these results suggest that AP protein response, as attested by SAA and CRP blood level changes, is not affected by a low T helper: T suppressor ratio (Marhaug et al. 1986).

Immunomodulation and AP Protein Response

There has been much research in the field of the biological properties of AP proteins which could play an important role in immunostimulation. CRP complexed or treated in a suitable way may bind to lymphocytes characterized by membrane Fc receptors such as B, T, non-B and non-T cells, but the significance of these observations to the host defence remains unclear. The ability of CRP to modulate in vitro T-cell activation as explored by stable E-rosettes needs further investigation. In addition, human CRP, following its autoaggregation or its binding with polyanion–polycation complexes and cation or phosphorylcholine-containing ligands, activates the classical complement pathway. The complement activation induced by CRP is as efficient as an antibody-induced activation (Kaplan and Volanakis 1974; Gewurz et al. 1982). These biological data are of importance, since they suggest that the involvement of CRP in resistance against infection could be significant. In addition, some experimental results argue in favour of this idea: treatment of mice with purified CRP from human pleural and peritoneal fluids protects animals against fatal infection with *Streptococcus pneumoniae* (Mold et al. 1981). However, CRP is able to protect mice against infection only when the AP protein binds to the bacteria, suggesting that CRP behaves like an opsonin, working even in the absence of complement (Nakayama et al. 1983). This opsonizing effect of CRP has been investigated in vivo in mice using the clearance of polysaccharide C-coated isologous erythrocytes. One can question the compatibility of experimental laboratory conditions with physiological AP response, since mouse AP protein response is quite different to human AP protein response.

Human AGP is able to suppress the murine lymphocyte mitogenic response to PHA, Con-A, LPS and allogeneic cells, antibody response to sheep erythrocytes, and the induction of cell mediated lympholysis of allogeneic target cells (Chiu et al. 1977; Whicher 1984). These effects are not influenced by desialyation but are augmented by the removal of both the terminal sialic acid and the penultimate galactose residue.

While the liver is the primary site of AGP synthesis and secretion during AP response, it has been demonstrated that AGP can be secreted by lymphocytes of T-cell lineage after activation in mixed lymphocyte culture. In our experiment using stable E-rosettes we demonstrated the capability of purified AGP to suppress T-cell activation with dose-dependent effects corresponding to AGP levels below those observed during AP response (Laurent et al. 1986). Thus, AGP seems to be a powerful modulator of lymphocyte activities. In addition, *Plasmodium falciparum* multiplication in vitro is decreased by 80% in the presence of concentrations of AGP corresponding to the level found in the blood of malaria patients. Some experimental observations strongly suggest that AGP is able to protect against infective agents that recognize and interact with the sialic moiety of AGP, such as herpes, influenza and measles viruses.

The fact that SAA is synthesized at the time of the onset of the acute phase response, with CRP and ACH, as well as at the time of antigenic stimulation, suggests that it might have an important function in immune regulation. In order to test this hypothesis, experiments were run to investigate antibody response both in vivo and in vitro in the presence of SAA. While the suppression of antibody response to a T-dependent antigen (SRBC) was reproducible in a dose-dependent fashion, non-suppression of a T-independent antigen (DNP ficoll) was shown. This suggests that the effect was not directly upon B lymphocytes (Benson et al. 1986). The physiological significance of immunomodulatory activity of human SAA on antibacterial and antiviral host-defences remains unclear.

The relationship of protease-antiprotease systems to lymphocyte response is supported by several lines of indirect evidence, including the modulation of lymphocyte response by the addition of proteases or protease inhibitors, the binding of protease inhibitors and the demonstration of surface-associated proteolytic activity in mouse B- and T-cells and human peripheral blood lymphocytes (Bata et al. 1980). The inhibiting effect of α1-AT on T-lymphocyte activation as investigated by stable E-rosettes is in agreement with previous data (Laurent et al. 1986). Macrophages are able to synthesize and secrete α2-macroglobulin (which is not an AP protein in man) and α1-AT (White et al. 1981). The inhibitory activity of macrophage-derived antiproteases on macrophage neutral proteinases and, further, the modulation of macrophage activation are questionable. Anyway, plasma and cellular inhibitors regulate proteolysis extracellularly and also promote endocytosis of inhibitor–proteinase complexes (Kaplan 1981) associated with expression on macrophage membranes of receptors for uptake of $\alpha 2$-macroglobulin complexes. Since activated macrophages seem to play a key role in the induction and regulation both of acute phase and immune response, modulation of macrophage activities by AP proteins, such as proteinase inhibitors, should be investigated actively.

Table 12.4. Biological properties of AP proteins

CRP	Complement system activation Opsonin Natural killer inhibition Lymphocyte activation?
SAA	Inhibition of antibody response Cholesterol transport/metabolism?
AGP	Inhibition of T lymphocyte activation Stimulation of fibroblastic proliferation
AT, ACH	Inhibition of collagenase, elastase Inhibition of cathepsin G Macrophage, lymphocyte activation?
HPT	Haemoglobin binding Inhibition of cathepsin B, H, L
FIB	Coagulation

Since these biological activities of various AP proteins may play a significant role in the host's defence reactions against infectious agents, it is relevant to investigate in experimental in vivo models the relationship between AP response as induced by non-biodegradable, non-diffusible and non-antigenic substances and the effect on host resistance to bacterial or parasitic infections. An inflammatory reaction was induced by subcutaneous injection of calcium phosphate gel containing magnesium silicate. Control animals were injected with pyrogen-free saline. Three to 17 days after the beginning of the systemic inflammation, animals were challenged with *Schistosoma mansoni* or 5 LD_{50} of *Listeria monocytogenes* or virulent *Salmonella typhimurium* or *Candida albicans*. It was found that mice had a higher resistance to infectious agents than control animals without acute inflammation (Fauve 1987). Furthermore, the efficiency of acute inflammation in improving host defences was found to be better than that of compounds well-known as immunostimulants, e.g. lipopolysaccharide, *Corynebacterium parvum* and BCG.

These results strongly suggest that AP response is associated with an improvement in host defences against microbial infections. In fact, these experimental data seem to disagree with clinical observations, since in most cases after major surgical trauma, or thermal injury in man, high susceptibility to infection is described (Laurent et al. 1986). We therefore have to ask ourselves at least two questions:

1. Are AP proteins reliable and useful markers of host defence efficiency against infection? In other words, could we use AP protein measurement in blood to detect patients characterized by a high risk of infection before surgery? The response to this first question is no. The pre-operative blood values of AP protein provide no information to detect patients at risk of post-operative infections. Of

all the biological variables measured before surgery, the state of anergy to intradermally applied recall antigens (standardized multiple puncture device to test delayed-type hypersensitivity) proved the most reliable predictor of post-operative infection (Kress et al. 1987).

2. Do some typical patterns exist for AP protein response during sepsis which can help clinicians to make a prognosis? The response in this case is yes since the highest values of CRP are detected in the group of patients who present a bad outcome associated with the paradoxical lowest values of AGP as described above. In addition, patients who demonstrate sustained depression of C4 level, below 0.15 g l^{-1} associated with CRP value up to 0.10 g l^{-1} with no tendency to return to baseline level are at a high risk of fatal septicaemia (Laurent et al. 1986).

References

Bata J, Cordier G, Revillard JP, Bonneau M, Latour M (1980) Transplantion and clinical immunology. Excerpta Medica, Amsterdam, pp 59–70

Benson MD, Dwulet FE, Kluve-Beckerman B, Aldo-Benson M (1986) Structure and function of SAA. In: Bienvenu J, Grimaud JA, Laurent PE (eds) Marker proteins in inflammation, vol 3. Walter de Gruyter, Berlin, pp 129–137

Chiu KM, Mortensen RF, Osmand AP, Gewurz H (1977) Interactions of alpha-1 acid glycoprotein with the immune system. Immunology 32:997–1005

Clarke D, Cost K (1983) Use of serum C reactive protein in differentiating septic from aseptic meningitis in children. J Pediatr 102:718–720

Dinarello CA (1986) The biological effects of purified and recombinant human interleukin-1. In: Bienvenu J, Grimaud JA, Laurent PE (eds) Marker proteins in inflammation, vol 3. Walter de Gruyter, Berlin, pp 17–26

Fauve RM (1987) Immunostimulants of the future. In: Azuma I, Jollès G (eds) Immunostimulants: now and tomorrow. Japan Sci Soc Press, Tokyo/Springer-Verlag, Berlin, pp 225–234

Fisher CL, Gill C, Forrester MG, Nakamura R (1976) Quantitation of "Acute phase proteins" postoperatively. Amer J Clin Pathol 66:840–846

Gewurz H, Mold C, Siegel J, Fiedel B (1982) C-reactive protein and the acute phase response. Year Book Medical Publishers, Chicago, pp 345–372

Kampschmidt RF, Pulliam LA, Merriman CR (1978) Further similarities of endogenous pyrogen and leukocytic endogenous mediator. Amer J Physiol 224:C118

Kaplan J (1981) Polypeptide-binding membrane receptors: analysis and classification. Science 212:14

Kaplan MH, Volanakis JE (1974) Interaction of C reactive protein complexes with the complement system. I: Consumption of human complement associated with reaction of C reactive phosphatides, lecithin and sphingomyelin. J Immunol 112:2135–2147

Koj A (1974) In: Allison AC (ed) Structure and function of plasma proteins. Plenum Press, London, pp 72–131

Koj A (1983) Metabolic studies of acute phase proteins. In: Mariani G (ed) Pathophysiology of plasma protein metabolism. Macmillan, London, pp 221–248

Kress HG, Gehrsitz P, Elert O (1987) Predictive value of skin testing, neutrophil migration and C-reactive protein for post operative infections in cardiopulmonary by-pass patients. Acta Anaesthesiol Scand 31:397–404

Kushner I (1982) The phenomenon of the acute phase response. Ann NY Acad Sci 389:39–48

Laurent PE, Marichy J (1982) In: Bienvenu A, Laurent P, Suskind (eds) Marker protein in inflammation. Walter de Gruyter, Berlin and New York, pp 497–502

Laurent PE, Marichy J, Buffet G, Gros P, Beal B (1986) Burn patients as typical model of acute phase response. I. In: Peeters (ed) Protides of the biological fluids, vol 34. Pergamon Press, Oxford, pp 561–565

Laurent PE, Gros P, Beal B, Gaspard A, Marichy J (1986) Burn patients as typical model of acute phase response. II: Acute phase proteins and their related immunomodulating activities. In: Peeters H (ed) Protides of the biological fluids, vol 34. Pergamon Press, Oxford, pp 567–572

Manroe BL, Rosenfeld CR, Weinberg AG, Browne R (1977) The differential leukocyte count in the assessment and outcome of early-onset neonatal group B streptococcal disease. J Pediatr 91: 632–637
Marhaug G, Stenson MØ, Husby G, Kolmannskog S, Flaegstad T, Stokland T, Husebekk A (1986) Clinical acute phase pattern of serum amyloid A. In: Peeters H (ed) Protides of the biological fluids, vol 34. Pergamon Press, Oxford, pp 375–378
McCarthy PL, Frank AL, Ablow RC, Masters SJ, Dolan SJ (1978) Value of C reactive protein test in the differentiation of bacterial and viral pneumonia. J Pediatr 92:454–456
Meliconi R, Parracino O, Schiattone ML, Chiodo F, Facchini A, Miglio F, Gasbarrini G (1986) Acute phase proteins (APP) in acute and chronic liver disease. In: Peeters H (ed) Protides of the biological fluids, vol 34. Pergamon Press, Oxford, pp 589–592
Mold C, Nakayama S, Holzer TJ, Gewurz H, du Clos TW (1981) C-reactive protein is protective against *Streptococcus pneumoniae* infection in mice. J Exp Med 154:1703–1706
Nakayama S, Gewurz H, Holzer T, du Clos TW, Mold C (1983) The role of the spleen in the protective effect of C-reactive protein in *Streptococcus pneumoniae* infection. Clin Exp Immunol 54:319–326
Owen JA (1967) Effect of injury on plasma proteins. Adv Clin Chem 2:1–41
Peltola H (1982) C-reactive protein for rapid monitoring of infections of the central nervous system. Lancet i:980–983
Philip AGS, Hewitt JR (1980) Early diagnosis of neonatal sepsis. Pediatrics 65:1036–1041
Powenda MC, Moyer ED (1981) In: Powenda MC, Canonico (eds) Infection: the physiologic and metabolic responses of the host. Elsevier/North-Holland, Amsterdam, pp 272–295
Powenda MC, Cockerell GL, Pekarek RS (1973) Amino acid and zinc movement in relation to protein synthesis early in inflammation. Amer J Physiol 225:399–405
Sabel KG, Wadsworth C (1979) C-reactive protein (CRP) in early diagnosis of neonatal septicemia. Acta Paediatr Scand 68:825–831
Sann L, Bienvenu F, Bienvenu J, Bourgeois J, Bethnod M (1984) Evolution of serum prealbumin, C reactive protein and orosomucoid in neonates with bacterial infection. J Pediatr 105: 977–981
Shabino ChL (1983) C-reactive protein and bacterial tracheitis. J Pediatr 103:1010
Speer CH, Bruns A, Gahr M (1983) Sequential determination of CRP, alpha 1 antitrypsin and haptoglobin in neonatal septicemia. Acta Paediatr Scand 72:679–683
Wager O, Jannson E (1957) C-reactive protein in serous meningitis and paralytic poliomyelitis. Ann Med Exp Fenn 35:352–356
Wannemacher RW, Pekarek RS, Beisel WR (1972) Mediator of hepatic amino acid flux in infected rats. Proc Soc Exp Biol Med 139:128–132
Wannemacher RW, Pekarek RS, Thompson WL, Curnow RT, Beall FA, Zenser TV, De Rubertis FR, Beisel WR (1975) A protein from polymorphonuclear leukocytes (LEM) which affect the rate of hepatic amino-acid transport and synthesis of acute-phase globulins. Endocrinology 96:651–661
Whicher JT (1984) Functions of acute phase proteins in the inflammatory response. In Arnaud P, Bienvenu J, Laurent PE (eds) Marker proteins in inflammation, vol 2. Walter de Gruyter, Berlin New York, pp 89–98
White R, Habicht GS, Godfrey HP, Janoff A, Barton E, Fox C (1981) Secretion of elastase and alpha 1 macroglobulin by cultured murine peritoneal macrophage: studies on their interaction. J Lab Clin Med 97:718–724

Chapter 13

C-Reactive Protein: Clinical Aspects

M. H. van Rijswijk and M. A. van Leeuwen

Introduction

The acute phase reaction is a rather complex generalized reaction to localized tissue damage. An inflammatory reaction will result in a number of local events comprising vasodilatation, increase of vascular permeability and oedema, platelet-aggregation, production of fibrin, leukocyte accumulation and the liberation of lysosomal enzymes, as well as a number of low-molecular weight substances such as histamine, kinins and oxygen radicals. Such a local reaction culminates in tissue damage and initiates the systemic "acute phase" response, which seems to be particularly designed to protect the organism against the potentially destructive action of inflammatory products. In other words, it is an attempt to confine the damage to the local area as much as possible. The systemic acute phase reaction comprises a number of fundamental metabolic changes. A key role in the balance between a local inflammatory reaction and the acute phase response is played by interleukin-1 (IL-1), acting as the main mediator, both for the local sequelae of inflammation and for its systemic effects (Dinarello 1984). IL-1 is synthesized by mononuclear phagocytes at the site of inflammation. Many of the effects of IL-1 have already been described under different names, indicating different biological effects (Table 13.1). At present, it is generally accepted that these different effects are all induced by a number of closely related polypeptides, in particular by IL-1. Recently it has been proposed that a number of IL-1 effects including the generation of the acute phase response, are mediated by IL-6 (hybridoma growth factor).

The Acute Phase Proteins

The acute phase proteins are synthesized by the liver during the acute phase reaction. An increase in the synthesis of acute phase proteins coincides with a

Table 13.1. Local and systemic effects of IL-1

A. Systemic effects
Induction of the synthesis of acute phase proteins (hepatocyte stimulating factor)
Induction of fever (via PGE_2) (leukocyte endogenous mediator/endogenous pyrogen)
Recruitment of PMNs from bone marrow
Increase in PMN adhesiveness
Induction of muscle catabolism (via PGE_2) (cachectin/tumor necrosis factor)
B. Local effects
Induction of cartilage degradation (catabolin)
Induction of fibroblast proliferation (platelet derived growth factor)
Activation of lymphocytes (lymphocyte activating factor)
Induction of bone resorption by osteoclasts (osteoclast activating factor)
Production of PGE_2, procollagenase, plasminogen activator (mononuclear cell factor)

The different names under which interleukin-1 effects have been described earlier are given in brackets.

decrease in the synthesis of other plasma proteins, e.g. albumin. The acute phase reaction may be considered as a component of the non-specific general reaction of the organism to a (localized) inflammatory process. Such an inflammatory reaction may be initiated by stimuli of different origin: chemical or physical agents, infections, allergic or immunological reactions, ischaemic necrosis, malignant tumours. Regarding the capacity of a number of acute phase reactants to eliminate potentially hazardous substances, the acute phase reaction apparently has an important function in the protection of the integrity of the organism against the deleterious effects of inflammatory products. Ceruloplasmin, for example, is a scavenger of oxygen free radicals, whereas α1-antitrypsin (AT) is an inhibitor of proteolytic enzymes.

The increase in the synthesis of acute phase proteins by hepatocytes is preceded by an increase of mRNA for these proteins induced by IL-1 (Dinarello 1984). The recently-acquired knowledge concerning the different functions of IL-1 supports clinical experience which suggest that the acute phase reaction is a good indicator of the activity and the extent of an inflammatory process.

The Erythrocyte Sedimentation Rate as an Indirect Indicator for the Acute Phase Reaction

The erythrocyte sedimentation rate (ESR) has been used as an indicator of inflammatory activity for many years. The ESR is mainly dependent on the plasma fibrinogen level and may therefore be considered to be an (indirect) indicator of the extent of the acute phase reaction. Other factors which may influence the ESR are immunoglobulin concentrations, dimension, shape and number of the erythrocytes, and food intake (Talstad and Haugen 1979; Mallya et al. 1982a). The slow dynamics of fibrinogen synthesis, its long half-life and the influence of the blood coagulation process on fibrinogen levels (Koj 1974, pp. 73–125) are also factors which must be considered when using the ESR as an indicator of inflammatory activity.

C-reactive Protein as a Direct Indicator of the Acute Phase Reaction

In the 1950s measurement of the level of C-reactive protein (CRP) by the capillary precipitation test or by the latex agglutination method was introduced as an alternative to the ESR. A real breakthrough, however, was hampered by the fact that the methods mentioned above are semi-quantitative and insensitive.

The introduction of quantitative and sensitive assays has evoked a renewed interest for the measurement of CRP. Arguments in favour of measurement of CRP, rather than the ESR are: (i) the CRP level is a direct indicator for the acute phase reaction in contrast to the ESR, which is an indirect indicator; (ii) measurement of CRP is not influenced by the shape or number of the erythrocytes, immunoglobulin levels or food intake; (iii) CRP has a much faster kinetic profile and a larger dynamic range than fibrinogen: under normal conditions CRP is only present in very low concentrations (less than 3 $\mu g\ ml^{-1}$); 4–6 hours after the beginning of an inflammatory reaction, the CRP level starts to rise, reaching peak values of 100–1000 times the starting value within 24–48 hours; its half time is about 12 hours (Pepys 1981a; Kushner et al. 1978; Chelladurai et al. 1982); (iv) CRP may be accurately quantified in the higher range whereas the ESR has a maximum of 150 mm.

Assays for CRP

The most commonly used methods of CRP measurement are the latex agglutination assay (Deyo et al. 1980), radial immunodiffusion (Mancini et al. 1965), electroimmunodiffusion according to Laurell (Pepys et al. 1978), rate-nephelometry (Gill et al. 1981) and the enzyme-immunoassay (Highton and Hessian 1984). The latex assay is obsolete, but the latter four assays are to a large extent comparable with regard to the quality of results (van Leeuwen et al. 1986). The differences between these assays are shown in Table 13.2.

Table 13.2. Comparison of the different CRP assays

	RID	EID	ICS	ELISA
Lower level of detection ($\mu g\ ml^{-1}$)	3	1	6	0.5
Duration of the assay (hours)	36	24	1	4
Coefficient of variation (%)				
Intraassay	4		2	5
Interassay	5		3	5

RID: radial immunodiffusion; EID: electroimmunodiffusion; ICS: rate nephelometer; ELISA: enzyme immunoassay.

CRP Levels in Different Diseases

In a number of diseases the course of the CRP levels in relation to the disease process has been investigated. Both in cross-sectional and in longitudinal studies, there appears to exist a good correlation between the level of CRP on the one hand, and the activity and the extent of the disease on the other hand. Exceptions are diseases such as SLE (Pepys et al. 1982, Van Leeuwen et al. 1986), scleroderma (Whicher et al. 1982), Sjögren (Moutsopoulos et al. 1983), dermatomyositis (Haas et al. 1982), and ulcerative colitis (Fagan et al. 1982), in which generally discrepantly low CRP levels are found. Such a discrepancy does not reside in a deficient synthesis of acute phase proteins, as patients with these types of diseases are perfectly able to mount an adequate acute phase response in the case of another inflammatory stimulus, such as bacterial infection. Apparently the stimulus for an acute phase response is decreased or even absent because of different mechanisms of inflammation in these diseases.

Infections

In general, bacterial infections cause a massive increase in CRP level, with a good correlation between the severity of the infection and the CRP level. After the institution of adequate antibiotic treatment, CRP levels will start to drop within 24 hours. Viral and fungal infections are usually accompanied by a less pronounced rise in CRP level (Schofield et al. 1982). Measurement of CRP levels is particularly useful for the (differential) diagnosis and treatment of infections of the central nervous system (Peltola 1982), septicaemia in neonates (Sabel and Wadsworth 1979; Alt et al. 1984), infections during pregnancy (Evans et al. 1980), as well as infectious complications following bone marrow transplantation (Walker et al. 1984), leukaemia (Schofield et al. 1982) and SLE (Pepys et al. 1982).

Myocardial Infarction

In patients with myocardial infarction, the CRP level will start to rise within a few hours. The CRP peak value correlates with the enzymatically determined infarction size, estimated from the area of CPK and beta-hydroxy-butyric acid (Smith et al. 1977). Persistent elevation or a second rise of the CRP level is indicative of a complication (de Beer et al. 1982a).

Surgery

Within 6 hours of uncomplicated surgical intervention, an increase in CRP level will occur, compared to pre-operative values. Persistent elevation or a second increase of the CRP level is frequently associated with a post-operative complication such as an infection or thromboembolic process. In a number of cases a deviation of the normal post-operative CRP response may occur before clinical signs of a complication become manifest (Fisher et al. 1976; Schentag 1984).

Renal Transplantation

In a number of studies it has been demonstrated that the course of the CRP level may indicate allograft rejection. Of course an increase in CRP level will also occur in the case of an infectious complication, and in the first three days after transplantation, CRP levels may be increased due to the surgical procedure. There is no relation between serum urea or creatinine concentration and CRP level. Neither is the serum CRP level influenced by haemodialysis (Freed et al. 1984; White et al. 1981; Groth et al. 1984).

Malignant Tumours

Many malignancies will be accompanied by increased CRP levels, in particular primary carcinomas of the lung, the bladder and the kidney, in which highly elevated levels of CRP are often found (Drahovsky et al. 1981; Weinstein et al. 1984). Elevated CRP levels are also encountered in Hodgkin's disease and in non-Hodgkin lymphomas, depending on the stage and the activity of the disease (Pepys et al. 1978; Child et al. 1980).

Rheumatic Diseases

In most of the inflammatory rheumatic diseases, such as rheumatoid arthritis (Amos et al. 1977; van Rijswijk et al. 1979; Mallya et al. 1982b), juvenile chronic polyarthritis (Pepys 1981b), ankylosing spondylitis (Cowling et al. 1980), reactive arthritis, psoriatic arthritis (Laurent et al. 1981), polymyalgia rheumatica (Eshagian and Goeken 1980), Behcet's disease (Lehner and Adinolfi 1980), Wegener's disease (Hind et al. 1984a) and polyarteritis nodosa (Hind et al. 1984b), the CRP level is a sensitive and objective indicator of disease activity.

Functions of CRP

From in vitro experiments it has become clear that the CRP molecule, on the one hand, has Ca-dependent binding sites for phosphorylcholine (Gotschlich et al. 1982) as well as for a number of polyanions like nucleic acids (Robey et al. 1984), whereas on the other hand CRP may bind in a Ca-independent way to polycations like histones (DiCamelli et al. 1980; Siegel et al. 1975). Phosphorylcholine is part of the C-polysaccharides in the cell wall of many bacteria, fungi and parasites, and it is a component of the phospholipids in the human cell membranes. Although it has been shown that CRP may bind to intact bacteria, the binding of CRP to phospholipids in cell membranes and to intracellular polycations and polyanions will occur mainly in the case of cell damage (Narkates and Volanakis 1982). After binding of CRP to one of the above mentioned substances has taken place, the complex may evoke a number of secondary effects such as precipitation and agglutination, complement activation by the classical pathway (Siegel et al. 1975; Claus et al. 1977), opsonization (Nakayama et al. 1982), thrombocyte aggregation and activation (Fiedel et al. 1982). In this respect CRP displays similarities with immunoglobulins. In vitro studies have shown that CRP complex may bind to lymphocytes bearing Fc-receptors for IgG (James et al. 1983), and also that aggregated CRP may bind to low-density-lipoproteins (LDL) as well as to very-low-density lipoproteins (VLDL) (de Beer et al. 1982b). The implications of these findings for the in vivo role of CRP are still unclear. A major function of CRP could be to bind products of pathogenic micro-organisms or components of damaged autologous cells in order to inactivate these substances and/or to promote their clearance from the circulation.

Conclusion

The acute phase reaction is, in most circumstances, a good indicator of (local) inflammatory activity and/or tissue damage. CRP is a direct and quantitative measure for the acute phase reaction and, because of its fast kinetics, provides adequate information on the actual situation. The ESR, on the other hand, is an indirect measure of the acute phase reaction, shows a much slower reaction to changes of inflammatory activity and is influenced by a number of other factors.

From studies on the "behaviour" of CRP it has become clear that diseases may differ with regard to the extent to which they induce an acute phase response. As an incidental measurement, the CRP level may add to the diagnostic procedure in selected cases, e.g. in the differentiation between a bacterial and a viral infection or between a bacterial infection and an exacerbation of diseases like SLE or leukaemia. In the case of an extremely elevated CRP level ($>100\ \mu g\ ml^{-1}$) the possibility of a bacterial infection should always be considered. In clinical practice CRP is particularly useful when serial measurements are performed. The course of the CRP level may be useful for monitoring the effect of treatment and for the early detection of post-operative complications or intercurrent infections. It will be apparent that measurement of CRP is mainly a tool in specialist practice.

The relationship between CRP and the local production and effects of IL-1 on the one hand and the possible functional role of CRP in the inflammatory process on the other hand, adds a dimension to the clinical use of CRP as a parameter of inflammatory activity.

References

Alt R, Erny P, Messer J, Willard D (1984) Infections bacteriennes neonatales. Etude cinétique de la C-reactive protein et de l'orosomucoide. Presse Med 13:1373–1376

Amos RS, Constable TJ, Crockson RA, Crockson AP, McConkey B (1977) Rheumatoid arthritis: relation of serum C-reactive protein and erythrocyte sedimentation rate to radiographic changes. Br Med J i:195–197

Chelladurai M, Macintyre SS, Kushner I (1982) Estimation of in vivo rates of C-reactive protein synthesis from turnover studies in rabbits. Ann NY Acad Sci 389:437–438

Child JA, Spati B, Illingworth S et al. (1980) Serum beta 2 microglobulin and C-reactive protein in the monitoring of lymphomas. Cancer 45:318–326

Claus DR, Siegel J, Petras K, Osmand AP, Gewurz H (1977) Interactions of C-reactive protein with the first component of human complement. J Immunol 119:187–192

Cowling P, Ebringer R, Cadwell D, Ishii M, Ebringer A (1980) C-reactive protein, ESR, and klebsiella in ankylosing spondylitis. Ann Rheum Dis 39:45–49

de Beer FC, Hind CRK, Fox KN, Allan RM, Maseri A, Pepys MB (1982a) Measurement of serum C-reactive protein concentration in myocardial ischaemia and infarction. Br Heart J 47:239–243

de Beer FC, Soutar AK, Baltz ML, Trayner IM, Feinstein A, Pepys MB (1982b) Low density lipoprotein and very low density lipoprotein are selectively bound by aggregated C-reactive protein. J Exp Med 156:230–242
Deyo RA, Pope RM, Persellin RH (1980) Interference by rheumatoid factor with the detection of C-reactive protein by the latex agglutination method. J Rheumatol 7:279–287
DiCamelli R, Potempa LA, Siegel J, Suyehira L, Petras K, Gewurz H (1980) Binding reactivity of C-reactive protein for polycations. J Immunol 125:1933–1938
Dinarello CA (1984) Interleukin-1 and the pathogenesis of the acute phase response. New Engl J Med 311:1413–1418
Drahovsky D, Dunzendorfer U, Ziegenhagen G, Drahovsky M, Kellen JA (1981) Re-evaluation of C-reactive protein in cancer sera by radioimmunoassay and radial immuodiffusion. Oncology 38:286–291
Eshagian J, Goeken JA (1980) C-reactive protein in giant cell (cranial, temporal) arteritis. Ophthalmology 87:1160–1166
Evans MI, Hajj SN, Devoe LD, Angerman NS, Moward AM (1980) C-reactive protein as a predictor of infectious morbidity with premature rupture of membranes. Am J Obstet Gynecol 138:648–652
Fagan EA, Dyck RF, Maton PN et al (1982) Serum levels of C-reactive protein in Crohn's disease and ulcerative colitis. Eur J Clin Invest 12:351–359
Fiedel BA, Simpson RM, Gewurz H (1982) Activation of platelets by modified C-reactive protein. Immunology 45:439–447
Fisher CL, Gill C, Forrester MG, Nakamura R (1976) Quantitation of 'Acute-phase proteins' postoperatively. Amer J Clin Pathol 66:840–846
Freed B, Walsh A, Pietrocola D, MacDowell R, Laffin R, Lempert N (1984) Early detection of renal allograft rejection by serial monitoring of serum C-reactive protein. Transplantation 37:215–218
Gill CW, Bush WS, Burleigh WM, Fisher CL (1981) an evaluation of a C-reactive protein assay using a rate immunonepholo-metric procedure. Amer J Clin Pathol 75:50–55
Gotschlich EC, Liu TY, Oliveira E (1982) Binding of C-reactive protein to C-carbohydrate and PC-substituted protein. Ann NY Acad Sci 389:163–169
Groth J, Kaden J, Hoffman P, Muller K, Mrochen H, Mebel M (1984) C-reactive protein and clinical kidney transplantation. Nephron 37:236–239
Haas RH, Dyck RF, Dubowitz V, Pepys MB (1982) C-reactive protein in childhood dermatomyositis. Ann Rheum Dis 41:483–485
Highton J, Hessian P (1984) A solid-phase enzyme immunoassay for C-reactive protein: clinical value and the role of rheumatoid factor. J Immunol Methods 68:185–192
Hind CRK, Winearls CG, Lockwood CM, Rees AJ, Pepys MB (1984a) Objective monitoring of activity in Wegener's granulomatosis by measurement of serum C-reactive protein concentration. Clin Nephrol 21:341–345
Hind CRK, Savage CO, Winearls CG, Pepys MB (1984b) Objective monitoring of disease activity in polyarteritis by measurement of serum C-reactive protein concentration. Br Med J 288:1027–1030
James K, Baum L, Adamowski C, Gewurz H (1983) C-reactive protein antigenicity on the surface of human lymphocytes. J Immunol 131:2930–2934
Koj A (1974) Acute-phase reactants. In: Allison AJ (ed) Structure and function of plasma proteins, vol I. Plenum Press, London, New York, pp 73–125
Kushner I, Broder ML, Karp D (1978) Control of the acute phase response. Serum C-reactive protein kinetics after acute myocardial infarction. J Clin Invest 61:235–242
Laurent MR, Panayi GS, Shephard P (1981) Circulating immune complexes, serum immunoglobulins, and acute phase proteins in psoriasis and psoriatic arthritis. Ann Rehum Dis 40:66–69
Lehner T, Adinolfi M (1980) Acute phase proteins, C9, factor B, and lysozyme in recurrent oral ulceration and Behcet's syndrome. J Clin Pathol 33:269–275
Mallya RK, Berry H, Mace BEW, De Beer FC, Pepys MB (1982a) Diurnal variation of erythrocyte sedimentation rate related to feeding. Lancet i:389–390
Mallya RK, de Beer FC, Berry H, Hamilton EDB, Mace BEW, Pepys MB (1982b) Correlation of clinical parameters of disease activity in rheumatoid arthritis with serum concentration of C-reactive protein and erythrocyte sedimentation rate. J Rheumatol 9:224–228
Mancini G, Carbonara AD, Heremans JF (1965) Immunochemical quantitation of antigens by single radial immunodiffusion. Immunochemistry 2:235–254
Moutsopoulos HM, Elkon KB, Mavridis AK, Acritidis NC, Hughes GRV, Pepys MB (1983) Serum C-reactive protein in primary Sjogren syndrome. Clin Exp Rheumatol 1:57–58
Nakayama S, Mold C, Gewurz H, Du Clos TW (1982) Opsonic properties of C-reactive protein in vivo. J Immunol 128:2435–2438

Narkates AJ, Volanakis JE (1982) C-reactive protein binding specificities: artificial and natural phospholipid bilayers. Ann NY Acad Sci 389:172–182

Peltola HO (1982) C-reactive protein for rapid monitoring of infections of the central nervous system. Lancet i:980–982

Pepys MB (1981a) C-reactive protein fifty years on. Lancet i:653–657

Pepys MB (1981b) Serum C-reactive protein, serum amyloid P-component and serum amyloid A protein in autoimmune disease. Clin Imm Allergy 1(I):77–101

Pepys MB, Dash AC, Markham RE, Thomas HC, Williams BD, Petrie A (1978) Comparative clinical study of SAP (amyloid P component) and C-reactive protein in serum. Clin Exp Immunol 32:119–124

Pepys MB, Lanham JG, de Beer FC (1982) C-reactive protein in systemic lupus erythematosus. Clin Rheum Dis 8(1):91–103

Robey FA, Jones KD, Tanaka T, Liu T-Y (1984) Binding of C-reactive protein to chromatin and nucleosome core particles. J Biol Chem 259:7311–7316

Sabel KG, Wadsworth C (1979) C-reactive protein (CRP) in early diagnosis of neonatal septicemia. Acta Paediatr Scand 68:825–831

Schentag JJ, O'Keeffe D, Marmion M, Wels PB (1984) C-reactive protein as an indicator of infection relapse in patients with abdominal sepsis. Arch Surg 119:300–304

Schofield KP, Voulgari F, Gozzard DI, Leyland MJ, Beeching NJ, Stuart J (1982) C-reactive protein concentration as a guide to antibiotic therapy in acute leukaemia. J Clin Pathol 35:866–869

Siegel J, Osmand AP, Wilson MF, Gewurz H (1975) Interactions of C-reactive protein with the complement system. II. C-reactive protein-mediated consumption of complement by poly-L-lysine polymers and other polycations. J Exp Med 142:709–721

Smith SJ, Bos G, Esseveld MR, Van Eyk HG, Gerbrandy J (1977) Acute-phase proteins from the liver and enzymes from myocardial infarction; a quantitative relationship. Clin Chim Acta 81:75–85

Talstad J, Haugen HF (1979) The relationship between the erythrocyte sedimentation rate (ESR) and plasma proteins in clinical materials and models. Scand J Clin Lab Invest 39:519–524

van Leeuwen MA, van Rijswijk MH, Marrink J, Westra J, De Jong HJ (1986) CRP measurements in rheumatic disorders. In: Peeters H (ed) Protides of the biological fluids, vol 34. Pergamon Press, Oxford, pp 315–318

van Rijswijk MH, Ruinen L, Marrink J, De Blecourt JJ (1979) Some aspects of SAA and CRP measurements in patients with rheumatoid arthritis. In: Feltkamp TEW, Van der Korst JK (eds) Disease evaluation and patient assessment in rheumatoid arthritis. Staphleu's Scientific Publishing Company, Alphen a/d Rijn, Brussels, pp 145–150

Walker SA, Rogers TR, Riches PG, White S, Hobbs JR (1984) Value of serum C-reactive protein measurement in the management of bone marrow transplant recipients. J Clin Pathol 37:1018–1026

Weinstein PS, Skinner M, Sipe JD, Lokich JJ, Zamcheck N, Cohen AS (1984) Acute-phase proteins or tumour markers: The role of SAA, SAP, CRP and CEA as indicators of metastasis in a broad spectrum of neoplastic diseases. Scand J Immunol 19:193–198

Whicher J, Bell A, Unwin J, Martin M, Dieppe P (1982) The prostaglandin induced acute-phate response is defective in endotoxin tolerant mice and in humans with scleroderma. Ann NY Acad Sci 389:474–475

White J, Meyer E, Hardy MA (1981) Prediction of onset and termination of renal allograft rejection by serum levels of C-reactive protein. Transplant Proc 13:682–685

Chapter 14

Pathogenesis of AA Amyloidosis

G. Husby

Definition and Classification of Amyloidosis

Amyloidosis is a heterogeneous group of disorders caused by extracellular deposition of a characteristic proteinaceous material, the amyloid substance, in various organs and tissues (Glenner 1980; Husby and Sletten 1986a, b). The pathophysiological consequences of amyloid deposition depend on its tissue and organ distribution, and the resulting symptoms and signs are partly caused by functional disturbances due to displacement of normal tissue structures, and partly by organ enlargement (Husby 1983). Benditt and Eriksen (1964) suggested for the first time that amyloidosis was heterogeneous also with respect to the nature of the amyloid deposits. It was, however, difficult to isolate the fibrils from affected tissues because of their very low solubility in physiological conditions. This property of amyloid was thought to be responsible for the irreversible and often progressive course of the disease, which in many cases leads to death within months or years of diagnosis (Husby 1983, 1985). Important progress was made when Pras et al. (1968) published their water extraction method for the purification of amyloid fibrils. They confirmed suggestions made more than a century earlier, that amyloid was principally of protein nature, although Virchow in 1854 had created the term amyloid (starch-like) because the deposits stained with iodine in a manner similar to that of starch or cellulose. The water extraction method permitted chemical and immunological analyses of isolated amyloid fibrils that had been solubilized by treatment with agents such as sodium hydroxide, guanidine or urea (Husby and Sletten 1986a). This opened up a new area for the understanding of the structural properties of amyloid.

Gel filtration of solubilized amyloid fibrils obtained from different clinical and experimental types of amyloidosis has in many instances revealed an elution pattern consisting of two main protein peaks (Husby 1980). One of these represents a material with very high molecular weight, the so-called void volume material. A similar material is also present in corresponding preparations of normal tissues and appears to contain a variety of macromolecules, among them fibronectin and a protein related to reticulin (Husby and Sletten 1977; Scott et al. 1983). Its role in amyloid formation is not established, and recent studies (van Andel et al. 1986) suggested that at least part of it is not integrated in the fibrils but rather co-isolated with them. Further studies, however, are needed to confirm this.

The second peak represents a distinct protein subunit with much lower molecular weight ranging from about 4000 to 30 000 Da in different amyloid preparations (Glenner 1980; Husby and Sletten, 1986a). It is the recognition and characterization of this low molecular weight amyloid protein subunit that has enabled a classification of amyloidosis by the nature of the respective fibril protein (Benditt and Eriksen 1971; Glenner 1980; Husby 1980; Husby and Sletten 1986a). It has thus been shown that several apparently non-related proteins can constitute this subunit in different cases of amyloidosis. However, a common feature of these proteins appears to be β-pleated sheets, a structure which is present in all amyloid fibrils so far studied (Glenner 1980). A steadily increasing number of such fibril proteins are characterized by their amino acid sequence (Table 14.1), and in some cases complementary DNA probes have also been established. The different amyloid proteins are often related to specific clinical forms of amyloidosis. Indeed, many types of amyloid disease can now be defined by structural analysis of the fibril proteins (Table 14.1) and/or the genes coding for them (Husby and Sletten 1986b).

Certain serum proteins are precursors for the different fibril proteins in the various systemic forms of amyloidosis (Glenner 1980; Husby and Sletten 1986b). Two important types of fibril protein related to systemic amyloidosis are the

Table 14.1. Human systemic amyloid fibril proteins verified by amino acid sequencing[a]

Clinical type	Amyloid protein	Related protein
Idiopathic (primary) amyloidosis		
Systemic	AL (Aκ, Aλ)	IgL, (Vκ, Vλ)
Localized	AL (Aκ, Aλ)	IgL, (Vκ, Vλ)
Myeloma-associated amyloidosis	AL (Aκ, Aλ)	IgL, (Vκ, Vλ)
Reactive (secondary) amyloidosis	AA	SAA
Heredofamilial amyloidosis		
Recessive, autosomal		
Familial Mediterranean fever (FMF)	AA or AA variant	SAA
Dominant, autosomal		
Familial amyloid nephropathy with urticaria and deafness (Muckle-Wells syndrome)	AA[b]	SAA
Familial amyloid polyneuropathy (FAP)	Prealbumin variants	Prealbumin
Familial amyloid cardiomyopathy (FAC)	Prealbumin variant	Prealbumin
Senile amyloidosis	Prealbumin (variant)	Prealbumin
Amyloidosis associated with chronic haemodialysis	β2-microglobulin	β2-microglobulin

[a] The nomenclature and abbreviations are according to those proposed by Benditt et al. (1980) with small modifications.
[b] Only residues 2–30 are sequenced.

amyloid L (AL) and amyloid A (AA) proteins (Table 14.1). Protein AL, which consists of homogeneous (monoclonal) immunoglobulin kappa or lambda light chains or the aminoterminal fragments thereof is seen in idiopathic and myeloma-associated amyloidosis (AL amyloidosis) (Glenner 1980; Husby 1983). Protein AA is associated with secondary amyloidosis, familial Mediterranean fever (FMF), familial nephropathic amyloidosis with febrile urticaria – the Muckle-Wells syndrome (Muckle 1979; Linke et al. 1983) and spontaneous and experimental amyloidosis in animals (Glenner 1980; Husby and Sletten 1986b). These forms of amyloidosis could therefore also be called AA amyloidosis. The pathogenetic aspects of these forms will be focused on in the following. In addition, genetic variants of prealbumin are associated with the autosomal dominant familial amyloid polyneuropathies (FAP) and familial amyloid cardiomyopathies (FAC) (Husby et al. 1985; Benson 1988; Nordlie et al. 1988), and also with systemic senile amyloidosis (Husby and Sletten 1986a; Gorevic et al. 1989). Normal β2-microglobulin makes up the fibrils in amyloidosis associated with long term haemodialysis (Gejyo et al. 1985).

However, there are clear exceptions to these clinical–chemical correlations. For example, we have characterized by complete amino acid sequence an AA protein from a patient with monoclonal gammopathy (Møyner et al. 1980), and others have detected AA amyloid in a patient with "primary" systemic amyloidosis without any underlying inflammatory or malignant disorder (Pras et al. 1980). Such cases clearly illustrate the advantage of classifying amyloidosis by the fibril protein.

Serum and Tissue Amyloid A Proteins

The unique protein AA was first described by Benditt and co-workers (1971). The precursor for tissue AA is serum amyloid A (SAA) as evidenced by experiments performed in our laboratory (Husebekk et al. 1985). We introduced human HDL–SAA complexes from acute phase sera to mice during the induction of amyloidosis with endotoxin. Control animals received endotoxin only. Human AA detected with specific antisera (not cross-reacting with mouse AA) was present in amyloid fibrils isolated from the animals who had received human SAA, while only mouse AA was found in the fibrils obtained from the control animals. The mouse thus converted human SAA to AA and incorporated it in its own amyloid fibrils, and the precursor–product relationship between SAA and AA was thereby established. We have reproduced these experiments using different, human SAA-rich HDL preparations to ensure the significance of our findings (Husebekk et al. 1987), which have been questioned by other workers (Baltz et al. 1986a). Furthermore, recent experiments by Tape and co-workers (1988) confirm the precursor–product relationship between SAA and tissue AA.

Structure of SAA and AA

Human SAA consists of 104 amino acid residues corresponding to a molecular weight of about 11 500 Da (Table 14.2) and is larger than, but otherwise essentially identical to, protein AA in its primary structure (Sletten and Husby 1974; Parmelee et al. 1982; Sletten et al. 1983). Protein AA is most probably formed by proteolytic cleavage of SAA at different positions varying from position 45 to 83 in different cases, leaving the N-terminus largely intact (Husby and Sletten, 1986a). AA is thus a different size N-terminal fragment of SAA lacking a C-terminal portion of the latter protein, and the sequence 46–104 is apparently not critical for fibrillogenesis. Also, different size AA proteins are often found in the same organ (Husby et al. 1988), although, the most common human AA consists of 76 amino acid residues, thus lacking the 28 C-terminal amino acids of SAA.

The pronounced hydrophobicity of the sequence consisting of the 11 first N-terminal amino acids is the most striking feature of SAA and AA (Tables 14.2, 14.3), while residues 25–49 of the two proteins and 76–104 of SAA have unspecified function (Parmelee et al. 1982; Husby and Sletten 1986b; Husebekk et al. 1986a; Turnell et al. 1986a,b). The sequence between residues 33–51 and 78–103 are highly conserved in evolution, pointing to an important biological function (Husby et al. 1988). However, apart from its being amyloid precursor, the functional significance of SAA is largely unknown. Computer analyses and wet gel X-ray studies of SAA (Turnell et al. 1986a,b) confirm the hydrophobicity at the N-terminal residues 1–11. These residues form an α-helix and are probably involved in lipid binding, thus supporting previous suggestions (Parmelee et al. 1982) that this part of the molecule may be involved in the complexing of SAA with high density lipoprotein (HDL). Also residues 50–74 appear to form an α-helix possibly involved in lipid binding (Parmelee et al. 1982), but this sequence is obviously not essential for fibril formation.

Chemical analysis of SAA (Table 14.2) and AA including amino acid sequence studies have clearly shown that both proteins are polymorphic (Husby et al. 1988). Complementary DNA probes for murine and human SAA are now available, and recent studies have shown that at least three SAA genes exist in both species, thus confirming the structural studies of the SAA protein products (Sipe et al. 1985; Yamamoto et al. 1986; Kluve-Beckerman et al. 1986; Meek and Benditt 1986; Woo et al. 1987; Sack 1988). We have also found polymorphism of SAA in the mink (Syversen et al. 1987) and horse (Sletten et al. unpublished work), which indicates that this is a general feature of the protein.

It has been suggested (Westermark et al. 1987) that the size of the AA molecule may influence the tissue distribution of AA amyloid, possibly due to the different electric charges of the various SAA degradation products. However, it could also be that differential enzymatic handling of SAA at the local site may produce different sized AA molecules. This is consistent with the observation of Prelli et al. (1987) of two AA molecules consisting of 45 and 82 amino acids, respectively, in amyloid obtained from a human thyroid, both having Ala-Ala at the C-terminus. This is the first hint that specific proteases may operate in AA amyloidosis, since the C-terminus of other different sized AA molecules described does not indicate any specific enzymatic handling of SAA.

Induction and Production of the Acute Phase Protein SAA

SAA belongs to a group of serum proteins called acute phase reactants (Husby and Natvig 1974; Pepys and Baltz 1983). The acute phase reflects a systemic

Table 14.2. Amino acid sequence of human (H) and mink (M) SAA[a]

					5					10					15					20
H	Arg	Ser	Phe	Phe	Ser	Phe	Leu	Gly	Glu	Ala	Phe	Asp	Gly	Ala	Arg	Asp	Met	Trp	Arg	Ala
											Ile									
M		PCA	Trp				Phe				Val	Gln			Trp			Tyr		
					25					30					35					40
H	Tyr	Ser	Asn	Met	Arg	Glu	Ala	Asn	Tyr	Ile	Gly	Ser	Asp	Lys	Tyr	The	His	Ala	Arg	Gly
M										Lys	Asn									
					45					50		*Ala*			55		*Ala*	*Ile*		60
H	Asn	Tyr	Asp	Ala	Ala	Lys	Arg	Gly	Pro	Gly	Gly	Val	Trp	Ala	Ala	Glu	Val	Leu	Ser	Asn
M				Gln								Ala				Lys	Val	Ile		Asp
					65					70					75					80
H	Ala	Arg	Glu	Asn	Ile	Glu	Arg	Phe	Phe	Gly	His	Gly	Ala	Glu	Asn	Ser	Leu	Ala	Asp	Glu
								Ile				*Ile*								
M					Gln			Val	Thr	Asp	Leu	Phe	Lys	Gly	Asp		Lys			
					85					90					95					100
H	Ala	Ala	Asn	Glu	Trp	Gly	Arg	Ser	Gly	Lys	Asp	Pro	Asn	His	Phe	Arg	Pro	Ala	Gly	Leu
M																		Pro		
				104																
H	Pro	Glu	Lys	Tyr-COOH																
M		Asp		-COOH																

[a] Sletten et al. (1983), Syversen et al. (1987).

Table 14.3. Partial N-terminal amino acid sequence of human, mink and murine SAA and AA

					5					10				
Human SAA/AA	Arg	Ser	Phe	Phe	Ser	Phe	Leu	Gly	Glu	Ala	Phe	Asp	Gly	Ala
Mink SAA		PCA	Trp				Phe				*Ile* Val	Gln		
Mink AA		PCA	Trp				Phe				Val	Gln		
Murine SAA[a]	Gly						*Ile* Val					Gln		
Murine AA[b]	Gly						Ile					Gln		

[a] Anders et al. (1977).
[b] Hoffmann et al. (1984).

response to infection, tissue injury or cell death involving a variety of cellular and biochemical events including a rise in the levels of the acute phase proteins in serum (Kushner 1982). SAA and C-reactive protein are the most characteristic human proteins in this respect (Marhaug et al. 1986a,b). The serum concentration of SAA increases up to a thousand-fold or more within 24–48 hours of adequate inflammatory stimuli. Indeed, we have measured serum concentrations of up to 1280 mg l^{-1} in patients with inflammation, while the lowest value we have recorded is 0.02 mg l^{-1}; this represents a difference of 64 000 times. The mean concentration in normal human sera as measured by radioimmunoassay in our laboratory is 1 mg l^{-1} (Marhaug 1983).

Clinically, SAA is a sensitive indicator of inflammation in diseases known to stimulate the acute phase response (McAdam et al. 1978, 1982; Marhaug 1983; Pepys and Baltz 1983). For example, we have shown that measurement of SAA is of significance for the early detection of infectious complications of cystic fibrosis (Marhaug et al. 1983), opportunistic infections in AIDS (Husebekk et al. 1986b) and has prognostic value with regard to determining the amount of tissue necrosis in myocardial infarction (Marhaug et al. 1986a,b).

Signals for SAA induction are mediated by cytokines such as interleukin-1 (McAdam et al. 1982), tumor necrosis factor (Rokita et al. 1987; Woo et al. 1987a) and interleukin-6 (Castell et al. 1988), released by activated mononuclear phagocytes, i.e. macrophages and monocytes, and other cell types. Although the mechanism by which these mediators stimulate SAA production has not been clarified (Pepys and Baltz 1983), activation of mononuclear phagocytes appears to be a key phenomenon in the acute phase response, including the induction, secretion and hence the serum concentration of SAA.

It is generally accepted that the liver (i.e. the hepatocyte) is the chief producer of SAA (Husby and Sletten 1986a). This most probably still holds true, although expression of murine SAA genes has been demonstrated in various extrahepatic sites following stimuli with inflammatory agents such as casein and bacterial endotoxin (lipopolysaccharide, LPS) (Meek and Benditt 1986; Rokita et al. 1987). However, of the three murine SAA gene products only SAA1 and SAA2 proteins have actually been shown to be produced and secreted to the circulation, and extrahepatic expression of these two genes is generally much lower than that of the liver. The murine SAA3 gene protein product has thus far not been demonstrated, although the gene is expressed in many tissues following LPS administration. It is interesting that peritoneal macrophages express the SAA3 gene at the same magnitude as the liver following LPS administration (Meek and Benditt 1986; Rokita et al. 1987). Perhaps all SAA3 remains in the intracellular compartment.

Since the spleen, which does not produce SAA1 or SAA2 or express these genes following LPS (Meek and Benditt 1986), is nevertheless the preferential organ for experimental murine AA amyloidosis, and since murine AA does not derive from SAA3, it appears clear that the chief source of the amyloid protein is hepatic SAA released to the circulation.

Apo SAA and "Apo AA"

It as now established that the bulk of SAA is complexed to high density lipoprotein (HDL) in serum (Benditt and Eriksen 1977; Skogen et al. 1979; Marhaug et al. 1982). Indeed, our own studies (Marhaug et al. 1982) as well as those of others (Clifton et al. 1985) have shown that SAA can constitute up to 87% of the total apolipoproteins of HDL in human acute phase sera, while only minute amounts are present in HDL in the normal state.

By immunoblotting experiments using monospecific antisera to human SAA/AA we demonstrated small amounts of protein AA-like molecules among the HDL apolipoproteins in acute phase sera containing large amounts of SAA (Husebekk et al. 1985, 1987), suggesting that SAA had been converted to AA when present in HDL in the circulation. This is in line with the findings that enzymes present on the surface of circulating monocytes (Lavie et al. 1978) as well as in serum (Skogen and Natvig 1981) are capable of transforming SAA to AA in vitro. It also shows that the 28 C-terminal amino acids of SAA which are not present in AA are not essential for the binding to HDL, in concert with recent computer analyses (Turnell et al. 1986a) showing that the strongest lipid binding property resides among the 11 first N-terminal residues of SAA.

Whether the existence of already-formed AA molecules in the circulation is of importance for amyloid formation is not clear at the present. However, we also studied the saturation kinetics between HDL and SAA/AA by adding different amounts of these proteins to purified normal HDL in vitro (Husebekk et al. 1987). The results showed that SAA has higher affinity for HDL than AA. The latter protein may therefore be more readily accessible for fibril formation. The lower affinity of AA for HDL is not readily explained by the lack of the 28 amino acid C-terminal sequence of SAA, since residues 75–104 in SAA, as mentioned, contain no lipid-binding sequence (Parmelee et al. 1982; Sletten et al. 1983; Turnell et al. 1986a). A more likely explanation may be that AA is formed by subspecies of SAA with low affinity for HDL. It could therefore be that apo SAA and "apo AA" molecules that have the lower affinity for HDL are those that form amyloid, and that "apo AA", although present in small amounts in the circulation, is of significance for amyloidosis.

Using the human–mouse model for AA amyloidosis we also showed that HDL–SAA complexes formed in vitro by adding purified SAA to normal HDL as well as similarly formed HDL–AA complexes, when given intravenously to the mice during amyloidogenesis, could be traced as human AA constituents of the amyloid fibrils formed in the mice (Husebekk et al. 1987, 1988a,b).

Displacement of SAA from HDL–SAA Complexes by Apo AI and Apo AII

It has been demonstrated that when SAA increases in serum during the acute phase, it displaces apo AI and apo AII or apo C proteins from the HDL particle (Park and Rudel 1985; Coetzee et al. 1986). We wanted to see if the reverse process, namely the displacement of SAA from acute phase HDL by apo AI or apo AII, can also take place (Husebekk et al. 1987). The results indeed showed that apo AI, but to an even greater extent apo AII, effectively displaced SAA from acute phase HDL when added in excess amounts in vitro (Husebekk et al. 1987, 1988a). High concentrations of these apoproteins in vivo may thus displace SAA and AA from HDL, and thus make the latter proteins available for amyloid formation. It is interesting that tissues like the liver and intestines that actually produce apo AI and apo AII, are also predilection sites for AA type amyloidosis.

Formation and Deposition of AA Amyloid

The recent developments in the molecular biology of amyloid-related proteins have provided new insights into possible pathogenetic mechanisms for amyloidosis. SAA is the precursor for tissue AA, and availability of the precursor must thus be one prerequisite for amyloidogenesis. It is well known that chronic inflammatory diseases associated with increased concentrations of acute phase proteins including SAA also underlie reactive amyloidosis (Husby 1985; Husby et al. 1988). It is also known that inflammatory agents like casein and LPS used to induce experimental amyloidosis also induce a rapid SAA response. However, only a minority of patients with inflammation develop amyloidosis, despite chronically raised SAA (Husby 1985) and there is no absolute correlation between SAA concentration and development of experimental amyloidosis in mice (Meek and Benditt 1986). Additional factors are therefore required for amyloidogenesis. In other words the aetiology of AA amyloidosis is multifactorial.

Besides the mere presence of SAA, the structure of this precursor protein appears to be of importance. As mentioned, SAA is heterogeneous (Table 14.2), and in 1977 our amino acid sequence studies of murine SAA revealed two residues, namely valine and isoleucine at position 7 (Table 14.3) possibly representing two isotypes of this protein (Anders et al. 1977). This heterogeneity is not evident in murine tissue AA (Hoffman et al. 1984) where only isoleucine has been found. Hoffman et al. (1984) confirmed that only one out of two isotypes of SAA, namely SAA2, forms amyloid, and the same group (Meek et al. 1986) further showed that although hepatic mRNA for SAA1 and SAA2 were expressed at the same magnitude, SAA2 was selectively removed from the circulation during amyloidogenesis, in contrast to equal clearance rates of SAA1 and SAA2 observed during the acute phase response not complicated by amyloidosis. We explored this observation further (Husebekk et al.1988a) and could show that mouse SAA2 indeed had lower affinity for HDL than SAA1 in vitro, which could facilitate the selective removal of SAA2. In addition, the expression of the "amyloid-prone" SAA2 gene has been shown to be defective in SJL mice that are resistant to amyloid induction (Yamomoto et al. 1986), further supporting the importance of this SAA isotype in amyloidogenesis.

The murine SAA3 gene, whose protein product is yet to be discovered (Meek and Benditt 1987), may nevertheless play a role in amyloidosis. The development of experimental amyloidosis in both the mouse (Rokita et al. 1987) and mink (Anders et al. 1976) can be divided into the pre-amyloid and amyloid phase, respectively, as originally described by Teilum (1964). Following administration of the inflammatory stimuli, the pre-amyloid phase is characterized by increased SAA concentrations, which gradually decrease towards the amyloid phase, paralleled by a similar decrease in the expression of mRNA for SAA1 and SAA2 (Rokita et al. 1987). In contrast, the SAA3 gene expression increased remarkably in peritoneal macrophages during the pre-amyloid phase, and this required persistent inflammatory stimulation. As AA amyloidosis is associated with long-standing inflammation, it may be concluded from these studies that SAA3 expression is involved in amyloidogenesis; perhaps by participating in the activation of macrophages required for the induction of hepatic production of amyloid precursor.

In order to examine whether these observations in murine AA amyloidosis are relevant to other species, we established an experimental model for AA amyloidosis in the mink induced with LPS (Husby et al. 1975). The kinetics of SAA in the pre-amyloid and the amyloid phases were strikingly similar to those of the mouse model (Anders et al. 1976). Furthermore the structure of AA (Waalen et al. 1980) and SAA (Syversen et al. 1987) in this species revealed a striking analogy to the situation in the mouse, with both valine and isoleucine indicating two SAA isotypes at position 10, while only valine was present at this position in AA (Table 14.3). Studies are presently in progress to elucidate the structure of the corresponding genes in the mink. Furthermore, our preliminary studies in the horse similarly indicate amyloidotic and non-amyloidotic molecular species of SAA (Sletten et al. unpublished work).

The importance of the N-terminal part of SAA in amyloidosis is clearly illustrated by recent experiments in the rat, an animal that probably never develops spontaneous or induced amyloidosis. SAA is not found in rat acute phase HDL (Baltz et al. 1987). A rat liver mRNA which increases 500-fold in the acute phase codes for an amino acid sequence which is 74% homologous to the 46 C-terminal residues of human SAA, thus lacking a large N-terminal segment of the protein (Liao and Stark 1986; Meek and Benditt 1988). This N-terminal sequence therefore appears to contain the region which both binds to HDL and is critical for amyloid formation.

What, then, is the situation in man? In contrast to mouse and mink, there is no obvious "amyloid-prone" sequence among the first 14 N-terminal amino acids of human SAA (Parmelee et al. 1982; Sletten et al. 1983). However, a heterogeneity at position 58 (leucine and isoleucine) in SAA (Sletten et al. 1983) does not occur in any human AA protein so far sequenced (see Table 14.2). However, residue 58 of SAA clearly belongs to a part of the molecule which is not necessarily critical for fibril formation. On the other hand, it has been shown that the inherited, type AA amyloid associated with familial Mediterranean fever (FMF) which is made up by an AA protein consisting of 76 amino acids, has threonine at residue 69 (Levin et al. 1972), whereas all AA proteins related to secondary amyloidosis have phenylalanine at this position. The latter is also the case with SAA from two human individuals (see Table 14.2) as well as a pool of human sera (Parmelee et al. 1982; Sletten et al. 1983). This substitution in FMF amyloid involving all three nucleotides, which has been confirmed by the structure of a corresponding human gene (Woo et al. 1987a), may be important for this inherited type of amyloidosis. However, AA protein from two other FMF patients reported (Prelli et al. 1987; Dr. E. Erken, personal communication) does not possess this substitution, since both have phenylalanine at position 69. This may reflect the marked variability in SAA genes (Sack 1988) and clinical phenotype (Pras 1986) of FMF patients of different ethnic origins. Furthermore, the AA molecule is also heterogeneous (Levin et al. 1972; Sletten and Husby 1974), showing that multiple phenotype of SAA molecules can make up amyloid fibrils (Prelli et al. 1987; Husby et al. 1988). Interestingly, the amino acid sequence deduced from the structure of a human SAA gene recently determined (Kluve-Beckerman et al. 1986) differed from any human AA protein so far studied, supporting the existence of "non-amyloidotic" SAA gene products also in humans.

Incomplete Degradation of SAA

SAA has a half-life of approximately 24 hours in humans (McAdam et al. 1978). Enzymes bound to white blood cells (Lavie et al. 1978), serum proteases (Skogen et al. 1981; Bausserman et al. 1987) and Kupffer cells in the liver (Fuks and Zucker-Franklin 1985) degrade SAA in vitro, which is compatible with both hepatic and extra-hepatic catabolism of SAA in vivo. The degradation is inhibited by SAA (Bausserman et al. 1987), supporting the observation of Prelli et al. (1987) that it takes place by a specific mechanism. An intermediate fragment in the catabolism of SAA is an AA-like molecule (Lavie et al. 1978; Skogen et al. 1981). Insufficient breakdown and removal of SAA or this AA-like fragment may lead to its deposition in amyloid (Lavie et al. 1978). In support of this is the experimental evidence of defective reticuloendothelial (i.e. Kupffer cell) function in pre-amyloidotic mice (Fuks and Zucker-Franklin 1985).

The question remains, however, whether limited degradation of SAA is a prerequisite for amyloid formation. Apparently, intact SAA makes up the fibrils in amyloid isolated from man (Isobe et al. 1980) and duck (Ericksson et al. 1987) and is also present together with different size AA molecules in many amyloid preparations (Husby et al. 1988; Husebekk et al. 1988b). Intriguing in this respect is the theory of Tape et al. (1988) that the enzymatic cleavage of SAA is a postfibrillar event. In newly formed amyloid, therefore, the fibrils would consist of intact SAA, in accord with the above observations. Another possibility that is not yet ruled out is that such degradation takes place ex vivo, eventually during the isolation procedure (Prelli et al. 1987). At any rate, the availability of intact or partly degraded SAA for fibril formation may have several explanations, for example, enzyme resistant (amyloidogenic) SAA, defective enzymes, presence of inhibitor(s), or an over-load of a normal catabolic system by too much SAA.

Amyloid Enhancing Factor and Glycosaminoglycans

Amyloid enhancing factor (AEF) is a poorly defined material which probably consists of both protein and carbohydrate, is induced in the spleen, liver and kidney during persistent inflammation, and is probably synthesized and secreted by reticulo-endothelial cells in these organs (Axelrad and Kisilevsky 1980; Hol et al. 1986; Kisilevsky 1987). AEF consistently precedes the occurrence of amyloid induced in these organs in the mouse (Kisilevsky 1987) and hamster (Niewold et al. 1987). Intravenously administered AEF has been shown to shorten the induction time of experimental amyloidosis from weeks to 24–48 hours, possibly by altering the catabolism of SAA, because AEF is not capable of inducing SAA (Deal et al. 1982). AEF is closely associated with the amyloid deposits. Indeed, it has been postulated that AEF is integrated in the fibrils themselves (Niewold et al. 1987). Furthermore, AEF has also been demonstrated in human organs laden with AA or AL as well as prealbumin-related amyloid, and AEF extracted from these amyloids enhances experimental murine AA amyloidosis (Varga et al. 1986).

Of interest is the observation that AEF enhances the murine SAA3 gene expression by peritoneal macrophages (Rokita et al. 1987), and the authors suggest that SAA3 might itself be AEF, or that it participates in amyloidosis together with AEF.

Glycosaminoglycans (GAGs) have been proposed to account for the carbohydrate moiety in AEF (Snow et al. 1987). GAGs occur in the tissues in close temporal and morphologic relationship to amyloid deposition (Snow et al. 1987). Recent experiments in our laboratory (Magnus et al. 1989) showed by direct chemical evidence that GAGs identified as chondroitin sulphate, dermatan sulphate and heparin/heparan sulphate are specifically co-purified with amyloid fibrils, in support of the suggestion of Niewold et al. (1987) and Snow et al. (1987) that GAGs (or proteoglycan) constitute the amyloid fibrils together with protein. The large negative charge of GAGs indicates an involvement in precursor protein folding and incorporation in the fibril (Snow et al. 1987).

Schubert et al. (1988) recently observed that the amyloid "β-protein" which make up the cerebral amyloid fibrils in Alzheimer's disease, Down's syndrome and other encephalopathies is possibly derived from a heparan sulphate proteoglycan core. The possible participation of proteoglycan in amyloidogenesis is further supported by the observed amino acid sequence homology between link proteins of the proteoglycan complex and immunoglobulin domains present in both kappa and lambda light chains, known to be precursors of AL amyloid (Bonnet et al. 1986).

Protein AP: The Amyloid "P Component"

Protein AP is an α-glycoprotein which is invariably present in amyloid deposits, regardless of the chemical nature of amyloid fibrils and the clinical type of amyloidosis (Holck et al. 1979; Pepys and Baltz 1983; Hawkins et al. 1988). AP is not a part of the amyloid fibrils, but is closely bound to them in a calcium-dependent fashion (Holck et al. 1979; Pepys and Baltz 1983). A normal plasma pentraxin protein, SAP, is identical to protein AP in structure and binding properties (Andersson and Mole 1982; Pepys and Baltz 1983). Studies using radiolabelled SAP injected intravenously (Baltz et al. 1986b) showed that SAP is the precursor of tissue AP. Furthermore, radionuclide imaging could be used in the diagnosis of various types of human amyloidosis (Hawkins et al. 1988). The authors showed that although the serum concentration of SAP is not increased in the acute phase or in patients with amyloidosis, the synthesis of the protein may be increased 100-fold in amyloidotic patients, pointing to a specific role of SAP in amyloidogenesis.

Mouse, but not human, SAP behaves like an acute phase reactant (Pepys and Baltz 1983). Otherwise, the biological significance of AP/SAP is largely unknown, although purified human AP has been shown to inhibit proteolytic activity of elastase in vitro (Li et al. 1984). This may have implications in amyloidogenesis, since AP could inhibit the complete enzymatic breakdown of amyloid precursor protein at the site of fibril deposition. AP may also inhibit SAA degradation indirectly via its ability to bind glycosaminoglycans (Woo et al. 1987b). These researchers described a DNA polymorphic site, 5′ to the SAP gene which was significantly associated with AA amyloidosis in juvenile rheumatoid arthritis, suggesting an active role of AP in amyloidogenesis.

An animal model for AA amyloidosis, the Syrian hamster, clearly illustrates a role for SAP in amyloidosis. Female protein (FP) is a hamster pentraxin and an analogue of human SAP (Coe and Ross 1985; Dowton and Colten 1986). The FP gene is regulated by mediators of inflammation, but also by sex steroid hormones, hence unstimulated female hamsters have high serum FP concentration which decreases with inflammation (i.e. "negative" acute phase protein). In contrast, serum FP is low in normal males, but increases in the acute phase (Coe and Ross 1985). Unstimulated female hamsters frequently develop amyloidosis, which is a significant cause of early death. Male hamsters are able to develop casein-induced amyloidosis, but less readily than the females, and indeed the female sex hormone, diethylstilboestrol, both increases serum FP and enhances experimental amyloidosis in the males.

Summary of Some Current Hypotheses Regarding AA Amyloid Formation

In all forms of systemic amyloidosis the fibril precursor is a serum protein, which in AA amyloidosis is SAA (Husebekk et al. 1985). Amyloidosis may be caused by excess amounts of precursor protein, as the result of increased production and/or decreased clearance. However, only a minor proportion of patients with inflammation develop AA amyloidosis in spite of chronically raised SAA (Husby 1985). On the other hand, individuals with presumably low concentrations of SAA may develop AA amyloidosis. Additional factors are obviously needed for the formation of AA amyloidosis. Genetically determined "amyloid-prone" SAA molecules is one such factor, together with diverse (Yamamoto et al. 1986) or altered (Rokita et al. 1987) expression of the SAA genes. Genes other than those coding for SAA may also be important, for example those coding for and regulating SAP and its analogue FP. In addition, the strong resistance of the A/J mouse strain to amyloid induction has been shown to be accounted for by a single gene (Wohlgethan and Cathcart 1980). Still another factor is AEF: virtually all

CBA/J mice receiving stimulus for persistent inflammation plus AEF develop amyloidosis within two days. This also suggests a proper combination of different amyloid-promoting factors is operating in this model (Kisilevsky 1987).

It is this author's belief that the relative importance of the various aetiological and pathogenetic factors operating in AA amyloidosis may differ in different cases of this disorder. The ongoing research, particularly that in the field of molecular biology of amyloidosis which has received increasing interest in recent years, will no doubt shed new light on the mechanisms for this often lethal disorder.

Acknowledgements. The author wants to express his sincere gratitude to the co-authors of the papers from our groups referred to in this overview, both for pleasant collaboration, and for valuable discussions regarding this work: Robin F. Anders, Anne Husebekk, Jeanette Magnus, Gudmund Marhaug, Jacob B. Natvig, Knut Nordstoga, Knut Sletten and Bjørn Skogen. Ms Sigrid Bolle, Ms Marit Espejord and Ms Wenche Johnsen are acknowledged for secretarial and technical assistance. This work was supported by the Norwegian Council for Science and the Humanities, the Norwegian Women's Public Health Association and Norsk Revmatikerforbund.

References

Axelrad MA, Kisilevsky R (1980) Biological characterization of amyloid-enhancing factor. In: Glenner GG, Costa PP, deFreitas AF (eds) Amyloid and amyloidosis. Exerpta Medica, Amsterdam, pp 527–533

Anders RF, Nordstoga K, Natvig JB, Husby G (1976) Amyloid related serum protein SAA in endotoxin-induced amyloidosis of the mink. J Exp Med 143:678–683

Anders RF, Natvig JB, Sletten K, Husby G, Nordstoga K (1977) Amyloid related serum protein SAA from three animal species: comparison with human SAA. J Immunol 118:229–234

Andersson JK, Mole JE (1982) Large scale isolation and partial primary structure of human plasma amyloid P-component. Ann NY Acad Sci 389:216–234

Baltz ML, Rowe IF, Caspi D, Turnell WG, Pepys MB (1986a) Is the serum amyloid A protein in acute phase plasma high-density lipoprotein the precursor of AA fibrils? Clin Exp Immunol 66:701–708

Baltz ML, Caspi D, Evans DJ, Rowe IF, Hind CRK, Pepys MB (1986b) Circulating serum amyloid P component is the precursor of amyloid P component in tissue amyloid deposits. Clin Exp Immunol 66:691–700

Baltz ML, Rowe IF, Capsi D, Turnell WG, Pepys MB (1987) Acute phase high-density lipoprotein in the rat does not contain serum amyloid A protein. Biochem J 242:301–303

Bausserman LL, Saritelli AL, van Zuiden P, Collaher CJ, Herbert PN (1987) Degradation of serum amyloid A by isolated perfused rat liver. J Biol Chem 262:1583–1589

Benditt EP, Eriksen N (1964) Starch gel electrophoretic analysis of some proteins extracted from amyloid. Arch Pathol 78:325–330

Benditt EP, Eriksen N (1971) Chemical classes of amyloid substance. Amer J Pathol 65:231–252

Benditt EP, Eriksen N (1977) Amyloid protein SAA is associated with high density lipoprotein from human serum. Proc Natl Acad Sci USA 74:4025–4028

Benditt EP, Eriksen M, Hermodsen MA, Ericsson LH (1971) The major proteins of human and monkey amyloid substance: common properties including unusual N-terminal amino acid sequences. FEBS Lett 19:169–173

Benson MD (1988) Hereditary amyloidosis – disease entity and clinical model. Hosp Pract 165–181

Bonnet F, Périn JP, Lorenzo F, Jollés J, Jollés P (1986) An unexpected sequence homology between link proteins of the proteoglycan complex and immunoglobulin-like proteins. Biochim Biophys Acta 873:152–155
Castell JV, Gómez-Lechón MJ, David M, Hirano T, Kishimoto T, Heinrich PC (1988) Recombinant human interleukin-6 (IL-6/BSF-2/HSF) regulates the synthesis of acute phase proteins in human hepatocytes. FEBS Lett 232:347–350
Clifton PM, Mackinnon AM, Barter PJ (1985) Effects of serum amyloid A protein (SAA) on composition, size, and density of high density lipoproteins in subjects with myocardial infarction. J Lipid Res 26:1389–1398
Coe JE, Ross MJ (1985) Hamster female protein, a sex-limited pentraxin, is a constituent of Syrian hamster amyloid. J Clin Invest 76:66–74
Coetzee GA, Strachan AS, van der Westhuysen DR, Hoppe HC, Jeenan MS, de Beer SC (1986) Serum amyloid A-containing human high density lipoprotein. J Biol Chem 261:9644–9651
Cohen AS (1967) Amyloidosis. N Engl J Med 277:522–530, 574–583, 628–638
Deal CL, Sipe JD, Tatsuta E, Skinner M, Cohen AS (1982) The effect of amyloid enhancing factor (AEF) on the acute phase serum amyloid A (SAA) and serum amyloid P (SAP) response to silver nitrate. Ann NY Acad Sci 439–441
Dowton SB, Colten HR (1986) Sites and regulation of biosynthesis of SAA. In: Marrink J, van Rijswijk MH (eds) Amyloidosis. Martinus Nijhoff, Dordrecht Boston Lancaster, pp 107–113
Ericsson LH, Eriksen N, Walsh KA, Benditt EP (1987) Primary structure of duck amyloid protein A. FEBS Lett 218:11–16
Fuks A and Zucker-Franklin D (1985) Impaired Kuppfer cell function precedes the development of secondary amyloidosis. J Exp Med 161:1013–1028
Gejyo F, Yamada T, Odani S et al. (1985) A new form of amyloid protein associated with chronic hemodialysis was identified as β2-microglobulin. Biochem Biophys Res Commun 129:701–706
Glenner GG (1980) Amyloid deposits and amyloidosis. N Engl J Med 302:1283–1292
Gorevic PD, Prelli FC, Wright J, Pras M, Frangione B (1989) Systemic senile amyloidosis: identification of a new prealbumin (transthyretin) variant in cardiac tissue: structural similarity to one form of familial amyloidotic polyneuropathy. J Clin Invest 83:836–843.
Hawkins PN, Lavender JP, Myers MJ et al. (1988) Diagnostic radionuclide imaging of amyloid: biological targeting by circulating human serum amyloid P component. Lancet i:1413–1418
Hoffman JS, Ericsson LH, Eriksen N, Walsh KA, Benditt EP (1984) Murine tissue amyloid protein AA. NH_2-terminal sequence identity with only one of two serum amyloid protein (ApoSAA) gene products. J Exp Med 159:641–646
Hol PR, Snel FWJJ, Niewold TA, Gruys E (1986) Amyloid enhancing factor (AEF) in the pathogenesis of AA-amyloidosis in the hamster. Virchows Arch B 52:273–281
Holck M, Husby G, Sletten K, Natvig JB (1979) The amyloid P component (protein AP): an integral part of the amyloid substance? Scand J Immunol 10:55–60
Husby G (1980) A chemical classification of amyloidosis – correlation with different clinical types of amyloidosis. Scand J Rheumatol 9:60–64
Husby G (1983) Immunoglobulin-related (AL) amyloidosis. Clin Exp Rheumatol 1:353–358
Husby G (1985) Amyloidosis and rheumatoid arthritis. Clin Exp Rheumatol 3:173–180
Husby G, Natvig JB (1974) A serum component related to non-immunoglobulin amyloid protein AS. A possible precursor of the fibrils. J Clin Invest 53:1054–1061
Husby G, Sletten K (1977) Structural similarities between a protein extracted from normal human tissue and a component of amyloid fibrils. Acta Pathol Microbiol Scand Sect C 85:153–160
Husby G, Sletten K (1986a) Chemical and clinical classification of amyloidosis 1985. Scand J Immunol 23:253–265
Husby G, Sletten K (1986b) Amyloid proteins. In: Marrink J, van Rijswijk MH (eds) Amyloidosis. Martinus Nijhoff, Dordrecht Boston Lancaster, pp 23–34
Husby G, Natvig JB, Sletten K, Nordstoga K, Anders RF (1975) An experimental model in mink for studying the relation between amyloid fibril protein AA and the related serum protein SAA. Scand J Immunol 4:811–816
Husby G, Ranløw PJ, Sletten K, Marhaug G (1985) The amyloid in familial amyloid cardiomyopathy of Danish origin is related to prealbumin. Clin Exp Immunol 60:207–216
Husby G, Husebekk A, Marhaug G, Skogen B, Sletten K (1988) Serum amyloid A (SAA): the precursor of protein AA in secondary amyloidosis Adv Exp Med Biol 243:185–192
Husebekk A, Skogen B, Husby G, Marhaug G (1985) Transformation of amyloid precursor SAA to protein AA and incorporation in amyloid fibrils in vivo. Scand J Immunol 21:283–287
Husebekk A, Husby G, Sletten K, Marhaug G (1986a) Characterization of amyloid protein AA and its serum precursor SAA in the horse. Scand J Immunol 23:703–709

Husebekk A, Permin H, Husby G (1986b) Serum amyloid protein A (SAA)—an indicator of inflammation in AIDS and AIDS-related complex (ARC). Scand J Infect Dis 18:389–394
Husebekk A, Skogen B, Husby G (1987) Characterization of amyloid proteins AA and SAA as apolipoproteins of HDL. Displacement of SAA from the HDL-SAA by apo AI and apo AII. Scand J Immunol 25:375–381
Husebekk A, Skogen B, Husby G (1988a) Replacement of SAA from the SAA-HDL complex by Apo AI and Apo AII. Relevance for amyloid formation? In: Isobe T et al. (eds) Amyloid and amyloidosis. Plenum, New York London, pp 51–56
Husebekk A, Husby G, Sletten K, Skogen B, Nordstoga K (1988b) Characterization of bovine amyloid proteins SAA and AA. Scand J Immunol 27:739–743
Isobe T, Husby G, Sletten K (1980) Characterization of an amyloid protein AA similar to SAA. In: Glenner GG, Costa P, Freitas F (eds) Amyloid and amyloidosis. Excerpta Medica, Amsterdam, pp 331–336
Kisilevsky R (1987) From arthritis to Alzheimer's disease: current concepts on the pathogenesis of amyloidosis. Can J Physiol Pharmacol 65:1805–1815
Kluve-Beckerman B, Long GL, Benson MD (1986) DNA sequence evidence for polymorphic forms of human serum amyloid A (SAA). Biochem Genet 24:795–803
Kushner I (1982) The phenomenon of the acute phase response. Ann NY Acad Sci 389:39–48
Lavie G, Zucker-Franklin D, Franklin EC (1978) Degradation of serum amyloid A protein by surface-associated enzymes of human blood monocytes. J Exp Med 148:1020–1031
Levin M, Franklin EC, Frangione B, Pras M (1972) The amino acid sequence of a major non-immunoglobulin component of some amyloid fibrils. J Clin Invest 51:2773–2776
Li JJ, McAdam PW (1984) Human amyloid P component: an elastase inhibitor. Scand J Immunol 20:219–226
Liao WS, Stark GS (1986) Cloning of rat cDNAs for eight acute phase reactants and kinetics of induction of mRNAs following acute inflammation. Adv Inflammation Res 10:220–222
Linke RP, Heilmann KL, Nathrath WBJ, Eulitz M (1983) Identification of amyloid A protein in a sporadic Muckle–Wells syndrome. N-terminal amino acid sequence after isolation from formalin-fixed tissue. Lab Invest 48:698–704
Mangus JH, Husby G, Kolset SO (1989) Glycosaminoglycans are present in purified AA type amyloid fibrils associated with juvenile rheumatoid arthritis. Ann Rheum Dis 48:215–219
Marhaug G (1983) Three assays for the characterization and quantitation of human serum amyloid A. Scand J Immunol 18:329–338
Marhaug G, Sletten K, Husby G (1982) Characterization of amyloid related serum protein SAA complexed with serum lipoproteins (apo SAA). Clin Exp Immunol 50:382–393
Marhaug G, Permin H, Husby G (1983) Amyloid-related serum protein (SAA) as an indicator of lung infection in cystic fibrosis. Acta Paediatr Scand 72:861–866
Marhaug G, Hårklau L, Olsen B, Husby G, Husebekk A, Wang H (1986a) Serum amyloid A protein in acute myocardial infarction. Acta Med Scand 220:303–306
Marhaug G, Østensen M, Husby G et al. (1986b) Clinical acute phase pattern of serum amyloid A. In: Peeters H (ed) Protides of the biological fluids, vol 34. Pergamon Press, Oxford, pp 375–378
McAdam KPWJ, Elin RJ, Sipe JD, Wolff SM (1978) Changes in human serum amyloid-A and C-reactive protein after etiocholanolene-induced inflammation. J Clin Invest 61:390–394
McAdam KPWJ, Li J, Knowles J, Foss NT, Dinarello CA, Rosenwasser LJ, Selinger MJ, Kaplan MM, Goodman R (1982) The biology of SAA: identification of the inducer, in vitro synthesis, and heterogeneity demonstrated with monoclonal antibodies. Ann NY Acad Sci 389:126–136
Meek RL, Benditt EP (1986) Amyloid A gene family expression in different mouse tissues. J Exp Med 164:2006–2017
Meek RL, Benditt EP (1988) Rat liver and lung express serum amyloid A related mRNAs. In: Isobe T et al. (eds) Amyloid and amyloidosis. Plenum, New York London, pp 283–288
Meek RL, Hoffmann JS, Benditt EP (1986) Amyloidogenesis. One serum amyloid A isotype is selectively removed from the circulation. J Exp Med 163:499–510
Muckle TJ (1979) The "Muckle–Wells" syndrome. Br J Dermatol 100:87–92
Møyner K, Sletten K, Husby G, Natvig JB (1980) An unusually large (83 amino acid residues) amyloid fibril protein AA from a patient with Waldenstrøm's macroglobulinaemia and amyloidosis. Scand J Immunol 11:549–554
Niewold TA, Hol PR, van Andel ACJ, Lutz ETG, Gruys E (1987) Enhancement of amyloid induction by amyloid fibril fragments in hamster. Lab Invest 56:545–549
Nordlie M, Sletten K, Husby G, Ranløv (1988) A new prealbumin variant in familiar amyloid cardiomyopathy of Danish origin. Scand J Immunol 27:119–122

Park JS, Rudel LL (1985) Alteration of high density lipoprotein subfraction distribution with induction of serum amyloid A protein (SAA) in the non-human primate. J Lip Res 26:82–91
Parmelee DC, Titani K, Ericsson LH, Eriksen N, Benditt EP, Walsh KA (1982) Amino acid sequence of amyloid related apoprotein (apoSAA) from human high-density lipoprotein. Biochemistry 21:3298–3303
Pepys M, Baltz ML (1983) Acute phase proteins with special reference to C-reactive protein and related proteins (pentraxins) and serum amyloid A protein. Adv Immunol 34:141–212
Pras M (1986) The hereditary amyloidoses. In: Marrink J, van Rijswijk MH (eds) Amyloidosis. Martinus Nijhoff, Dordrecht Boston Lancaster, pp 185–193
Pras M, Schubert M, Zucker-Franklin D, Rimon A, Franklin EC (1968) The characterization of soluble amyloid prepared in water. J Clin Invest 47: 924–933
Pras M, Zaretzky NA, Frangione B, Franklin EC (1980) AA protein in a case of "primary" or "idiopathic" amyloidosis. Amer J Med 68:291–294
Prelli F, Pras M, Frangione B (1987) The degradation and deposition of amyloid AA fibril is tissue specific. Biochemistry 26:8251–8256
Rokita H, Shirahama T, Cohen AS, Meek RL, Benditt EP, Sipe J (1987) Differential expression of the amyloid SAA3 gene in liver and perinoneal macrophages of mice undergoing dissimilar inflammatory processes. J Immunol 139:3849–3853
Sack GH Jr (1988) Serum amyloid A (SAA) gene variations in familial Mediterranean fever. Mol Biol Med 5:61–67
Schubert D, Schroeder R, LaCorbiere M, Saitoh T, Cole G (1988) Amyloid β protein precursor is possibly a heparan sulfate proteoglycan core protein. Science 241:223–226
Scott DL, Marhaug G, Husby G (1983) Comparative studies of the high molecular weight amyloid fibril proteins and similar components from normal tissues. Clin Exp Immunol 52:693–701
Sipe JD, Colten HR, Godberger G et al. (1985) Human serum amyloid A (SAA): biosynthesis and postsynthetic processing of preSAA and structural variants defined by complementary DNA. Biochemistry 24:2931–2935
Skogen B, Natvig JB (1981) Degradation of amyloid proteins by different serine proteases. Scand J Immunol 14:389–396
Skogen B, Børresen AL, Natvig JB, Berg K, Michaelsen TE (1979) High-density lipoproteins as carrier for amyloid-related protein SAA in rabbit serum. Scand J Immunol 10:39–45
Sletten K, Husby G (1974) The complete amino acid sequence of non-immunoglobulin amyloid fibril protein AS in rheumatoid arthritis. Eur J Biochem 41:117–125
Sletten K, Westermark P, Natvig JB (1976) Characterization of amyloid fibril proteins from medullary carcinoma of the thyroid. J Exp Med 143:993–997
Sletten K, Marhaug G, Husby G (1983) The covalent structure of amyloid related serum protein SAA from two patients with inflammatory disease. Hoppe–Zeyler's Z Physiol Chem 364:1039–1046
Snow AD, Willmer J, Kisilevsky R (1987) A close structural relationship between sulfated proteoglycans and AA amyloid fibrils. Lab Invest 57:687–698
Syversen V, Sletten K, Marhaug G, Husby G, Lium B (1987) The amino acid sequence of serum amyloid A (SAA) in mink. Scand J Immunol 26:763–767
Tape C, Tan R, Nesheim M, Kisilevsky R (1988) Apo SAA and AA: precursor and product revisited. In: Isobe T et al. (eds) Amyloid and amyloidosis. Plenum, New York London, pp 57–62
Teilum G (1964) Pathogenesis of amyloidosis. The two-phase cellular theory of local secretion. Acta Pathol Microbiol Scand 61:21–45
Turnell W, Sarra R, Glover ID et al. (1986a) Secondary structure prediction of human SAA1. Presumptive identification of calcium and lipid binding sites. Mol Biol Med 3:387–407
Turnell W, Sarra R, Baum JO, Caspi D, Baltz ML, Pepys MB (1986b) X-ray scattering and diffraction by wet gels of AA amyloid fibrils. Mol Biol Med 3:409–424
Van Andel ACJ, Niewold TA, Lutz BTG, Messing MWJ, Gruys E (1986) The significance of non-protein AA material in water-soluble bovine AA-amyloid fibrils. In: Marrink J, van Rijswijk MH (eds) Amyloidosis. Martinus Nijhoff, Dordrecth Boston Lancaster, pp 169–175
Varga J, Flinn MS, Shirahama T, Rodgers OG, Cohen AS (1986) The induction of accelerated murine amyloid with human splenic extract. Probable role of amyloid enhancing factor. Virchows Arch B 51:177–185
Waalen K, Sletten K, Husby G, Nordstoga K (1980) The primary structure of amyloid fibril protein AA in endotoxin-induced amyloidosis of the mink. Eur J Biochem 104:407–412
Westermark GT, Westermark P, Sletten K (1987) Amyloid fibril protein AA. Characterization of uncommon subspecies from a patient with rheumatoid arthritis. Lab Invest 57:57–64
Wohlgethan JR, Cathcart ES (1980) Amyloid resistance in A/J mice. Studies with a transfer model. Lab Invest 42:663–667

Woo P, Sipe J, Dinarello CA, Colten HR (1987a) Structure of a human serum amyloid A gene and modulation of its expression in transfected L cells. J Biol Chem 262:15790–15795

Woo P, O'Brien J, Robson M, Ansell BM (1987b) A genetic marker for systemic amyloidosis in juvenile arthritis. Lancet ii:767–769

Yamamoto KI, Shiroo M, Migita S (1986) Diverse gene expression for isotypes of murine serum amyloid A protein during acute phase reaction. Science 232:227–229

Chapter 15

Serum Amyloid P Component: A Specific Molecular Targeting Vehicle in Amyloidosis

P. N. Hawkins and M. B. Pepys

Introduction

Amyloidosis is a disorder of protein metabolism characterized by abnormal extracellular protein material which is deposited in organs and tissues. Amyloid deposits may be focal, localized, or systemic in distribution. When focal or small in amount, amyloid may be clinically silent and only an incidental finding at autopsy. Indeed such deposition in the brain, heart, seminal vesicles and joints seems to be an almost universal accompaniment of ageing. However, systemic amyloidosis or local deposition on a major scale disrupts the normal structure and function of the affected tissues or organs and is usually associated with serious morbidity and mortality. There is no known effective therapy for systemic amyloidosis, and since it rarely regresses, the usual natural history is inexorable progression toward organ failure and death.

Amyloid deposits are composed largely of protein fibrils, the nonglycosylated peptide subunits of which are of different chemical constitution in different forms of the disease. Sulphated glycosaminoglycans are intimately associated with the fibrils in all forms of amyloid deposit, and there is in addition another non-fibrillar minor constituent, the normal plasma glycoprotein, amyloid P component (AP). Indeed, the observation that AP isolated from amyloid material under the electron microscope had a unique pentameric structure was the first evidence for the existence of what is now known as the pentraxin family of plasma proteins. This family includes C-reactive protein (CRP), the classical acute phase reactant of man, and stably conserved homologous proteins throughout the vertebrate phylum. Description of the calcium-dependent ligand binding properties of serum amyloid P component (SAP) provided an explanation for the general presence of amyloid P component in amyloid deposits, but its contribution to the pathogenesis, and the possible significance of the presence of a related (or identical) protein in normal human connective tissues remain obscure.

The normal physiological functions of the pentraxins are not yet known, despite description of many ligand binding interactions and biological properties in experimental systems. However, a novel "iatrogenic" function for SAP has been developed, which has culminated in the use of radiolabelled pure SAP as a method of imaging and possibly quantifying human amyloid deposits in vivo.

Serum Amyloid P Component

Serum amyloid P component is composed of 10 identical glycosylated polypeptide subunits that are noncovalently associated in two pentameric disc-like rings interacting face to face (Osmand et al. 1977; Pepys et al. 1977; Pinteric et al. 1976). It is identical to amyloid P component (Pepys and Baltz 1983; Prelli et al. 1985; Mantzouranis et al. 1985), with a direct precursor–product relation between SAP and AP, the latter being derived by deposition of the former from the circulation (Baltz et al. 1986a). SAP is distinct, both in its structure and its amino acid sequence, from all other human proteins with the exception of CRP, the classical acute phase protein. CRP and SAP share 51% strict residue for residue identity, and when conservative substitutions are taken into account the overall homology is 66% (Mantzouranis et al. 1985; Woo et al. 1985). No polymorphism of SAP or AP has yet been firmly established.

Serum amyloid P component and CRP are clearly members of the same, unique protein "superfamily", known as pentraxins. The genes for both proteins are located on chromosome 1 (Mantzouranis et al. 1985; Whitehead et al. 1983) and have presumably been derived by a process of gene duplication. This duplication must have occurred at an early stage in vertebrate evolution, if not before, as proteins homologous with human SAP and/or CRP are present in all vertebrates in which they have been sought (Baltz et al. 1982a; Pepys et al. 1978; Robey et al. 1983), and, indeed, a protein homologous with CRP is even present in an invertebrate, *Limulus polyphemus*, the horseshoe crab (Robey and Liu 1981). Such stable conservation during evolution is a strong pointer that these proteins have important physiological functions, although they have yet to be fully characterized.

Ligand Binding by SAP/AP

Serum amyloid P component and AP bind calcium ions and display specific calcium-dependent ligand binding properties. This characteristic was first described in the case of human SAP with respect to agarose (Pepys et al. 1977), and SAP of other species has been isolated and identified on the basis of the same binding specificity. Subsequently, SAP was shown to undergo calcium-dependent binding to AA- and AL-type amyloid fibrils in vitro (Pepys et al. 1979), and the occurrence of this interaction in vivo accounts for the presence of AP in amyloid deposits. However, SAP has presumably not been conserved in evolution in order to bind either to agarose, a linear galactan hydrocolloid from marine algae, or to amyloid fibrils.

Aggregated human SAP expresses new reactivities that are not expressed by native, nonaggregated SAP. These reactivities are directed toward two normal plasma proteins, fibronectin and C4-binding protein (de Beer et al. 1981). When suitably immobilized SAP is exposed to whole normal human serum, it selectively binds these two proteins in a strictly calcium-dependent reaction. A pair of SAP molecules, for example, bound by an immobilized IgG anti-SAP molecule binds a single molecule of fibronectin. SAP within plasma or in serum is not bound to either fibronectin or C4-binding protein and apparently exists freely as single, uncomplexed molecules (Baltz et al. 1982b).

The chemical nature of the autologous physiological or even pathophysiological ligand(s) for SAP is not known. SAP does not just bind to any deposit in vivo containing calcium or to any polyanion capable of binding calcium; it displays considerable specificity. In the case of agarose, the pyruvate moiety, a variable trace constituent, is essential for SAP binding, and even specific methylation of its carboxyl hydroxyl group is sufficient to abolish uptake of SAP (Hind et al. 1984c). The free 4,6-cyclic pyruvate acetal of methyl β-D-galactose (MOβDG), the form in which pyruvate exists in agarose, has been synthesized (Hind et al. 1984c). It inhibits or reverses all the known binding reactions of SAP (Hind et al. 1985, 1984b), including the binding of AP deposited with amyloid fibrils in vivo. Such activity is strictly dependent on the structure and conformation of MOβDG, presumably reflecting the specific nature of the binding site in the SAP molecule.

Serum amyloid P component in normal human serum also binds selectively to bacteria that contain the 4,6-cyclic pyruvate ketal of galactose in their coat but does not bind to those containing the 2,3-cyclic form, noncyclic ketals, or cyclic ketals of glucose (Hind et al. 1985). SAP is thus a circulating lectin, and it also binds to zymosan (yeast cell wall preparation) and various other poorly characterized mannans and galactans (Pepys et al. 1979). However, MOβDG itself is not found in mammalian tissues, and although it is the best characterized ligand for SAP so far, the precise carbohydrate specificity of SAP remains to be determined.

Binding of SAP to Chromatin

Of considerable interest has been the recent study of the specific interaction between SAP and nuclear chromatin. Robey et al. (1984) have reported that CRP binds to chromatin with great avidity and suggested that this represents a significant biological role for the protein. SAP was not studied. However, we have recently shown that under conditions of physiological ionic strength SAP binds to both single and double stranded DNA as well as to nucleosome core particles, to long chromatin and to long chromatin stripped of H1 histone (Pepys and Butler 1987). In our studies CRP did not bind to any significant degree to any of these ligands and from the milieu of whole normal or acute phase human serum, SAP was the only protein which underwent specific calcium-dependent binding to DNA. With isolated pure proteins, binding of both pentraxins was considerably enhanced at low ionic strength (34 mM). The specific interaction of SAP with DNA was conserved among species and was also noted using sera from the mouse, rat, guinea pig, cow and plaice. In contrast, CRP isolated from rat, rabbit, plaice and *Limulus*, as well as hamster female protein, did not bind DNA, thus resembling the behaviour of human CRP.

The finding that SAP is the single major calcium-dependent DNA binding protein in whole normal serum, that this interaction occurs in a physiological milieu and that it is conserved across species, argues that it may represent an important biological phenomenon. Furthermore, we have recently found deposition of SAP in vivo on extracellular aggregates of chromatin in lesional skin of certain patients with systemic lupus erythematosus, indicating that SAP may indeed participate in the handling of such material in vivo (Breathnach et al. 1989b).

SAP and Amyloidosis

SAP has been detected immunochemically in all forms of amyloidosis, regardless of the site in the body in which it occurs (Breathnach et al. 1981a; Pepys et al. 1982; Pitkanen et al. 1983; Rowe et al. 1984), though modified immunohistochemical techniques have been required to detect its presence in certain tissues, particularly when formalin-fixed (Coria et al. 1988). AP is identical to and derived from SAP, which evidently has the capacity for calcium-dependent binding to specific ligands present in all different types of amyloid fibrils. This is readily demonstrable in vitro (Pepys et al. 1979) and it has been firmly established that circulating SAP is laid down as AP in amyloid deposits in vivo (Baltz et al. 1986a).

Amyloid P component is a minor constituent of amyloid deposits, the mass present being in the range of 5%–15% of the mass of fibrils present (Pepys et al. 1979). It is intimately associated with the fibrils but is not part of their structure. The role of AP in amyloidosis is not known. One possibility is that its presence there is an epiphenomenon, reflecting the coincidental presence of a ligand to which it binds on the amyloid fibrils, and that having bound it has no functional effects. Alternatively AP may contribute in some way to the deposition and persistence of amyloid fibrils and hence of the deposits themselves.

The nature of the ligand on amyloid fibrils to which SAP binds is not known, but possible candidates are a protein structural conformation shared by all the different protein subunits which can form amyloid fibrils (Hind et al. 1984c), or, perhaps more likely, the sulphated glycosaminoglycans which are intimately associated with them. Recent work has demonstrated that SAP specifically binds heparan sulphate and dermatan sulphate in a calcium-dependent interaction (Hamazaki 1987). Once bound to the fibrils in vivo it has been postulated that the bound SAP may serve a "protective" role in amyloid, shielding the fibrils in some way from degradation, possibly by acting as an enzyme inhibitor or by acting as an autologous "native" protein coat masking the altered nature of the fibrils and thereby preventing them from stimulating or activating phagocytic cells (Pepys et al. 1982; Baltz et al. 1986b). One way of testing this hypothesis would be to dissociate AP from the fibrils in vivo, and low molecular weight sugar ligands have been identified and synthesized which can do this in vitro (Hind et al. 1984a,c), but analogues which function in vivo have not yet been developed. Once they are obtained they or their congeners or derivatives may have potential as therapeutic agents.

Tissue Amyloid P Component

It is intriguing that a protein which is immunochemically indistinguishable from circulating SAP is present as a normal connective tissue matrix protein in human tissues. It is an integral constituent of the glomerular basement membrane (Dyck et al. 1980), where it is covalently linked to collagen and/or other matrix glycoproteins and it is also found in non-covalent association with the peripheral microfibrillar mantle of elastic fibres throughout the body (Breathnach et al. 1981b; 1989a). This normal tissue amyloid P component is not present during foetal life or at birth but it accumulates during early childhood and is constantly present thereafter (Khan and Walker 1984). It is not known whether the tissue amyloid P component expresses the ligand binding properties of serum amyloid P component but if it does this could constitute a focal mechanism for the localization and deposition of its ligands in association with connective tissue. It is noteworthy that in many tissues early deposition of amyloid fibrils takes place in close relation to elastic fibres (Yanagihara et al. 1985).

SAP as a Targeting Vehicle in Amyloidosis

Regardless of whether serum amyloid P component plays a part in the pathogenesis of amyloidosis, its specific deposition from the circulation into amyloid deposits can be used as a means for transporting agents coupled to it into them. This was demonstrated for the first time with ^{125}I-labelled mouse serum amyloid P component in experimental mouse AA amyloidosis (Baltz et al. 1986a). After intravenous injection into the mice, ^{125}I–SAP localized specifically to the amyloid deposits, and was detected by organ counting and by autoradiography after killing the animals.

In these studies only a small proportion of the injected dose localized, though this was highly significant compared to control animals. The plasma clearance of ^{125}I-mouse SAP was similar in amyloid and control mice, and appeared to have a circulating half-time in the order of 7.2–8.0 hours. In view of later data it seems likely that the lack of an accelerated plasma clearance in the presence of amyloidosis was due to the very small proportion of the injected SAP which localized. Human SAP, either in purified form or in whole normal human serum, was also used in these studies, and was shown by immunohistochemistry to localize specifically in the amyloid deposits, though the proportion of the injected material was not quantitated.

Scintigraphic Imaging of Amyloid Deposits In Vivo

Clinical diagnosis of amyloidosis is dependent on histological examination of tissue biopsy material or resected specimens. Generally, these are very small samples and cannot be used to determine the extent and distribution of amyloid deposits within the tissue or organ in question or in the body as a whole. In vivo radionuclide imaging techniques, based on the presence of abundant calcium in soft tissue amyloid deposits (Kula et al. 1977), have been attempted with variable success, but they are not specific and have limited sensitivity (Yood et al. 1981; Johnston et al. 1982). Such studies have generally used ^{99m}Tc-pyrophosphate, as in bone scanning, and have in particular been evaluated in cardiac amyloid, with echocardiography having been found superior by several groups (Wizenberg et al. 1982; Falk et al. 1983; Eriksson et al. 1984; Hind et al. 1986).

A more specific approach has been to direct radiolabelled monoclonal antibodies to the AA protein subunits of the fibrils in experimentally induced murine reactive systemic amyloidosis (Marshall et al. 1986). This proved effective, though the procedure would be severely limited in a clinical setting by a number of difficulties, especially the need to develop specific antibodies against each type of amyloid fibril: AA, AL (derived from monoclonal immunoglobulin light chains and requiring individual anti-idiotype antibody for each), prealbumin variants, and others, the chemistry of which is not even known in many cases. Injection of heterologous immunoglobulins would also pose problems.

A more logical, specific and widely applicable approach to this situation has been made by our group using radioiodine-labelled SAP, both in experimental murine amyloidosis and more recently in man.

Radiolabelled-SAP Studies in Experimental Murine Amyloidosis

Experimental AA amyloidosis may be induced in susceptible strains of laboratory mice either by repeated daily subcutaneous injections of casein, causing a chronic inflammatory reaction that results in amyloid deposition after several weeks (Janigan 1965), or within 24 hours by producing a brief but marked acute phase response, with subcutaneous silver nitrate, following a single intravenous injection of amyloid enhancing factor (AEF) (Axelrad et al. 1982). This powerful substance is produced from crude extracts of homogenized amyloidotic organs, or in lesser quantities from normal tissues, though the mechanism by which AEF accelerates amyloid deposition is not yet known despite much effort to characterize its chemical properties. In our laboratory the AEF method of amyloid induction generally results in lesser and more variable deposits than the casein method.

The first radiolabelled-SAP imaging experiments used ^{131}I-labelled human SAP in mice with casein induced amyloid (Caspi et al. 1987). Selective concentration of radioactivity was seen by gamma camera imaging within the abdomens of the amyloid mice compared to no significant retention of the tracer in the controls. Though images at 24 and 48 hours were diagnostic for amyloid, resolution was poor and individual organs could not be identified. However, resolution was later improved enormously by using ^{123}I as the label (Hawkins et al. 1988a), this isotope having ideal qualities for imaging purposes, being a medium energy pure gamma emitter. Excellent results were obtained with both heterologous human SAP and isologous mouse SAP, and these are described in detail below.

Mouse Imaging Studies Using ^{123}I-Human SAP

After intravenous injection of ^{123}I-human SAP into normal, control mice, gamma camera images showed the presence of activity compatible with its distribution in

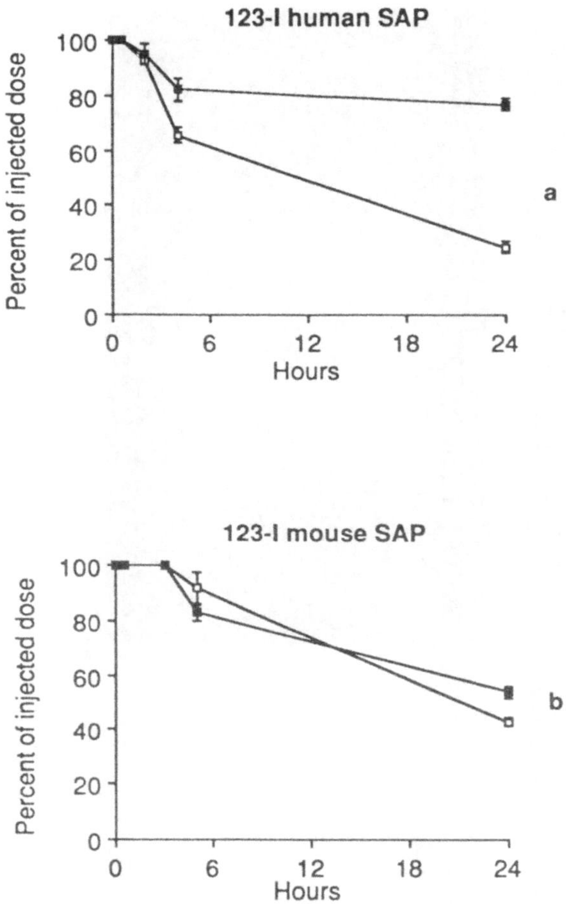

Fig. 15.1a,b. Whole body clearance of ^{123}I-SAP in amyloidotic (■) and control (□) mice. Each point represents the mean ±SD of three animals. **a** ^{123}I-human SAP; **b** ^{123}I-mouse SAP.

the blood pool. There was rapid whole body clearance so that only 65% of injected counts remained after 4 hours and 25% or less at 24 hours (Fig. 15.1a). Over this period there was some uptake of activity into the thyroid glands, which were not blocked beforehand, and minor retention of iodinated material in the upper abdomen and the bladders of some animals. Control mice killed at 24 hours had a significant proportion of the total body counts in their stomachs, presumably as a result of swallowing excreted material during grooming, combined with gastrointestinal stasis due to repeated ether anaesthesia. There was no evidence of any significant localization of activity elsewhere in the control animals.

In contrast, mice that had received repeated casein injections, and that all had amyloidosis, yielded distinctly different images within 30 min of injection of ^{123}I-human SAP (Fig. 15.2). There was major, specific localization of activity in their spleens and livers, no localization elsewhere, and significant whole body retention of counts compared with the controls. These results were confirmed by the counts detected in the individual organs removed 24 hours after injection

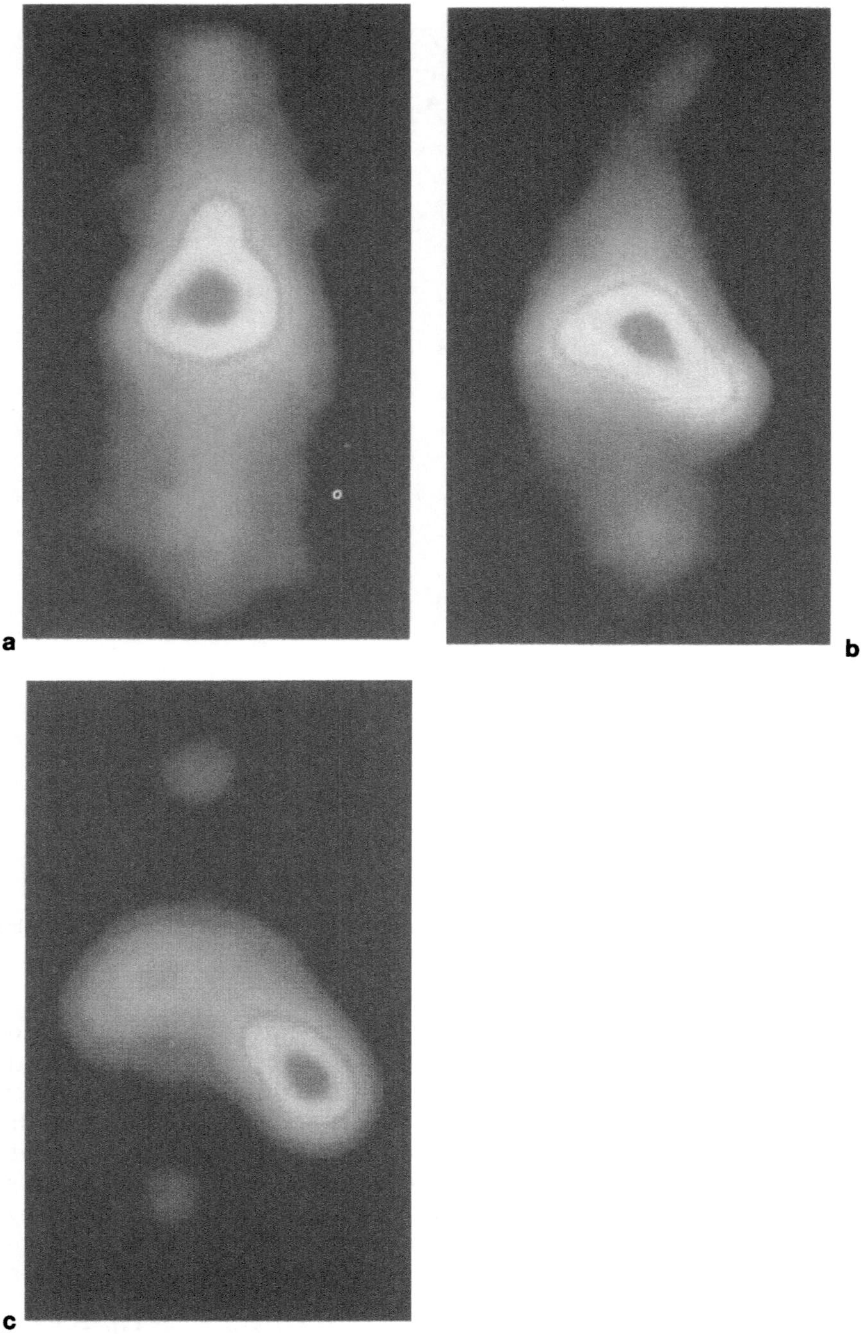

Fig. 15.2.

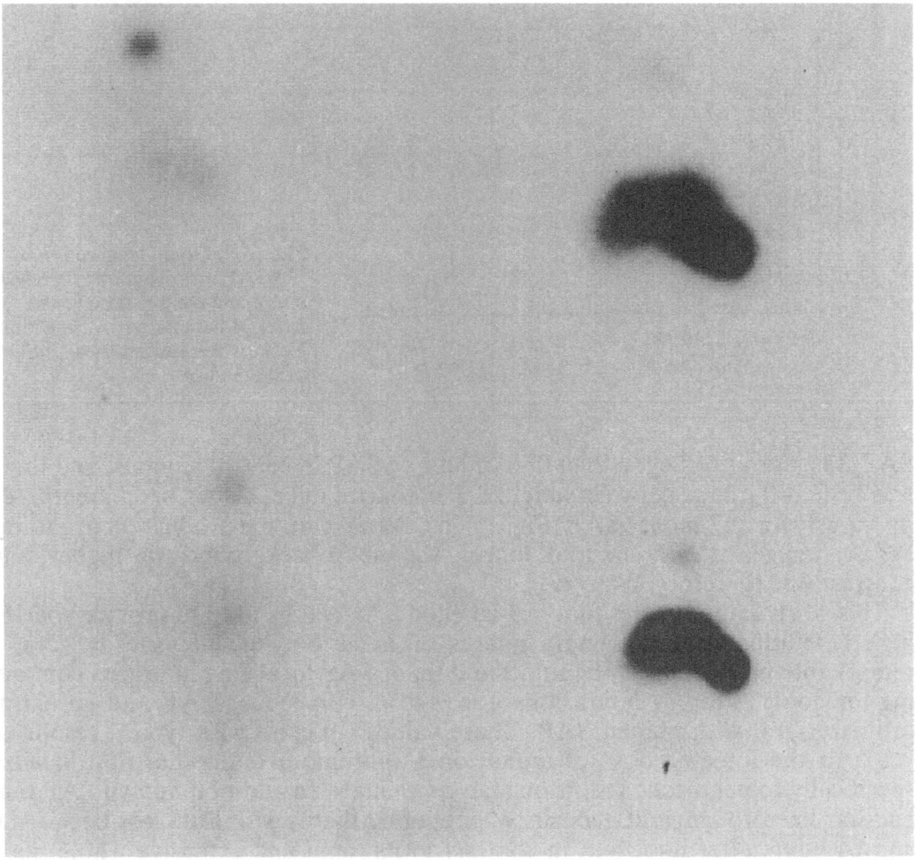
d

Fig. 15.2a–d. Ventral scintigrams of mice after i.v. injection of ^{123}I-human SAP. Serial images of a single mouse with experimental amyloidosis **a** immediately after injection, **b** at 0.5 hours and **c** at 24 hours. Activity has localized from the blood pool to the amyloid laden liver and spleen. **d** Composite image of four mice 24 hours after injection. Controls are on the *left*, casein-induced amyloid mice on the *right*. The latter show massive localization and persistence of activity in the liver and spleen with traces in the (unblocked) thyroid. The controls show thyroid activity only.

(Fig. 15.3), corresponding exactly with the presence of amyloid in histological sections. None of the control animals contained any amyloid, whereas the casein-injected mice all had similar, massive amyloid deposits in their spleens and less extensive deposits in their livers. There were only minute traces of amyloid in the kidneys of some of the test mice and none was found elsewhere.

Mouse Imaging Studies Using ^{123}I-Mouse SAP

Injection of labelled mouse SAP gave results similar to those with human SAP except that the proportion of the dose localized to amyloidotic organs was smaller. Furthermore, there was more rapid clearance of organ-localized activity than occurred with labelled human SAP. Whole body retention of ^{123}I-mouse

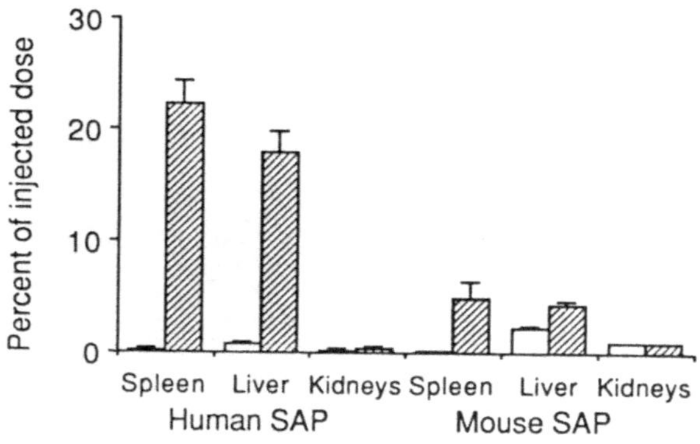

Fig. 15.3. Organ localization of ^{123}I-SAP in amyloidotic (*hatched bars*), and control mice (*open bars*) 24 hours after i.v. injection. Each bar represents mean ±SD of three mice.

SAP was more prolonged than that of human SAP in control animals, and there was little difference between amyloid and control mice, as has been mentioned previously for ^{125}I-mouse SAP (Fig. 15.1b). As a result, the production of gamma camera images at 24 hours took longer, the blood background was higher, and contrast was therefore less good.

This work demonstrates that ^{123}I-labelled SAP can be used to provide specific high resolution gamma camera images of tissue amyloid deposits in vivo, a remarkable proportion of the injected dose of SAP localizing in organs containing amyloid. After 24 hours, this was ~10% with mouse SAP and an extraordinary 40% with human SAP. These values compare with typical results of <1% of the injected dose of monoclonal antitumour antibodies that localize specifically to neoplastic lesions in man or animals. In addition human SAP that had localized to amyloid deposits was retained there, with little loss between 4 and 24 hours after injection, in contrast to its rapid and extensive whole body clearance in control mice. The whole body clearance of mouse SAP was slower than that of human SAP but was similar in amyloidotic and control mice. Together with the smaller proportion of the injected dose that localized to amyloid, this suggests that mouse SAP has a lower affinity for binding to isologous AA fibrils than does the heterologous protein, human SAP. The significance of this finding is unclear but in purely operational terms, it means that human SAP is a better targeting agent for mouse AA amyloid than is mouse SAP.

Localization of ^{125}I-Pentraxins in AEF-Induced Amyloid

The specificity of radiolabelled SAP localization as a method of detecting amyloid has been further examined by using the homologous human and *Limulus* CRP as control proteins (Fig. 15.4). Autopsy organ counting of radioactivity was performed 24 hours after injection of various ^{125}I-pentraxins. The CRP molecules closely resemble their SAP counterparts in terms of structure but differ with respect to their calcium-dependent ligand-binding specificity. Both human and mouse ^{125}I-SAP showed highly significant specific uptake into amyloidotic compared with normal, control livers and spleens, and the proportions of the injected doses that localized were similar to those seen in the above imaging experiments. Labelled human CRP showed marginally greater uptake into

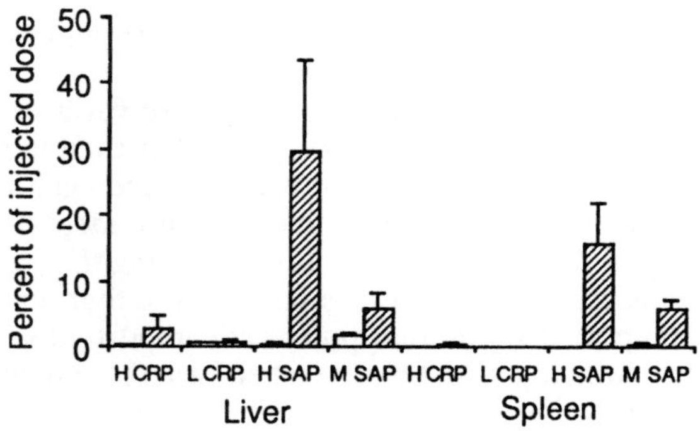

Fig. 15.4. Organ localization of ^{125}I-pentraxins in AEF-induced amyloidotic (*hatched bars*), and control mice (*open bars*) 24 hours after i.v. injection. Each bar represents the mean ±SD of five mice.

amyloidotic than normal organs, but the quantities involved were extremely small; e.g. the amount localizing to the spleen was ~10% of that of mouse SAP and 2.5% of human SAP. *Limulus* CRP did not show any such localization at all.

^{125}I-SAP Localization as a Method of Quantitating Murine Amyloid Deposits

Although the localization of radiolabelled SAP into amyloid deposits following its introduction into the circulation has thus been shown to be a highly specific phenomenon, it remains unclear as to how quantitative such data may be, that is whether the quantity of SAP that localizes reflects the absolute amount of amyloid present. This question is difficult to answer precisely because of inherent problems in the accurate estimation of the amyloid content of tissues. Histological examination of Congo red-stained sections cannot reasonably sample more than a very small proportion of the affected organ, and even then the light microscopic evaluation of green birefringent areas may not reflect quantitatively the absolute amount of amyloid fibril deposits. Nevertheless, in the above experiments the mice with casein-induced amyloid all had similar and extensive Congophilic, green birefringent deposits in their spleens and livers and there was very little variation in the proportion of the injected dose of labelled SAP that localized. On the other hand, in the AEF-induced model there was much less amyloid in the tissues assessed histologically, and much greater variation between mice than in the casein-induced model. While in these mice the absolute specificity of SAP for amyloidotic organs was retained, there were only imperfect correlations between labelled SAP localized and histological assessment of the amount of amyloid present.

A radioimmunoassay for tissue AA was developed in the hope of overcoming the sampling and interpretation problems of histology. Although this was highly specific and gave no false-positive results it showed only partial correlation with Congo red staining and with quantity of SAP localized as a means of quantitating amyloid. Given the complexity of amyloid deposits, including fibrils of different ages, glycosaminoglycans, and SAP, there is no *a priori* reason to expect that there should necessarily be a close correlation between Congo red birefringence (the precise molecular basis of which is not known), AA antigen content, and SAP deposition. Indeed, appreciably more SAP seems to localize in the spleen than in the liver in association with apparently similar quantities of amyloid in the mouse model.

It is possible, therefore, that these variable correlations are due to inherent features of amyloid pathology, rather than merely the operational problems of amyloid quantitation. For example, the ligand on amyloid fibrils to which SAP binds has not yet been biochemically identified; it is not known whether it is exposed on all fibrils at all times during the pathogenesis of fibril deposition, or whether competitive ligands exist either in body fluids or on other tissue components. Another potentially important variable is local vascular permeability, which may determine when, where, and how much SAP (Mr 254 000) leaves the circulation to enter amyloidotic tissue. It will be important, therefore, to determine whether or not SAP localizes with the same kinetics and to the same extent in relation to similar amounts of amyloid fibrils in different tissue sites and at different stages of the disease.

However, there is recent evidence that mouse organ counting following injection of ^{125}I-human SAP can be a considerably more sensitive marker of amyloid deposition than is conventional Congo red histology (P. N. Hawkins and M. B. Pepys, unpublished observations).

Radiolabelled SAP Studies in Man

Preparation of ^{123}I-Human SAP for Clinical Studies

The logical and exciting progression from the work described above was to apply the same technique to man. Though simple in principle, inevitably, several problems were posed. Heading this list was the question of the source of SAP for clinical use, and initially the protein was prepared from each subject using 50 ml of autologous serum. It rapidly became evident that the yield of material of acceptable purity in such small scale preparations was inconsistent, and the procedure was technically difficult. Therefore the use of isologous SAP was considered. There is no evidence for the existence of structurally polymorphic forms of SAP in the normal population or among subjects who develop amyloidosis, though a restriction fragment length polymorphism in genomic DNA upstream of the human SAP gene has recently been described (Woo et al. 1987). However, the polymorphic site is not within the region coding for SAP and extensive studies of SAP from individuals of the three different genotypes described, and from a subject with AA amyloidosis have failed to reveal any phenotypic differences in structure or ligand binding properties (P. N. Hawkins and M. B. Pepys, unpublished observations). In addition it is known that normal serum levels, around 30 mg l^{-1}, are not significantly altered in amyloidosis. Therefore SAP was isolated from 5 litres of serum donated by a single healthy accredited volunteer, kindly provided by the National Blood Transfusion Service, and was purified to greater than 99% purity under strictly sterile conditions. Pure SAP was stored at −70° C until required when 1 mg aliquots were labelled with ^{123}I using the mild oxidant *N*-bromosuccinimide as described by Mather and Ward (1987). Very high efficiency labelling can be achieved with

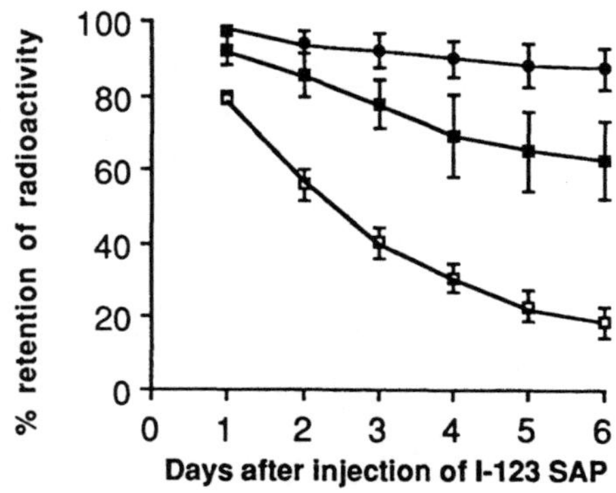

Fig. 15.5. Whole body retention of radioactivity following injection of ^{123}I-SAP in control subjects (□, *n* = 5), amyloid patients with (●, *n* = 2) and without (■, *n* = 5) hepatomegaly.

incorporation of 90%–95%. Functional integrity of the SAP is not affected by this procedure and the resulting material was repeatedly shown to be sterile and non-pyrogenic. Imaging studies with this tracer have also permitted for the first time direct measurement of the turnover of SAP in health and disease.

Imaging of Human Amyloid Deposits

Images obtained after intravenous injection of ^{123}I-SAP into normal subjects showed the distribution of activity confined to the blood pool. There was rapid whole body clearance of activity, with only 56% of injected counts remaining at 2 days and less than 20% at 6 days (Fig. 15.5). Minor uptake occurred around the salivary glands, the thyroid gland and the stomach mucosa (despite cold iodide "blockade") reflecting the normal localization of radioiodine (Fig. 15.6). In contrast, in all patients with known or suspected amyloidosis there was major specific localization of the ^{123}I-SAP, resulting in very high resolution scintigrams, permitting amyloid deposits to be identified within the liver, spleen, kidneys, heart, bone marrow and even the adrenal glands and carpal tunnel region (Figs. 15.7, 15.8).

In some sites, organ localization was seen within minutes of injection, this frequently being noted with hepatic amyloid, though in other sites it appeared to occur more slowly, for example in the heart. However, as the tracer appeared to persist once it had localized, most patients could be imaged at 24 hours. With persistence of the tracer within the amyloid, and further clearance of it from the normal "background" blood pool, the signal-to-noise ratio can be seen to improve with time, and indeed imaging at 48 hours could on occasions clarify the situation with borderline deposits, despite the isotope nearing the end of its useful life.

Human SAP Turnover Studies

The normal turnover of human SAP has now been established. It has a circulating half-time of 31 hours and a steady state circulating level of around 30 mg l^{-1}. It is thus being synthesized at a rate of around 1.0 mg hour^{-1} and the large molecules

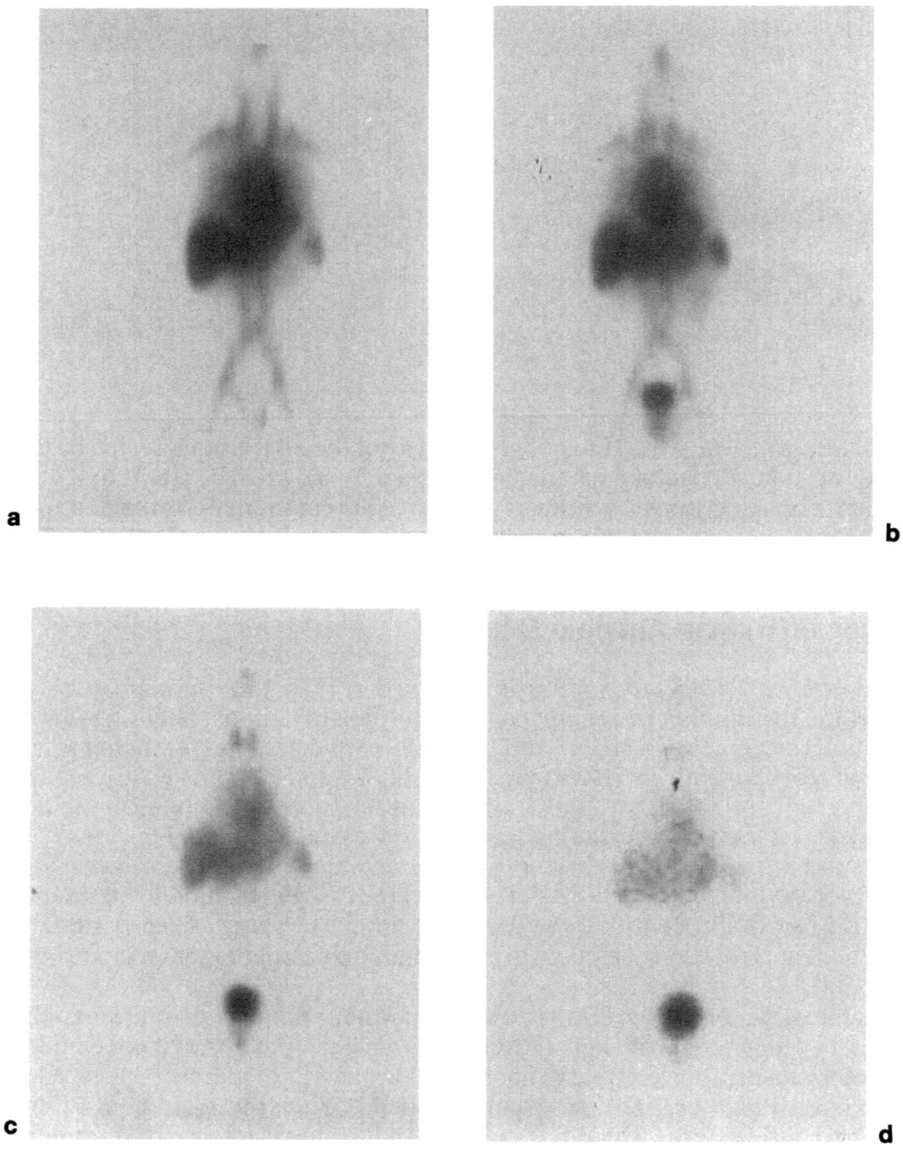

Fig. 15.6a–d. Anterior whole body images of a normal adult male after i.v. injection of ^{123}I-SAP: **a** 30 min; **b** 6 hours; **c** 24 hours; **d** 48 hours. Activity is confined to the blood pool with no selective localization. Rapid degradation of the tracer is associated with appearance of excreted products in the bladder.

remain predominantly intravascular. In amyloidosis SAP can apparently leave the circulation with great rapidity and become incorporated and persist within amyloid deposits (Fig. 15.9). Plasma clearance of injected ^{123}I-SAP is therefore increased, and as the radioactive protein which has localized to the amyloid is apparently protected from the normal rapid degradation pathways that exist

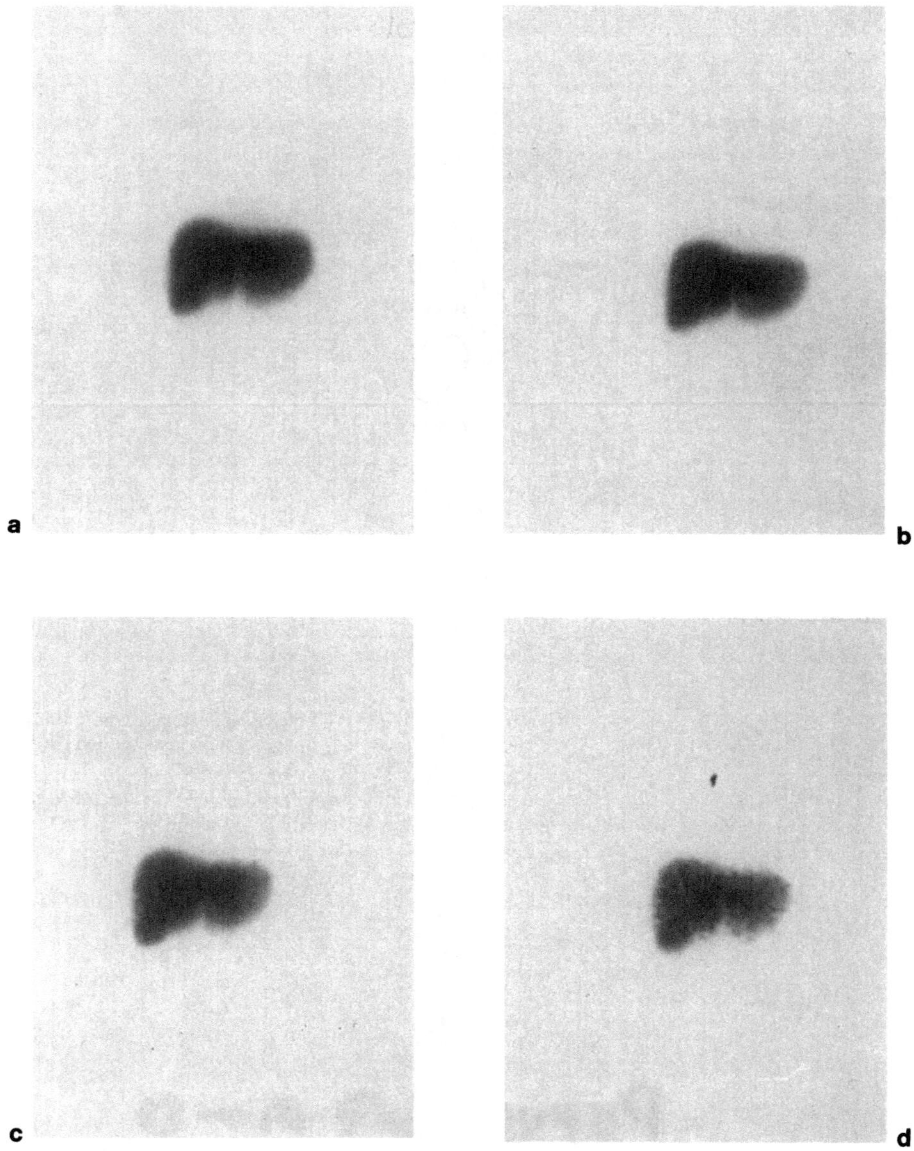

Fig. 15.7a–d. Anterior whole body images of a patient with AA amyloidosis complicating mesothelioma after i.v. injection of ^{123}I-SAP: **a** 30 min; **b** 6 hours; **c** 24 hours; **d** 48 hours. More than 95% of the tracer has localized to the hepatic (and splenic) amyloid deposits and persisted there throughout the study. The spleen is not seen in this project.

within the circulation, whole body retention of radioactivity is increased. This can be monitored either by whole body scanning or, probably more accurately, by counting the excretion of radioactivity in serial complete 24 hour urine collections, as there has been shown to be very little loss via other routes (Matthews 1957).

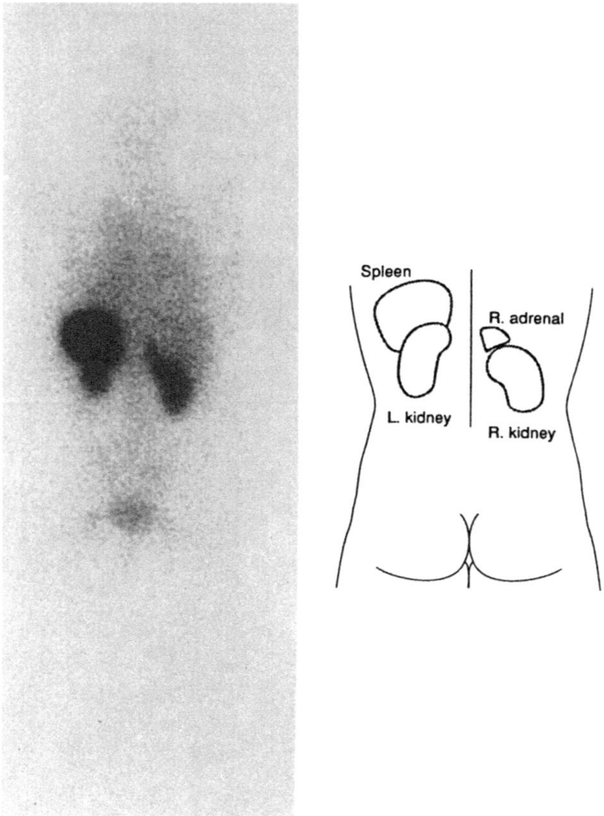

Fig. 15.8. Posterior whole body image of a 54 year old lady with AA amyloidosis and rheumatoid arthritis, 24 hours after i.v. injection of ^{123}I-SAP.

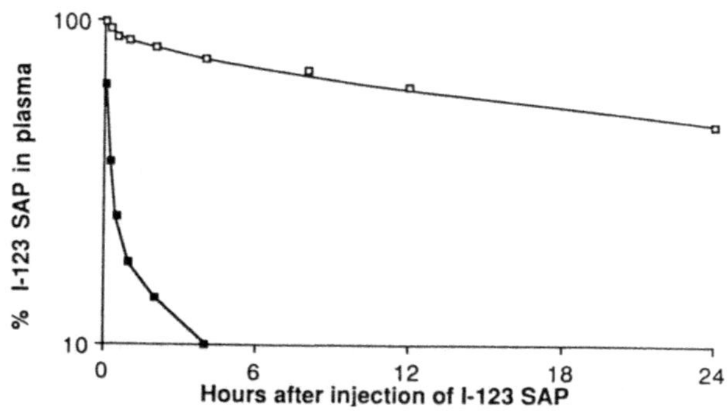

Fig. 15.9. Plasma clearance of ^{123}I-SAP in healthy control subjects (□, n = 5) and amyloid patients with hepatomegaly (■, n = 4). Each point represents the mean of each group.

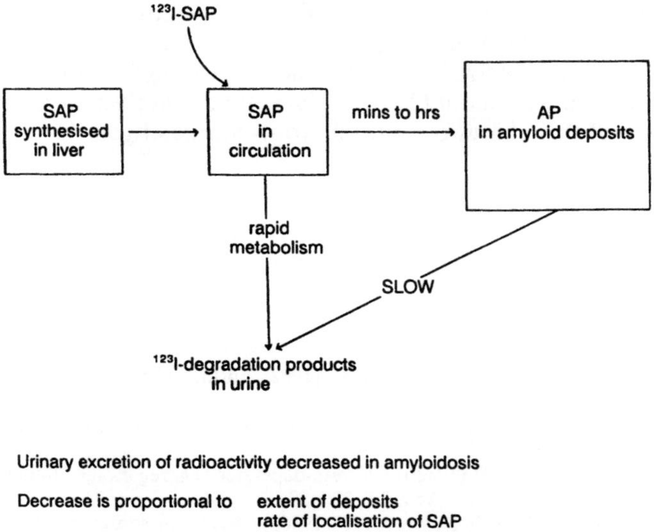

Fig. 15.10. Schematic representation of the abnormal metabolic pathway of SAP in amyloidosis.

The data from these metabolic studies indicates that there must exist an exchange of SAP between the circulation and the amyloid deposits, and also that the pool of SAP in the latter is several orders of magnitude greater than the former. This has been confirmed by eluting AP from whole amyloidotic organs obtained at autopsy in several patients (unpublished observations). Preliminary results from the turnover studies also suggest that the rate of synthesis is considerably increased in some amyloid patients, an area of research currently being assessed in more detail. A summary of SAP metabolism is given in Fig. 15.10.

Conclusions

The diverse, but phylogenetically conserved ligand binding properties of SAP have been discussed, both regarding a possible physiological role and the pathological situation in amyloidosis.

The latter has been used as the basis of a diagnostic clinical investigation (Hawkins et al. 1988b), and the availability of SAP as a highly specific targeting vehicle for amyloid deposits opens up the possibility of its use in therapy, for example, by coupling to SAP an agent capable of promoting the destruction and resorption of amyloid fibrils. The sustained persistence of SAP in amyloid deposits makes this a promising approach.

The normal function of SAP remains elusive, though a specific interaction with nuclear chromatic released from dead or dying cells is an attractive hypothesis. Resolution of the crystal structure of SAP, now at an advanced stage of progress (Wood et al. 1988 and this volume), should help to elucidate the molecular basis of the binding mechanism of this ubiquitous protein to its various ligands.

References

Axelrad MA, Kisilevsky R, Willmer J, Chen SJ, Skinner M (1982) Further characterisation of amyloid-enhancing factor. Lab Invest 47:139–146

Baltz ML, de Beer FC, Feinstein A et al. (1982a) Phylogenetic aspects of C-reactive protein and related proteins. Ann NY Acad Sci 389:49–75

Baltz ML, de Beer FC, Feinstein A, Pepys MB (1982b) Calcium-dependent aggregation of human serum amyloid P component. Biochim Biophys Acta 701:229–236

Baltz ML, Caspi D, Evans DJ, Rowe IF, Hind CRK, Pepys MB (1986a) Circulating serum amyloid P component is the precursor of amyloid P component in tissue amyloid deposits. Clin Exp Immunol 66:691–700

Baltz ML, Caspi D, Rowe IF, Hind CRK, Evans DJ, Pepys MB (1986b) Pathogenetic mechanisms and precursor–product relationships in murine amyloidosis. In: Glenner GG, Osserman EF, Benditt EP, Calkins E, Cohen AS, Zucker-Franklin D (eds) Amyloidosis. Plenum Press, New York, pp 101–108

Breathnach SM, Bhogal B, Dyck RF, de Beer FC, Black MM, Pepys MB (1981a) Immunohistochemical demonstration of amyloid P component in skin of normal subjects and patients with cutaneous amyloidosis. Br J Dermatol 105:115–124

Breathnach SM, Melrose SM, Bhogal B et al. (1981b) Amyloid P component is located on elastic fibre micofibrils of normal human tissues. Nature 293:652–654

Breathnach SM, Booker J, Ashworth J, Pepys MB, Hintner H (1989a) Tissue amyloid P component in normal human dermis is non-covalently associated with elastic fiber microfibrils. J Invest Dermatol (in press)

Breathnach SM, Kofler H, Sepp N et al. (1989b) Serum amyloid P component binds to cell nuclei in vitro and to in vivo deposits of extracellular chromatin in systemic lupus erythematosus. (submitted)

Caspi D, Zalzman S, Baratz M et al. (1987) Imaging of experimental amyloidosis with ^{131}I-serum amyloid P component. Arth Rheum 30:1303–1306

Coria F, Castano E, Prelli F et al. (1988) Isolation and characterization of amyloid P component from Alzheimer's disease and other types of cerebral amyloidosis. Lab Invest 58:454–458

de Beer FC, Baltz M, Holford S, Feinstein A, Pepys MB (1981) Fibronectin and C4-binding protein are selectively bound by aggregated amyloid P component. J Exp Med 154:1134–1149

Dyck RF, Lockwood M, Kershaw M et al. (1980) Amyloid P component is a constituent of normal glomerular basement membrane. J Exp Med 152:1162–1174

Eriksson P, Backman C, Bjerle P, Eriksson A, Holm S, Olofsson B-O (1984) Non-invasive assessment of the presence and severity of cardiac amyloidosis. A study in familial amyloidosis and polyneuropathy by cross sectional echocardiography and technetium-99m-pyrophosphate scintigraphy. Br Heart J 52:321–326

Falk RH, Lee VW, Rubinow A, Hood WB, Cohen AS (1983) Sensitivity of technetium-99m-pyrophosphate scintigraphy in diagnosing cardiac amyloidosis. Am J Cardiol 51:826–830

Hamazaki H (1987) Ca^{2+}-mediated association of human serum amyloid P component with heparan sulfate and dermatan sulfate. J Biol Chem 262:1456–1460

Hawkins PN, Myers MJ, Epenetos AA, Caspi D, Pepys MB (1988a) Specific localization and imaging of amyloid deposits in vivo using ^{123}I-labelled serum amyloid P component. J Exp Med 167:903–913

Hawkins PN, Myers MJ, Lavender JP, Pepys MB (1988b) Diagnostic radionuclide imaging of amyloid: biological targeting by circulating human serum amyloid P component. Lancet i:1413–1418

Hind CRK, Collins PM, Caspi D, Baltz ML, Pepys MB (1984a) Specific chemical dissociation of fibrillar and non-fibrillar components of amyloid deposits. Lancet ii:376–378

Hind CRK, Collins PM, Pepys MB (1984b) Calcium-dependent aggregation of human serum amyloid P component. Inhibition by the cyclic 4,6-pyruvate acetal of galactose. Biochem Biophys Acta 802:148–150

Hind CRK, Collins PM, Renn D et al. (1984c) Binding specificity of serum amyloid P component for the pyruvate acetal of galactose. J Exp Med 59:1058–1069

Hind CRK, Collins PM, Baltz ML, Pepys MB (1985) Human serum amyloid P component, a circulating lectin with specificity for the cyclic 4,6-pyruvate acetal of galactose. Biochem J 225:107–111

Hind CRK, Gibson DG, Lavender JP, Pepys MB (1986) Non-invasive techniques for demonstrating cardiac involvement in the acquired forms of systemic amyloidosis. In: Glenner GG, Osserman EF, Benditt EP, Calkins E, Cohen AS, Zucker-Franklin D (eds) Amyloidosis. Plenum Press, New York, pp 617–620

Janigan DT (1985) Experimental amyloidosis. Studies with a modified casein method, casein hydrolysate and gelatin. Am J Pathol 47:159–171

Johnston JB, Rayner HL, Travenen L, Greenberg D (1982) Skeletal muscle uptake of Tc 99m pyrophosphate in amyloidosis. Am J Hematol 13:247–251

Khan AM, Walker F (1984) Age related detection of serum amyloid P in the skin. J Pathol 143:183–186

Kula RW, Engel WK, Line BR (1977) Scanning for soft-tissue amyloid. Lancet i:92–93

Mantzouranis EC, Dowton SB, Whitehead AS, Edge MD, Bruns GAP, Cotton HR (1985) Human serum amyloid P component. cDNA isolation, complete sequence of pre-serum amyloid P component, and localization of the gene to chromosome 1. J Biol Chem 280:7752–7756

Marshall J, McNally W, Muller D et al. (1986) In vivo radioimmunodetection of amyloid deposits in experimental amyloidosis. In: Glenner GG, Osserman EF, Benditt EP, Calkins E, Cohen AS, Zucker-Franklin D (eds) Amyloidosis. Plenum Press, New York, pp 163–174

Mather SJ, Ward BG (1987) High efficiency iodination of monoclonal antibodies for radiotherapy. J Nucl Med 28:1034–1036

Matthews CME (1957) The theory of tracer experiments with ^{131}I-labelled plasma proteins. Phys Med Biol 2:36–53

Osmand AP, Friedenson B, Gewurz H, Painter RH, Hofmann T, Shelton E (1977) Characterization of C-reactive protein and the complement subcomponent Clt as homologous proteins displaying cyclic pentameric symmetry (pentraxins). Proc Natl Acad Sci USA 74:739–743

Pepys MB, Baltz KL (1983) Acute phase proteins with special reference to C-reactive protein and related proteins (pentaxins) and serum amyloid A protein. Adv Immunol 34:141–212

Pepys MB, Butler PJG (1987) Serum amyloid P component is the major calcium-dependent specific DNA binding protein of the protein. Biochem Biophys Res Commun 148:308–313

Pepys MB, Dash AC, Munn EA et al. (1977) Isolation of amyloid P component (protein AP) from normal serum as a calcium-dependent binding protein. Lancet i:1029–1031

Pepys MB, Dash AC, Fletcher TC, Richardson N, Munn EA, Feinstein A (1978) Analogues in other mammals and in fish of human plasma proteins C-reactive protein and amyloid P component. Nature 273:168–170

Pepys MB, Dyck RF, de Beer FC, Skinner M, Cohen AS (1979) Binding of serum amyloid P component (SAP) by amyloid fibrils. Clin Exp Immunol 38:284–293

Pepys MB, Baltz M, de Beer FC et al. (1982) Biology of serum amyloid P component. Ann NY Acad Sci 389:286–297

Pinteric L, Assimeh SN, Kells DIC, Painter RH (1976) The ultrastructure of Clt, a subcomponent of the first component of complement: an EM and ultracentrifuge study. J Immunol 117:79–83

Pitkanen P, Westermark P, Cornwell GG, Murdoch W (1983) Amyloid of the seminal vesicles. A distinctive and common localised form of senile amyloidosis. Amer J Pathol 110:64–69

Prelli F, Pras M, Frangione B (1985) The primary structure of human tissue amyloid P component from a patient with primary idiopathic amyloidosis. J Biol Chem 260:12895–12898

Robey FA, Liu T-Y (1981) Limulin: a C-reactive protein from *Limulus polyphemus*. J Biol Chem 256:969–975

Robey FA, Tanaka T, Liu T-Y (1983) Isolation and characterisation of two major serum proteins from the dogfish, *Mustelus canis*, C-reactive protein and amyloid P component. J Biol Chem 258:3889–3894

Robey FA, Jones KD, Tanaka T, Liu T-Y (1984) Binding of C-reactive protein to chromatin and nucleosome core particles. A possible physiologic role of C-reactive protein. J Biol Chem 259:7311–7316

Rowe IF, Jensson O, Lewis PD, Candy J, Tennent GA, Pepys MB (1984) Immunohistochemical demonstration of amyloid P component in cerebrovascular amyloidosis. Neuropathol Appl Neurobiol 10:53–61

Whitehead AS, Bruns GAP, Markham AF, Colten HR, Woods DE (1983) Isolation of human C-reactive protein complementary DNA and localization of the gene to chromosome 1. Science 221:69–71

Wizenberg TA, Muz J, Sohn YH, Samlowski W, Weissler AM (1982) Value of positive myocardial technetium-99m-pyrophosphate scintigraphy in the noninvasive diagnosis of cardiac amyloidosis. Amer Heart J 103:468–473

Woo P, Korenberg JR, Whitehead AS (1985) Characterization of genomic and complementary DNA sequence of human C-reactive protein, and comparison with the complementary DNA sequence of serum amyloid P component. J Biol Chem 260:13384–13388

Woo P, O'Brien J, Robson M, Ansell B (1987) A genetic marker for systemic amyloidosis in juvenile arthritis. Lancet ii:767–769

Woods SP, Oliva G, O'Hara BP et al. (1988) A pentameric form of human serum amyloid P component: crystallization, X-ray diffraction and neutron scattering studies. J Mol Biol 202:169–173

Yanagihara M, Kato F, Shikano Y, Fukushima N, Mori S (1985) Intimate structural association of amyloid and elastic fibers in systemic and cutaneous amyloidosis. J Cutan Pathol 12:110–116

Yood RA, Skinner M, Cohen AS, Lee VW (1981) Soft tissue uptake of bone seeking radionuclide in amyloidosis. J Rheumatol 8:760–766

Subject Index

ACH 157
α1-acid glycoprotein (AGP) 30, 151, 154–6
  rat and human regulatory regions 37–8
ACTH 12
Acute inflammatory reaction (AIR) 97
Acute phase proteins 1, 7–11, 151–62
  biological properties 156
  biosynthesis by liver cells 97–105
  classification 152
  clinical measurement of 151–60
  general pattern of response in infection 154–5
  monitoring disease activity using 155–6
  synthesis 161–2
Acute phase reactants 1, 21, 29, 30, 97, 151, 172
  Class I 33
  Class II 33
  expression during acute inflammatory reaction 101–3
  expression in normal unstimulated liver 98–101
  role of hepatocytes in in vivo biosynthesis 98–103
  role of sinusoidal cells in in vivo biosynthesis 103
Acute phase response (APR) 2, 6–11, 13, 137, 151, 161
  characterization 47
  definition 29, 47
  heterogeneity in 77
  in viral infection 155–6
  molecular biology of 32
  molecular mechanisms underlying 29–30
  physiological role 30–1
  promoters induced during 37–41
  species-specific 31
  systems for study of 31–2
  transcriptional regulations of genes 35–41
Acute phase responsive elements (APREs) 54
AIDS 156
Alzheimer's disease 179
Amino acid residues 48
Amino acid sequences
  amyloid A 174
  amyloid fibril proteins 170–1
  amyloid precursor protein 60
  Apo-SAA 138
  CRP 50, 134–5
  pentraxins 49, 134
  serum amyloid A (SAA) 172, 174
Amyloid A amyloidosis 169–85
  aetiology of 176
  current hypotheses 180–1
  experimental model for 177
Amyloid A proteins 137, 171
  amino acid sequences 174
  formation and deposition of 176–8
  serum. *See* Serum amyloid A (SAA)
  structure 171–2
  tissue 171
Amyloid deposits
  AEF-induced 196–7
  imaging of human 199
  quantitating murine 197–8
  scintigraphic imaging in vivo 191–2
Amyloid enhancing factor (AEF) 178–9, 192
Amyloid fibril proteins, amino acid sequences 170–1
Amyloid L protein 171
Amyloid P component (AP) 179–80, 187–8
  ligand binding 188–9
  tissue 191
Amyloid precursor protein 60
Amyloidosis 59, 187
  classification 169–71
  definition 169
  experimental murine 192
  pathogenetic mechanisms for 176
  SAA role in 176–8
  SAP as targeting vehicle in 191
  SAP role in 190
  see also Amyloid A amyloidosis
Angiotensin I 119
Angiotensin II 119
Angiotensinogen 118–19
Antichymotrypsin 109
α1-Antichymotrypsin (ACH) 71, 151
Antigen-3 (LFA-3) 6
Antithrombin III 112, 113, 115, 117

Antitrypsin 107–9, 111
cleaved, relaxed (R) form 110
deficiency of 111–12
microheterogeneity 119
native, stressed (S) form 110
Pittsburgh 113
reactive centre oxidation 113–14
α1-Antitrypsin (α1-AT) 151, 154, 162
Apolipoproteins 60, 145, 148
Apo AI 33, 55, 176
Apo AII 33, 55, 176
Apo AA 175
Apo-SAA 59–67, 137, 175
amino acid sequences 138
cellular association and degradation of 145–7
cellular interactions and metabolism of 61
function of 65, 147–8
in plasma 138
metabolism 142–3
phosphorylation 148
plasma clearance 144–5
structure 64, 138–40
synthesis 143–4

C2 85–96
constitutive expression of 89–90
interleukin-1 and interferon-γ regulated expression 90–3
regulated gene expression 90
synthesis and secretion of 89
C4 85–6, 91–2
Caffeine 76
Candida albicans 158
Cardiopulmonary bypass 117
Cercopithecus aethiops 144
Cerebrospinal fluid (CSF) 156
Ceruloplasmin 151, 162
Chloramphenicol acetyl transferase (CAT) 22, 23, 34, 35, 54, 91
Chromatin, SAP binding to 189–90
Chronic granulomatous disease 114
COLO-16 72
Colony-stimulating factors (CSF) 11, 12
Complement proteins, molecular regulation of 85–96
Conditioned medium from monocytes (MoCM) 32, 34, 35, 38, 39, 41
Corticosteroids 12–13, 70, 75–7
Cortosol binding globulin (CBG) 107, 115–17
Corynebacterium parvum 158
C-reactive protein (CRP) 10, 21, 29, 30, 47, 69–83, 123, 124, 151, 154–7, 174, 187
amino acid sequences 50, 134–5
as direct indicator of acute phase reaction 163–6
assays for 163
CD spectra 133
clinical aspects 161–8
dynamics of secretion by rabbit hepatocytes 78–9
functions of 165–6
gene expression 39–41
genetics 54–5
intron 53
levels in different diseases 163–5
orthorhombic crystal 132
regulation of biosynthesis 71–7
regulation of secretion 77–80
retention within endoplasmic reticulum 80
role of cofactors 75–7
signal transduction 75–7
subcellular localization of intracellular pool 79–80
3′ untranslated region 53
Cysteine proteinase inhibitor (CPI) 8
Cystic fibrosis 117
Cytokines
elements responding to 54
in acute phase protein induction 71–5

Dermatomyositis 164
Diacylglycerol (DG) 24
Diisopropyl-fluorophosphate (DEP) 145
DNA 26, 40, 41, 42, 55
Drosophila 51, 52

Emphysema
in antitrypsin deficiency 111
role of oxidation in pathogenesis 114
Endoplasmic reticulum 79–80
Endothelial cells 98, 146
Endotoxaemia 117
Endotoxin (LPS) 3
Erythrocyte sedimentation rate (ESR) 155, 162
Erythrocytes 162

Factor B 85–96
constitutive expression of 89–90
expression regulated by tumour necrosis factor and interleukin-6 93
interleukin-1 and interferon-γ regulated expression 90–3
regulated gene expression 90
synthesis and secretion 87–9
Familial amyloid cardiomyopathies (FAC) 171
Familial amyloid polyneuropathies (FAP) 171
Familial Mediterranean fever (FMF) 171
Female protein (FP) 180
Fever 6–7
Fibrinogen 31, 151, 162

Glucocorticoid responsive element (GRE) 38
Glycosaminoglycans (GAGs) 178–9
GMCSF 5

Haptoglobin gene promoter 38–9
Haptoglobin (HPT) 151, 154, 156
Heat shock elements 52–3
Hepatocyte stimulating factors 7, 32, 72
  HSF-I and HSF-II 35
Hepatocytes 30, 32, 97
  role in in vivo biosynthesis of APR 98–103
High density lipoprotein (HDL) 137–50, 172, 175, 176
  apo-SAA-enriched 141
  in plasma 138
  metabolic function 142–3
  structure of acute phase 140–2
HNF1 26

ICAM-1 6
Immunoglobulins 162
Immunomodulation 156–9
Immunostimulation 156
Infections 164
Inflammation 29, 31, 59, 109–11, 118–19, 152, 174, 176
Interferon-β2 (IFN-β2) 33
Interferon-γ (IFN-γ) 91
Interleukin-1 (IL-1) 31, 33–5, 37, 42, 54, 71–3, 90–3, 102, 153, 161
  and acute phase reactions 6–11
  and corticosteroid 12–13
  and other cytokines 11–13
  functions of 162
  history and identification 2
  in acute and chronic inflammatory disorders 13
  in acute phase response 1–20, 13
  in vivo action of 13
  induction of other cytokines by 11–12
  interaction with other cytokines 4
  local effects 162
  multiple systemic effects 4
  pleotropic actions 4–6
  receptor 3–4
  regulation 3
  sources 3
  structure 2–3
  synergy of action 11
  synthesis 3
  systemic effects 162
Interleukin-6 (IL-6) 11, 33–5, 37, 39, 42, 54, 73–5, 91, 93, 102, 161
Ito cells 98

Kaposi's sarcoma 156
Kupffer cells 98, 103, 178

Leukocytosis 7
Limulus 48, 49, 51, 196–7
Limulus polyphemus 70, 188
Lipid metabolism 65
Lipopolysaccharide (LPS) 31, 72, 174–5
Lipoproteins 60
Liver cells
  biosynthesis of acute phase proteins by 97–105
  composition of 97
Liver disease in antitrypsin deficiency 111–12
Liver-specific gene expression 32–5
Liver-specific promoters and enhancers 35–7
Low density lipoproteins (LDL) 165
LTB4 7
Lymphocyte activating factor 2

Major histocompatibility complex. *See* MHC
Malignant tumours 165
Meningitis 154
Methyl 4,6-O-(1-carboxyethylidene)-β-D-galactopyranoside (MOβDG) 127, 128, 189
MHC class I genes 86
MHC class II genes 86
MHC class III genes 85, 86, 90
Microbial growth 152–3
MLTF 26
MMTV 38
mRNA 30, 34, 52, 53, 61, 87–90, 100
  SAA specific 22
MRP-8 55
MRP-14 55
Myocardial infarction 164

Negative acute phase reactants 33
Nuclear transcription factor 25

Oligo-A stretch 53

Pentraxins 30, 47–57, 123
  in AEF-induced amyloid 196–7
  structure of 48–52
PEPCK enzyme 5, 10
Peritonitis 154
Phorbol-ester (TPA) 3
Phorbol 12-myristate 13-acetate (PMA) 22–5
Phosphatidylcholine (PC) 25
Plasma proteins 29
Plasma serine protease inhibitors. *See* Serpins
Plasmodium falciparum 157
Pneumocystis carinii 156
Positive acute phase reactants 29
Pre-RNA 52
Protein AP 179–80
Protein kinase C (PKC) 147–8
Protein synthesis 40
Pseudomonas aeruginosa 116
Purine-pyrimidine repeat region 53

Recombinant DNA 32
Renal transplantation 164–5
Rheumatic diseases 165
RNA 32, 40
RNA polymerase II 41

Salmonella typhimurium 158
Schistosoma mansoni 158
Scintigraphic imaging, amyloid deposits in vivo 191–2
Scleroderma 164
Septicaemia 117, 154
Serine protease inhibitors. *See* Serpins
Serpins 107–22
  cleaved tertiary structure 108
  evolutionary loss of SR change 118
  human 108
  inactivation by pathological proteases 116–17
  local inactivation of inhibitors 115
  modulation of hormone transport 115–16
  N-termina cleavage 118–19
  native tertiary structure 108–9
  neutrophil chemotaxis 115
  protease inhibition mechanism 112
  reactive centre 112–14
  shock syndromes 117–18
  specificity of inhibition 112–14
  SR conformational change 114–18
  structural modifications in inflammation 118–19
  structure 107–9
  synthesis in inflammation 109
Serratia marcescens 116
Serum amyloid A (SAA) 7, 8, 21–7, 30, 47, 59, 69–83, 151, 154, 171
  amino acid sequences 172, 174
  chemical analysis 172
  cytokine control of expression 23–6
  distribution 61
  function 61–2
  gene expression regulation by cytokines 21–7
  genes related to, in rats 63–4
  in amyloidosis 176–8
  in spleen 62
  in vitro expression of 21–2
  incomplete degradation of 178
  induction and production of acute phase protein 172–5
  mRNAs in mice 62–3
  PMA induction 22–3
  regulation of biosynthesis 71–7
  role in lipid metabolism 65
  role of cofactors 75–7
  signal transduction 75–7
  structure 171–2
  synthesis 62
Serum amyloid P (SAP) 8, 30, 47, 187–206
  amino acid sequences 134–5
  and amyloidosis 190
  as targeting vehicle in amyloidosis 191
  binding to chromatin 189
  CD spectra 133
  composition of 188
  crystal forms of 125
  genetics 54–5
  human turnover studies 199–203
  ligand binding 188–9
  mouse imaging studies using ^{123}I-human 192–5, 198–203
  mouse imaging studies using ^{123}I-mouse 192, 195–6
  pentameric disk of 128
  preparation for clinical studies 198–9
  prismatic crystals of 124
  quantitating murine amyloid deposits 197–8
  three-dimensional structure 123–36
  X-ray data sets 129
Shock syndromes 117–18
Sinusoidal cells, role in in vivo biosynthesis of APR 103
SLE 164
Slp 93
Staphylococcus aureus 116, 152
Streptococcus pneumoniae 47, 157
Substance P 11
Surgical intervention 164

TATA 41, 52
TFIID 41
Thyroxine 116
Thyroxine binding globulin (TBG) 107, 115–16
Transforming growth factor-β (TGF-β) 33
Tumour necrosis factor (TNF) 7, 11, 21, 33, 72, 86, 93, 102

Ulcerative colitis 164

Very low density lipoproteins (VLDL) 165

GPSR Compliance

The European Union's (EU) General Product Safety Regulation (GPSR) is a set of rules that requires consumer products to be safe and our obligations to ensure this.

If you have any concerns about our products, you can contact us on ProductSafety@springernature.com

In case Publisher is established outside the EU, the EU authorized representative is:

Springer Nature Customer Service Center GmbH
Europaplatz 3
69115 Heidelberg, Germany

Batch number: 10607330

Printed by Printforce, the Netherlands